Stolen from.
Don Jenkins

CHEMICAL INFORMATION SOURCES

Gary Wiggins
Head, Chemistry Library
Indiana University

wiggins@indiana.edu

McGraw-Hill, Inc.
New York   St. Louis   San Francisco   Auckland   Bogotá
Caracas   Hamburg   Lisbon   London   Madrid   Mexico
Milan   Montreal   New Delhi   Paris   San Juan   São Paulo
Singapore   Sydney   Tokyo   Toronto

CHEMICAL INFORMATION SOURCES

Copyright © 1991 by McGraw-Hill, Inc.
All rights reserved. Printed in the United States of America.
Except as permitted under the United States Copyright Act of 1976,
no part of this publication may be reproduced or distributed in
any form or by any means, or stored in a data base or retrieval
system, without the prior written permission of the publisher.

2 3 4 5 6 7 8 9 0 DOC DOC 9 5 4 3 2 1

P/N 070157-1
PART OF
ISBN 0-07-909939-4

The editor was Kirk Emry;
the production supervisor was Richard Ausburn.
R.R. Donnelley & Sons Company was printer and binder

Library of Congress Catalog Card Number: 90-61663.

CONTENTS

Preface

Chapter Outline

Tables   xvii

Figures   xxi

Acknowledgements   xxiii

Chapters
1. Introduction to Chemical Information Science   1
2. Database Directories, Guides to the Literature, and Search
   Strategies   21
3. General Techniques and Benefits of Computer-Based Searching   33
4. Searching by Author or Organization Names and by Known
   Citations   59
5. Searching by Subject   73
6. Patents   101
7. Online Chemical Dictionaries and Other Sources for Chemical
   Compound Searches   115
8. Structure Searching   139
9. Searching for Information Involving Chemical Measurements
   (Constitutional Chemistry)   161
10. Searching Which Involves Chemical and Physical Properties of
    Substances   181
11. Searching for the Synthesis or Reactions of Compounds (Reaction
    Chemistry)   211
12. Searching for Chemical Safety or Toxicology Information   245
13. Current Awareness, Research in Progress, Background Reading,
    and Document Delivery Services   263
14. The Personal Library and Science Writing Aids   285
15. Miscellaneous Information Sources   299

Appendices   315
  I. Index to Some Library of Congress Classification Numbers
     Relevant to Chemistry   317
 II. ACS Recommended Journals   321
III. Instructions for Using the CRSD Database with Pro-Cite   323
 IV. Subject Term Authority List for the Chemistry Reference Sources
     Database (CRSD)   333
     Index   341

iii

PREFACE

Chemical Information Sources is a textbook designed to give to the chemist, librarian, or student the command of the chemical literature which is needed to successfully solve most chemical information problems. A student who has taken a one-year beginning foreign language course has not really mastered the language. Likewise, one should not expect to be a polished chemical information specialist after completing this book. However, the material found herein will lay a solid foundation upon which to build toward mastery of the vast domain of **chemical information science**, the discipline which deals with the storage and retrieval of chemical information in all of its manifestations.[a] The reader of this book will learn basic chemical information sources and concepts which should be known by anyone who completes an undergraduate chemistry program certified by the American Chemical Society. An up-to-date discussion of the sources and approaches for solving chemical information problems is presented in a fashion which integrates computer-readable and printed sources of information throughout the book. This approach gives the reader a good overview of the range of sources available for solving chemical information problems.

There is a growing belief among chemical information specialists and chemistry librarians that instruction in the use of the chemical literature should take place in a formal course in the undergraduate curriculum. Nonetheless, a recent study of chemical education in the United States points out that "Formal training in chemical information retrieval is seldom included in the chemistry curriculum, yet every professional chemist must use a variety of complex printed sources as well as an increasing number of computerized databases."[1] Furthermore, the most recent American Chemical Society guidelines for professional education in chemistry note that "The increasing volume and complexity of the literature means that students can no longer acquire skills in information retrieval without some formal instruction."[2] In part, it is to provide a textbook for use in such formal classroom instruction that *Chemical Information Sources* was written. However, if the opportunity to study the material in a formal setting is not available, the book will still be a valuable source of information.

---

[a] When new or particularly important terms or concepts are used in the text, they are entered in **boldface** type and defined at that point.

The goal of the work is then to introduce a minimal set of chemical information sources and concepts. Those who complete *Chemical Information Sources* should:
- develop an understanding of the flow of information and documents and the role of libraries, information centers, and other information providers in that process
- learn to distinguish the various types of chemical information sources and to choose appropriate sources to solve specific chemical information problems
- learn to recognize the primary literature by format and to construct appropriate bibliographic citations[b]
- improve writing skills
- become aware of computer-based sources and techniques[c]
- learn to perform efficient searches for subjects and for authors
- understand the importance of indexing techniques in information retrieval
- understand the importance and nature of major chemical abstracting tools and those of related sciences
- be able to locate chemical and physical properties of substances, including spectra
- learn sources of methods for performing certain tasks in chemistry
- learn sources of assistance in designing and performing chemical syntheses
- understand the importance of patents for chemistry and learn sources of patent information.

There are actually two parts to *Chemical Information Sources*. The first is the printed book itself, where a number of basic sources, both printed and computer-readable, are described in detail. Most of these reference materials should be easily accessible through libraries which serve chemists. Almost all of the printed secondary sources recommended in *Guidelines and Suggested Title List for Undergraduate Chemistry Libraries* [1] have been included in the printed text.[d] In addition, I have selected a number of other works, especially those published in the 1980s. Computer-based services or databases which complement the printed sources are discussed alongside those works.

---

[b] Bibliographic references here and in the "Selected Readings" sections of the chapters are entered in the style preferred by the American Chemical Society. However, full titles of journal articles or chapter titles are included, information which is not normally required by *The ACS Style Guide* [110] (1986).

[c] There is definitely a bias in this book toward the versions of the *Chemical Abstracts* database found on the STN International system. Although other online search vendors and databases are discussed at appropriate places, the STN search techniques for the *Chemical Abstracts* database are emphasized throughout.

[d] *Guidelines and Suggested Title List for Undergraduate Chemistry Libraries*. From a list compiled by Judith A. Douville; Rev. ed.; Brasted, Robert C.; Clapp, Leallyn, Eds.; American Chemical Society: Washington, DC, 1982. [1]

   Works in the Chemistry Reference Sources Database which are included in the *Guidelines* can be found by searching **GL**. If a later edition of the work has appeared since the compilation of the *Guidelines*, only the more recent edition was entered in the database.

A supplement to the book is the Chemistry Reference Sources Database (CRSD), a computer-readable file of over 2150 records. Included is the complete reference collection of the Indiana University Chemistry Library, supplemented with entries from the corresponding chemistry libraries at Purdue University, the University of Illinois, and the University of Michigan. Additionally, records for other printed reference works, online databases, and software of interest to chemists were entered into the CRSD. Many secondary works and descriptions of databases or computer-based products which are not discussed in the textbook are included in the Chemistry Reference Sources Database. Hence, the database can be searched by subject to obtain additional sources of information if the basic reference works described in the printed text fail to provide an answer. To facilitate such searches, a "Subject Term Authority List" is included on the diskette as file INDEX.LST. The list may also be found as the final appendix in this book.

It is only in the Chemistry Reference Sources Database that full bibliographic information is found for the works discussed in the book. The number in brackets following a title in the book, for example, [557], easily leads to the full description of the item in the database. The record number could be found simply by entering **557**.[e] With the database management software package Pro-Cite, which labels the fields of a record, the search for the number can be limited to the field for the accession number. Details on searching CRSD with Pro-Cite are in the file CRSDNTRO on the CRSD diskette and in Appendix III in this book.

Another benefit of using the database is the ability to discover whether the item is held by the Indiana University Chemistry Library and is available for borrowing. Although many of the works are reference materials and hence are non-circulating, quite a number can be loaned. The presence of a Library of Congress call number in the record indicates that it may be borrowed. Details are included on the diskette.

It is hoped that *Chemical Information Sources* will foster an increase in the number of chemistry departments which offer formal courses in the subject, as well as prove useful to those who need to review the subject matter or update their knowledge in this area. I welcome comments and criticism of the work, especially suggestions for improvements in future editions.

I would like to thank the many people who have provided useful criticism and suggestions for improving this work. Although the list is far too long to acknowledge everyone here by name, I am especially grateful to Roger Beckman, Don McMasters, and Ned O. Tykomka (Indiana University); Michaeleen Trimarchi and Maimie V. Reitano (Eastman Kodak Company); Phae H. Dorman, Michael S. Feider, and Dave Fifield (The Dow Chemical Company); Richard Love and John Hearty (American Chemical Society); W.V. Metanomski (Chemical Abstracts Service); Anne Leinbach

---

[e] Input for searches of all databases is entered in **boldface** type throughout this book. Examples of searches in the Chemistry Reference Sources Database which utilize the controlled vocabulary in Appendix IV are shown in the format used in Pro-Cite. For example, **INDX="MOLECULAR SEQUENCE DATA-BANK"**.

and Barbara Temos (Institute for Scientific Information); and Tina Chrzastowski (University of Illinois). Finally, a very special thank-you goes to Esma Karic, the person who so accurately typed the manuscript in all of its many revisions and assisted in the preparation of the database. Without her exceptional talents, this book might never have been finished.

>Gary Wiggins
>Chemistry Library
>Indiana University
>Bloomington, IN 47405
>Bitnet address: WIGGINS@IUBACS
>FAX number: 812-855-6611
>January 1, 1990

REFERENCES

[1] *Tomorrow: The Report of the Task Force for the Study of Chemistry Education in the United States*; American Chemical Society: Washington, DC, 1984; p 45. [2]

[2] American Chemical Society Committee on Professional Training. *Undergraduate Professional Education in Chemistry: Guidelines and Evaluation Procedures*; American Chemical Society: Washington, DC, 1983, reprinted with Extension, 1988; p 13. [3]

CHAPTER OUTLINE

1	**INTRODUCTION TO CHEMICAL INFORMATION SCIENCE**	**1**
1.1	INTRODUCTION	1
1.2	THE SCIENCE OF CHEMISTRY AND ITS STRUCTURE	2
1.3	CLASSIFICATION VS. SUBJECT ACCESS	4
1.3.1	Library of Congress Call Numbers	
1.3.2	*Chemical Abstracts* Classification Scheme	
1.3.3	Subject Indexing	
1.4	THE PUBLICATION PROCESS: PRIMARY LITERATURE	7
1.4.1	Journals	
1.4.2	Conference Proceedings	
1.4.3	Government Reports, Dissertations, Patents	
1.4.4	Formats: Document Types	
1.5	THE SECONDARY LITERATURE	10
1.5.1	Treatises	
1.5.2	"Multigraphs" (Multiply-Authored Books) and Monographs	
1.5.3	Encyclopedias and Dictionaries	
1.5.4	Handbooks and Data Compilations	
1.5.5	Reviews and Review Serials	
1.5.6	Bibliographies	
1.5.7	Current Awareness and Retrospective Searches: Indexing and Abstracting Services	
1.6	THE RELATIONSHIP OF THE SECONDARY LITERATURE TO NEW PRIMARY LITERATURE	13
1.7	COMPUTER-READABLE SOURCES	14
1.7.1	Bibliographic Databases and Document Access	
1.7.2	Numeric Databases	
1.7.3	Text Databases	
1.7.4	Dictionary Databases	
1.7.5	Directory Databases	
1.8	OPTIONS FOR DATABASE SEARCHING	18
1.9	TERTIARY SOURCES: GUIDES TO THE LITERATURE	19
1.10	SUMMARY	19
1.11	SELECTED READINGS	20
	REFERENCES	20
2	**DATABASE DIRECTORIES, GUIDES TO THE LITERATURE, AND SEARCH STRATEGIES**	**21**
2.1	INTRODUCTION	21
2.2	USER AIDS FOR COMPUTER-READABLE DATABASES	22

2.3	DATABASE GUIDES AND ONLINE AIDS FOR DATABASE SELECTION	
2.4	COMPREHENSIVE CHEMISTRY GUIDES	23
2.5	SEARCH STRATEGY FORMULATION	25
2.5.1	Statement of the Information Need	29
2.5.2	Preliminary Selection of Search Terms and Other Search Keys	
2.5.3	Selection of Appropriate Secondary Sources	
2.5.4	Search of the Appropriate Sources	
2.6	SUMMARY	32
2.7	SELECTED READINGS	32
3	GENERAL TECHNIQUES AND BENEFITS OF COMPUTER-BASED SEARCHING	33
3.1	INTRODUCTION	33
3.2	COMMON FEATURES OF ONLINE RETRIEVAL SYSTEMS	34
3.2.1	Time Periods and Subjects Covered by the Databases	
3.2.2	Virtues and Benefits of Online Searching	
3.2.3	Costs of Online Searching	
3.2.4	Problems With Online Searching	
3.2.5	Other Factors	
3.3	EQUIPMENT AND SOFTWARE NEEDED FOR ONLINE SEARCHING	39
3.4	TELECOMMUNICATIONS WITH THE HOST COMPUTER	40
3.5	PROTOCOLS FOR SIGNING ON	40
3.6	FEATURES OF THE SYSTEM SEARCH SOFTWARE	43
3.6.1	Logging On and Logging Off	
3.6.2	Modes of Searching: Levels of Proficiency and Defaults	
3.6.3	Retrieval Software	
3.6.4	Records and Fields in the CA File; Displaying an Answer Set	
3.6.5	Assistance in Search Formulation and Retrieval	
3.6.5.1	*Inverted Dictionary Files (Indexes)*	
3.6.5.2	*The Basic Index*	
3.6.5.3	*Other Inverted Dictionary Files*	
3.6.6	Search Commands and Other Software Features	
3.7	SUMMARY	56
3.8	SELECTED READINGS	56
	REFERENCES	57
4	SEARCHING BY AUTHOR OR ORGANIZATION NAMES AND BY KNOWN CITATIONS	59
4.1	INTRODUCTION	59
4.2	*SCIENCE CITATION INDEX* AND OTHER ISI PRODUCTS	60
4.2.1	Citation and Author Searching in the *Science Citation Index*	
4.2.2	Citation and Author Searching in SciSearch	
4.2.3	Using the Corporate Source Approach in SciSearch and the *Science Citation Index*	

4.2.4	Author and Corporate Searches in Other ISI Products	
4.3	AUTHOR AND CORPORATE SEARCHES IN *CHEMICAL ABSTRACTS*	69
4.4	SUMMARY	72
4.5	SELECTED READINGS	72
5	**SEARCHING BY SUBJECT**	**73**
5.1	INTRODUCTION	73
5.2	SUBJECT SEARCHES	73
5.2.1	Truncation and Masking vs. Neighboring or Expanding	
5.2.2	Key-Word Searches	
5.2.3	The Permuterm Subject Index to *Science Citation Index*	
5.2.4	Subject Term Authority Lists	
5.2.5	*Chemical Abstracts* Subject Term Authority Lists: The Index Guide and Others	
5.3	THE STRUCTURE OF *CHEMICAL ABSTRACTS*, THE CA FILE, AND THEIR SUBJECT INDEXES	86
5.4	FORMATS: DOCUMENT TYPES	91
5.5	SEARCHES OF DATABASES AND REFERENCE WORKS DEVOTED TO A DOCUMENT TYPE	92
5.5.1	Dissertations	
5.5.2	Conference Proceedings	
5.5.3	Reports: NTIS	
5.6	BIBLIOGRAPHIES	94
5.7	CD-ROM DATABASES FOR BIBLIOGRAPHIC SEARCHING	95
5.8	TEXT DATABASES	95
5.8.1	Chemical Journals Online: The CJO File on STN	
5.8.2	Other Text Searching in Bibliographic and Primary Journal Text Databases	
5.9	SUMMARY	100
5.10	SELECTED READINGS	100
6	**PATENTS**	**101**
6.1	INTRODUCTION	101
6.2	DEFINITION OF A PATENT AND THE CRITERIA FOR JUDGING PATENTABILITY	102
6.3	THE PATENT PROCESS	103
6.4	THE VALUE OF PATENT INFORMATION: REASONS FOR CONDUCTING A PATENT SEARCH	105
6.5	THE MAKE-UP OF A PATENT SPECIFICATION	106
6.6	PATENT REFERENCE TOOLS AND PATENT AGENCIES	106
6.6.1	The United States Patent and Trademark Office	
6.6.2	Derwent Patent Publications, Databases, and Products	
6.6.3	INPADOC, the International Patent Documentation Center's Database	
6.6.4	IFI/Plenum's Patent Services	

6.6.5	Chemical Abstracts Service's Coverage of Patents	
6.6.6	Other Patent Reference Sources	
6.7	TRADEMARKS	112
6.8	SUMMARY	113
6.9	SELECTED READINGS	113
7	**ONLINE CHEMICAL DICTIONARIES AND OTHER SOURCES FOR CHEMICAL COMPOUND SEARCHES**	**115**
7.1	INTRODUCTION	115
7.2	COMPOUND NAME SEARCHES	116
7.2.1	*Chemical Abstracts* Index Guide	
7.2.2	*Chemical Abstracts* Chemical Substance Indexes	
7.2.3	Other Sources of Compound Names	
7.3	SEARCHING FOR COMPOUNDS WITH CODES	119
7.3.1	Molecular Formulas	
7.3.2	*Chemical Abstracts* Formula Indexes	
7.3.3	Ring Indexes	
7.3.4	Fragment Codes	
7.3.5	Wiswesser Line Notation and Other Linear Notations	
7.3.6	Connection Codes and CAS Registry Numbers	
7.4	ONLINE CHEMICAL DICTIONARIES	129
7.4.1	Online Chemical Dictionaries on DIALOG and ORBIT	
7.4.2	The National Library of Medicine's Online Chemical Dictionary	
7.4.3	The Chemical Information System's Online Chemical Dictionary	
7.4.4	STN International's Online Chemical Dictionary	
7.5	DISPLAY OPTIONS IN THE REGISTRY FILE	134
7.6	SEARCHING WITH CHEMICAL NAMES IN THE CA FILE	136
7.7	ELEMENT TERM (/ET) SEARCHES ON STN INTERNATIONAL	137
7.8	SUMMARY	137
7.9	SELECTED READINGS	137
8	**STRUCTURE SEARCHING**	**139**
8.1	INTRODUCTION	139
8.2	STRUCTURE SEARCHING ON STAND-ALONE COMPUTERS	140
8.2.1	Microcomputer-Based Systems	
8.2.2	Structure Searching Systems on Larger Stand-Alone Computers	
8.2.2.1	*MACCS*	
8.2.2.2	*OSAC*	
8.3	STRUCTURE SEARCHING ON REMOTE SYSTEMS	142
8.3.1	SANSS	
8.3.2	DARC and Markush Searches	
8.3.3	CAS ONLINE Registry File	
8.3.3.1	*General Features of Searching in the Registry File*	
8.3.3.2	*How to Create a Structure and Search in the Registry File*	
8.4	STRUCTURE INPUT TO THE REGISTRY FILE AND OTHER DATABASES VIA FRONT-END SOFTWARE	156

8.5	STRUCTURE SEARCHING OF BEILSTEIN	157
8.6	OTHER STRUCTURE SEARCHING SYSTEMS	158
8.7	SUMMARY	158
8.8	SELECTED READINGS	158
9	**SEARCHING FOR INFORMATION INVOLVING CHEMICAL MEASUREMENTS (CONSTITUTIONAL CHEMISTRY)**	**161**
9.1	INTRODUCTION	161
9.2	COLLECTIONS-I: CONTINUING METHODS SERIES, TREATISES, AND ENCYCLOPEDIAS	162
9.2.1	*Techniques of Chemistry* and *Physical Methods of Chemistry*	
9.2.2	*Treatise on Analytical Chemistry*	
9.2.3	*Encyclopedia of Industrial Chemical Analysis*	
9.2.4	*Comprehensive Analytical Chemistry*	
9.2.5	*Comprehensive Treatise of Electrochemistry*	
9.2.6	*Encyclopedia of Electrochemistry of the Elements*	
9.2.7	*Methods in Enzymology*	
9.2.8	*Methods of Enzymatic Analysis*	
9.3	COLLECTIONS-II: STANDARD METHODS, HANDBOOKS, AND SMALLER WORKS	167
9.3.1	Standard Methods and Pharmacopoeias	
9.3.2	Handbooks and Selected Smaller Works	
9.3.3	Dictionaries and Nomenclature Aids	
9.4	COLLECTIONS-III: SPECTRAL, STRUCTURAL, AND SEQUENCE ANALYSIS COLLECTIONS	170
9.4.1	Spectral Data Collections	
9.4.2	Crystallography	
9.4.3	Biochemical Compounds and Sequence Databases	
9.5	REVIEWS	175
9.5.1	Special Review Issues of *Analytical Chemistry*	
9.5.2	*Methods of Biochemical Analysis*	
9.5.3	Other Review Serials	
9.6	ABSTRACTING AND INDEXING JOURNALS	177
9.7	SUMMARY	178
9.8	SELECTED READINGS	178
10	**SEARCHING WHICH INVOLVES CHEMICAL AND PHYSICAL PROPERTIES OF SUBSTANCES**	**181**
10.1	INTRODUCTION	181
10.2	GUIDES TO THE LITERATURE	182
10.3	CRITICALLY EVALUATED DATA AND INFORMATION ANALYSIS CENTERS	182
10.4	LARGE DATA COMPILATIONS	183
10.4.1	*Landolt-Börnstein Numerical Data and Functional Relationships*	

10.4.2	*Beilstein Handbook of Organic Chemistry*	
10.4.2.1	*Coverage and Arrangement of Beilstein*	
10.4.2.2	*Access to the Information in Beilstein*	
10.4.2.3	*Beilstein Databases*	
10.4.3	*Gmelin Handbook of Inorganic Chemistry*	
10.4.3.1	*Coverage and Arrangement of Gmelin*	
10.4.3.2	*Access to the Information in Gmelin*	
10.5	SMALLER COLLECTED WORKS AND HANDBOOKS	200
10.5.1	Important Smaller Collected Works and One-Volume Printed Handbooks	
10.5.2	Printed Compilations Which Lead From a Known Value of a Property to a Compound	
10.6	INDEXES TO NUMERIC DATA	205
10.6.1	Comprehensive Indexes to Numeric Data Compilations	
10.6.2	Secondary Indexes to Data in the Primary Literature	
10.7	OTHER PHYSICAL PROPERTY AND RELATED DATABASES	207
10.8	MISCELLANEOUS SOURCES	209
10.9	SUMMARY	210
10.10	SELECTED READINGS	210
11	**SEARCHING FOR THE SYNTHESIS OR REACTIONS OF COMPOUNDS (REACTION CHEMISTRY)**	**211**
11.1	INTRODUCTION	211
11.2	GUIDES TO THE LITERATURE	212
11.3	SOME TRADITIONAL SOURCES OF INFORMATION	213
11.3.1	*Beilstein* and *Gmelin*	
11.3.2	*Methoden der Organischen Chemie (Houben-Weyl)*	
11.3.3	English-Language Treatises Devoted Largely to Reaction Chemistry	
11.3.3.1	*Inorganic Reactions and Methods*	
11.3.3.2	*Pergamon's Comprehensive Treatises*	
11.3.3.3	*Other Treatises*	
11.4	ABSTRACTING AND INDEXING SERVICES	223
11.4.1	Searching *Chemical Abstracts* and CAS ONLINE for the Preparation or Reactions of a Compound	
11.4.2	CASREACT	
11.4.3	*Index Chemicus* and *Current Chemical Reactions*	
11.4.4	The *Journal of Synthetic Methods* and *Theilheimer*	
11.4.5	Other Abstracting Services with Reaction Diagrams	
11.5	ANNUAL SURVEYS AND REVIEWS	236
11.5.1	*Organic Syntheses*	
11.5.2	*Organic Reactions* and Other Reviews of Organic Synthesis	
11.5.3	Other Surveys and Reviews	
11.6	COMPENDIA OF SYNTHETIC METHODS AND IMPORTANT SERIES	238

11.7	NAME REACTIONS	239
11.8	COMPUTER-BASED REACTION SEARCHING SYSTEMS FOR IN-HOUSE USE	240
11.8.1	REACCS	
11.8.2	ORAC	
11.8.3	SYNLIB	
11.9	SUMMARY	
11.10	SELECTED READINGS	242
	REFERENCES	243
12	**SEARCHING FOR CHEMICAL SAFETY OR TOXICOLOGY INFORMATION**	**245**
12.1	INTRODUCTION	245
12.2	MAJOR COLLECTIONS OF DATABASES CENTERED ON ENVIRONMENTAL OR TOXICOLOGICAL CONCERNS	247
12.2.1	NLM's Toxicology Information Program and the TOXicology Data NETwork (TOXNET).	
12.2.2	The Chemical Information System	
12.2.3	Other Vendors of Environmental and Toxicology Databases	
12.3	SOURCES OF INFORMATION ON CHEMICALS KNOWN (OR THOUGHT) TO BE TOXIC OR HAZARDOUS	252
12.3.1	Sources Available as Databases	
12.3.2	Sources Available Only in Printed Format	
12.4	SAFE HANDLING, USE, AND DISPOSAL OF CHEMICALS	257
12.4.1	Books	
12.4.2	Abstracts, Indexes, Databases	
12.5	CHEMICAL LAWS AND REGULATIONS: LEGAL REQUIREMENTS FOR THE CHEMICAL INDUSTRY	259
12.6	SUMMARY	260
12.7	SELECTED READINGS	261
13	**CURRENT AWARENESS, RESEARCH IN PROGRESS, BACKGROUND READING, AND DOCUMENT DELIVERY SERVICES**	**263**
13.1	INTRODUCTION	263
13.2	CURRENT AWARENESS SOURCES	264
13.2.1	Table-of-Contents Services	
13.2.2	Printed Table-of-Contents Journals from Commercial Publishers	
13.2.3	Databases for Current Awareness	
13.2.4	Selective Dissemination of Information (SDI)	
13.2.5	Standard Interest Profiles	
13.3	RESEARCH IN PROGRESS	274
13.4	SOURCES FOR BACKGROUND READING	275
13.4.1	Reviews	
13.4.2	Encyclopedias	
13.4.3	Dictionaries	

13.4.4	Treatises, Monographs, and Multigraphs	
13.4.5	Finding Out About Books	
13.4.6	Finding Out About Serials	
13.5	DOCUMENT DELIVERY OPTIONS	279
13.5.1	Commercial Document Delivery Services	
13.5.2	Author Reprints	
13.5.3	Translations	
13.6	SUMMARY	283
13.7	SELECTED READINGS	283
14	**THE PERSONAL LIBRARY AND SCIENCE WRITING AIDS**	**285**
14.1	INTRODUCTION	285
14.2	SOFTWARE FOR PERSONAL DATABASES AND WORDPROCESSING	286
14.3	AIDS TO WRITING	289
14.3.1	Laboratory Notebooks	
14.3.2	Style Guides	
14.3.3	"How-to-Write" Books	
14.4	BIBLIOGRAPHIC STANDARDS FOR SERIALS AND LIBRARY RECORDS	291
14.4.1	*CASSI* and Serial Abbreviations	
14.4.2	Library Forms of Serial Entries: *CASSI* vs. *AACR*	
14.5	BOOKS AND OTHER WORKS ON THE NOMENCLATURE OF CHEMICAL COMPOUNDS	293
14.6	OTHER STANDARDS FOR NUMERICAL DATA, SYMBOLS, AND TERMINOLOGY	295
14.7	SUMMARY	296
14.8	SELECTED READINGS	296
15	**MISCELLANEOUS INFORMATION SOURCES**	**299**
15.1	INTRODUCTION	299
15.2	HISTORY OF CHEMISTRY	299
15.3	BIOGRAPHICAL SOURCES	300
15.4	DIRECTORIES	302
15.4.1	Directories of People	
15.4.2	Directories of Organizations	
15.4.3	Directories of Suppliers of Chemicals and Chemical Laboratory or Plant Equipment	
15.5	TRADE LITERATURE	306
15.6	INFORMATION SOURCES FOR THE CHEMICAL INDUSTRY	307
15.7	STANDARDS AND SPECIFICATIONS	308
15.8	TEACHING OF CHEMISTRY	309
15.9	THE STUDY OF CHEMISTRY/CAREERS IN CHEMISTRY	312
15.10	SUMMARY	314
15.11	SELECTED READINGS	314

TABLES

Table 1.1	Subdivisions of the Discipline of Chemistry in Several Scientific Organizations	2
Table 1.2	Major Divisions of the Library of Congress Classification Schedule for Chemistry	4
Table 1.3	Major Subject Divisions of *Chemical Abstracts*	5
Table 2.1	Major U.S. Vendors of Databases Relevant to Chemistry	22
Table 3.1	Comparison of Commands Used on Various Online Searching Systems	37
Table 3.2	Telecommunications Companies for Online Searching in the United States	41
Table 3.3	Telecommunications Problems and Potential Solutions	42
Table 3.4	Selected Front-End or Gateway Systems for Online Searching	47
Table 4.1	Time Periods Covered in *Science Citation Index* Cumulative Volumes	60
Table 4.2	SciSearch on DIALOG and ORBIT	63
Table 5.1	Truncation and Masking Symbols on STN International	75
Table 5.2	Codes for Source Items in *Science Citation Index* and *SciSearch*	78
Table 5.3	Codes for Major Subject Sections of *Chemical Abstracts* in the CA File	88
Table 5.4	Results of File Crossover from the Registry File to the CA and CAOLD Files for Sets Containing Isatin and Related Compounds	90
Table 5.5	Codes Used in the Document-Type Field in the CA File on STN International and in the Printed *CA*	92
Table 5.6	Chemical Journals Online (CJO) Text Database Files on STN International	97
Table 5.7	Fields Included in the Basic Index of the CJACS File on STN International	98
Table 5.8	Display Options in the STN Chemical Journals Online (CJO) File	98

xvii

Table 6.1	Derwent Classification Sections for Chemical Patents in the WPI Database and the *Chemical Patents Index*	108
Table 7.1	Standard Subject Divisions for Qualified Chemical Substances in *Chemical Abstracts*	117
Table 7.2	Selected Printed Sources of Chemical Compound Names	118
Table 7.3	Inverted Files for the CAS ONLINE Registry File	132
Table 7.4	Compound Class Identifiers in the CAS ONLINE Registry File	133
Table 8.1	A Sample of Databases Which Can Be Searched With the CAS Registry Number	142
Table 8.2	Some Pre-Drawn Ring Systems for Use in Creating Structures in the CAS ONLINE Registry File	147
Table 8.3	Valid Node Symbols in the Registry File	148
Table 8.4	Specifications for Generic Group Symbols in Structure Building in the Registry File	149
Table 8.5	Shortcut Symbols for Use in Building Structures in the CAS ONLINE Registry File	149
Table 8.6	Bond Codes Used in the CAS ONLINE Registry File	150
Table 8.7	Outline of the Search Sequence in the Registry File	152
Table 8.8	Templates Available With STN Express	156
Table 9.1	Contents of *Techniques of Chemistry*	162
Table 9.2	Contents of *Physical Methods of Chemistry*	163
Table 9.3	Contents of *Wilson and Wilson's Comprehensive Analytical Chemistry*	164-165
Table 9.4	Contents of *Comprehensive Treatise of Electrochemistry*	166
Table 9.5	Subject Indexes to *Methods in Enzymology*	167
Table 9.6	Collections of Mass Spectra Covered by the *Eight Peak Index of Mass Spectra*	170
Table 9.7	Analytical Chemistry Review Serials	176
Table 9.8	Subject Sections of *Analytical Abstracts* on DIALOG	178
Table 10.1	Selected Systems of Units for Non-Electrical Physical Constants	183
Table 10.2	Arrangement of Compounds in the *Beilstein Handbook of Organic Chemistry*	186
Table 10.3	Time Periods Covered in the *Beilstein Handbook of Organic Chemistry*	186
Table 10.4	Recommended Labeling of *Beilstein* Volumes	187
Table 10.5	German Abbreviations and Terms for Physical Constants in *Beilsteins Handbuch der Organischen Chemie*	188-189
Table 10.6	Physical Property Data Searchable and/or Displayable in the Beilstein Database	196
Table 10.7	The *Gmelin Handbook of Inorganic Chemistry* System Numbers	199
Table 10.8	Search Examples from STN's GFI (Gmelin Formula Index) Database	201

Table 10.9	Indexes to the *Journal of Physical and Chemical Reference Data*	207
Table 10.10	Physical Chemistry Review Serials	210
Table 11.1	Selected German Words in *Beilstein* Relevant to Reaction Chemistry	214-215
Table 11.2	Contents of *Inorganic Reactions and Methods*	218
Table 11.3	Contents of *Comprehensive Inorganic Chemistry*	220
Table 11.4	Contents of *Comprehensive Organic Chemistry*	220
Table 11.5	Contents of *Comprehensive Heterocyclic Chemistry*	221
Table 11.6	Contents of *Comprehensive Coordination Chemistry*	222
Table 11.7	Contents of *Comprehensive Polymer Science*	222
Table 11.8	*Index Chemicus* Personal Databases Available for Microcomputers	232
Table 11.9	*Current Chemical Reactions* Personal Databases Available for Microcomputers	232
Table 11.10	Thematic Group Codes Used in the *Journal of Synthetic Methods*	234
Table 11.11	Reaction Codes in *Theilheimer's Synthetic Methods of Organic Chemistry*	235
Table 11.12	Selected Review Serials for Reaction Chemistry	238
Table 12.1	Toxic Chemical Information Needs	246
Table 12.2	Secondary Sources From Which NLM's TOXLINE and TOXLIT Database Citations Are Drawn	248
Table 12.3	Data Fields in the National Library of Medicine's Hazardous Substances Data Bank	249
Table 14.1	IUPAC Books on Chemical Nomenclature	294
Table 14.2	Selected Standards for Numerical Data, Symbols, and Terminology Published in the CODATA *Bulletin*	296
Table 14.3	Selected Standards for Numerical Data, Symbols, and Terminology Published in *Pure and Applied Chemistry*	296
Table 15.1	Chronological Coverage of Scientists in Poggendorff's *Bio-Bibliographical Handbook of the Exact Sciences*	301

FIGURES

Figure 1-1	Typical Abstracts for Journal Articles in the Printed *Chemical Abstracts*	5
Figure 1-2	Typical Report, Dissertation, and Patent Entries from *Chemical Abstracts*	9
Figure 1-3	Time Lag in Secondary Literature Coverage of New Knowledge	14
Figure 1-4	Time Lag in Current Awareness Services' Coverage of New Knowledge	15
Figure 2-1	Search on DIALINDEX for ONTAP Learning Files with Information on Flavonoids	24
Figure 3-1	Logging On and Off STN International's CAS ONLINE Academic Program via Telenet	44-45
Figure 3-2	Sample Records for Journal Articles from the CA File on STN International	48-49
Figure 3-3	Relationships Among the CAS ONLINE CA, CAOLD, and Registry Files	50
Figure 3-4	Posting of Accession Numbers in a Hypothetical Inverted Dictionary File	52
Figure 3-5	Inverted File Author Search in the CA File on STN International	53
Figure 4-1	Use of the Citation Index to *Science Citation Index*	62
Figure 4-2	Cited Reference Search in the ONTAP SciSearch File	65
Figure 4-3	Corporate Name Search in the ONTAP SciSearch File	66
Figure 4-4	Search by Corporate Name in the *Science Citation Index* Corporate Index	68
Figure 4-5	Use of the EXPAND Command in an Author Search for a Hyphenated Name in the STN CA File	70
Figure 4-6	Use of the EXPAND Command in the STN CA File for an Author Search Limited to a Particular Journal	71
Figure 5-1	Use of the EXPAND Command During a Subject Search for a Review in the CA File	76
Figure 5-2	Typical Subject Search in the *Science Citation Index*	79
Figure 5-3	Use of Truncation in a Subject Search of SciSearch	81
Figure 5-4	Contents of the *Chemical Abstracts* Index Guide for 1987-91	84
Figure 5-5	Sample Tables of Contents for Two Successive Weekly Issues of *Chemical Abstracts*	87
Figure 5-6	History of a Search Session in the CAS ONLINE Files on STN International	90

Figure 5-7	Isatin Reference in the CA File Found by Searching with CA Section Codes (Displayed in the ALL Format)	91
Figure 5-8	Text Record from the CJACS Database Displayed in the BIB Format	98
Figure 6-1	Patent Index Entry for a U.S. Patent Covered by *Chemical Abstracts*	111
Figure 7-1	SEE Reference for a Compound Name in the *Chemical Abstracts* Index Guide	117
Figure 7-2	Selected Molecular Formulas in Hill System Order and Alphabetized as in a Formula Index	121
Figure 7-3	*Chemical Abstracts* Formula Index Entry	121
Figure 7-4	Typical *Ring Systems Handbook* Entry	123
Figure 7-5	Fragment Code for Epichlorohydrin	124
Figure 7-6	Connection Code for Epichlorohydrin	126
Figure 7-7	CAS ONLINE Registry File Record Displayed in the IDE Format on a Type 2 (Graphics) Terminal	135
Figure 8-1	Chemical Information Systems, Inc. SANSS File Record	143
Figure 8-2	Structure Search for Isatin in the CAS ONLINE Registry File on a Type 3 (Text) Terminal	153-155
Figure 10-1	Use of SANDRA to Locate Isatin in the *Beilstein Handbook of Organic Chemistry*	190
Figure 10-2	Selected Pages of the Isatin Listing in the *Beilstein* Basic Series and the Third/Fourth Supplementary Series	192-195
Figure 10-3	Use of the Beilstein Database for Numeric Data Searching	197
Figure 10-4	Isatin Entry in the *Dictionary of Organic Compounds*	203
Figure 11-1	Use of the Beilstein Database in a Reaction Chemistry Search	216-217
Figure 11-2	Sample Record from the CASREACT Database	227
Figure 11-3	Typical Abstract from *Index Chemicus*	229
Figure 11-4	Typical Entry in *Current Chemical Reactions*	231
Figure 11-5	Typical Entry from the *Journal of Synthetic Methods*	233
Figure 12-1	OHM/TADS Entry for Methyl Ethyl Ketone	253
Figure 12-2	Methyl Ethyl Ketone Entry From the *Registry of Toxic Effects of Chemical Substances* (RTECS)	254
Figure 13-1	Table of Contents of a Journal Issue in Cauzin Softstrip Code	266-268
Figure 13-2	Sample *Chemical Titles* Entry	269
Figure 13-3	Sample *Current Contents* Entry	270
Figure 13-4	Sample CAS Individual Search Service Output	273
Figure 14-1	Bibliographic Portion of an Entry from *Chemical Abstracts Service Source Index* (CASSI)	292
Figure 15-1	Sample Entry From the *ACS Directory of Graduate Research*	303
Figure 15-2	Typical Entry From the ERIC (Education Resources Information Clearinghouse) Database	310

ACKNOWLEDGMENTS

I wish to thank the publishers and companies listed below for their permission to reproduce material found in the text.

American Chemical Society for quotations found in the Preface and for Figure 15-1 and Appendix II

Chapman and Hall for the reproduction of the entry in Figure 10-4 from the *Dictionary of Organic Compounds* **1982**, **3**, 3370

Chemical Abstracts Service for Figures 1-1, 1-2, 3-2, 3-3, 3-5, 4-5, 4-6, 5-1, 5-4, 5-5, 5-6, 5-7, 5-8, 6-1, 7-1, 7-3, 7-4, 7-7, 8-2, 11-2, 13-2, 13-4, 14-1, and Tables 3.1 and 10.8

Chemical Information Systems, Inc. for Figure 8-1

Derwent Publications, Ltd. for Figure 11-5

DIALOG Information Services, Inc. for Figure 2-1

Ellis Horwood, Ltd. for the quotation in Chapter 11

Elsevier Science Publishers for material paraphrased in Chapter 3

ERIC and ORBIT Search Service for Figure 15-2

Institute for Scientific Information for Figure 4-1, 4-2, 4-3, 4-4, 5-2, 5-3, 11-3, 11-4, 13-3

John Wiley & Sons, Inc. for the quotation in Chapter 12 taken from *Patty's Industrial Hygiene and Toxicology* **1982**, *2C*, 4733

Learned Information, Inc. for Table 12.1 which is adapted from: Benz, Joachim; Vogt, Kristina; Mücke, Wolfgang. "Strategy for Computer-Aided Searches for Information About Chemicals." *Online Review* **1989**, *13*(5), 390

Marcel Dekker, Inc. for Table 11.11, taken from: Woodburn, Henry M. *Using the Chemical Literature: A Practical Guide*; Marcel Dekker, Inc.: New York, 1974; p 134

New Science Publications for the paraphrase in Chapter 1 of Dr. David Jones' article which first appeared in *New Scientist* magazine, London, the weekly review of science and technology

Online, Inc. for the paraphrase of Lucinda D. Conger's article in Chapter 1.

Pergamon Press, Inc. for Figure 13-1

Sigma-Aldrich Corporation for the quotation in Chapter 12 from: *The Sigma-Aldrich Library of Chemical Safety Data* **1988**, *1*, vi

Springer-Verlag for Figures 10-1, 10-2, 10-3, 11-1, Tables 10.5 and 11.1, and for the reproductions from the *Landolt-Börnstein Comprehensive Index* **1987**, p 244 and p 256.

CHAPTER 1

INTRODUCTION TO CHEMICAL INFORMATION SCIENCE

1.1 INTRODUCTION

Chemical information science encompasses all of the activities formerly referred to as chemical documentation. Today it includes traditional library science areas which are used for chemical information storage and retrieval, as well as many computer-based techniques for retrieving chemical information. Thus, **chemical information science** embraces all of the intellectual endeavors and products which result in the codification of chemical knowledge, especially those techniques, products, and services which enhance the retrieval of information from the total storehouse of chemical knowledge.

There are certain information-gathering activities in which all scientists engage. In this book, we will cover the major reference sources which allow chemists to find the information they most commonly need, whether that information is in printed or computer-readable sources. Parallel, complementary reference sources exist in different scientific disciplines to solve similar information problems. Likewise, there are general reference works, some not limited to science, which contain information on chemical topics. In *Chemical Information Sources*, we will concentrate on works which are specific to chemistry, while covering a few of the most important general or interdisciplinary sources.

1.2 THE SCIENCE OF CHEMISTRY AND ITS STRUCTURE

Chemistry has been called "the central science," a phrase which emphasizes how close chemistry is to other scientific disciplines. Indeed, there is much overlap between chemistry and other subjects such as geology, physics, and biology. Thus, a person who is seeking an answer to a chemical question must decide how far into the knowledge base of other related disciplines it is necessary or desirable to go.

Knowing the structure of a scientific discipline is essential to retrieving infor-

TABLE 1.1
Subdivisions of the Discipline of Chemistry in Several Scientific Organizations

NSF	CAS	IUPAC	RSC	ACS
Analytical & Surface	SEE: Physical	Analytical	Analytical	Analytical Colloid & Surface
1	Biochemistry	—	—	2
Inorganic, Bioinorganic, & Organometallic	SEE: Physical	Inorganic	Inorganic	Inorganic
Organic & Macromolecular	Organic Macromolecular	Organic Macromolecular	Organic	Organic Macromolecular
Physical	Physical, Inorganic, & Analytical	Physical	Physical	Physical
1	Applied Chemistry & Chemical Engineering	Applied Chemistry	Industrial	Industrial & Engineering
—	—	Clinical	—	2
Chemical Instrumentation	—	—	—	—

1. Biochemistry and chemical engineering are not handled by the NSF Chemistry Division.
2. Only ACS divisions corresponding to those of the other organizations are listed. There is no clinical chemistry division and no biochemistry division within ACS, although chemists who work in those areas belong to related ACS divisions. The macromolecular section is presently a "secretariat".

mation from the body of recorded knowledge in that area. Pure chemistry has traditionally been divided into five areas: analytical chemistry, biochemistry, inorganic chemistry, organic chemistry, and physical chemistry. Many of the printed reference works in a library were developed by people who thought of themselves as analytical chemists or biochemists or physical chemists, etc. In the last few decades, however, the traditional divisions of chemistry have become somewhat blurred, resulting in such interdisciplinary areas as biogeochemistry and physical organic chemistry.

Let us look at how the discipline of chemistry has been subdivided by some major organizations which are involved with chemistry. These are the National Science Foundation (NSF), Chemical Abstracts Service (CAS), the International Union of Pure and Applied Chemistry (IUPAC), the Royal Society of Chemistry (RSC), and the American Chemical Society (ACS). (See Table 1.1.)

While the division of chemistry by the five organizations in Table 1.1 is similar, there are some differences. For example, the National Science Foundation links macromolecular chemistry with organic in one section. Three of the others have a separate section for macromolecular chemistry, but the Royal Society of Chemistry apparently subsumes macromolecular within organic chemistry without mentioning it by name.

Despite the seemingly distinct branches of chemistry noted above, interdisciplinary research, which blends into a single research project the knowledge gained over the centuries in the separate fields of chemistry, is becoming increasingly common. This has led to new ways of thinking about the science of chemistry which ignore some of the older, arbitrary divisions.

Chemistry is a science which deals with substances, either pure substances (containing only one kind of molecule) or those comprising a mixture (containing molecules of more than one type). Much of chemistry is concerned with chemical reactions which occur when the conditions are favorable and the molecules in a mixture react to form new molecules. Many of the distinctions between inorganic, organic, and biochemistry are diminished if we view chemistry in this light. There are two major areas in which many chemists who think of themselves as inorganic chemists, organic chemists, biochemists, etc. really work: constitutional chemistry and reaction chemistry. **Constitutional chemistry** is the area which is involved with finding out the chemical nature of a sample of matter. For a pure substance this is structural chemistry which investigates the types of atoms in a molecule, their spatial arrangement, and the types of bonds which hold the atoms together in the molecule. The constitutional chemistry of mixtures is usually dealt with in analytical chemistry. **Reaction chemistry** is concerned with making new substances by combining appropriate starting materials or treating them with different types of energy.[1]

Of course, many chemists are very much interested in both constitutional chemistry and reaction chemistry. Organic and inorganic compounds have geometries in common, forming a continuum from binary compounds through polymers. Reactivity can, in fact, be viewed as a consequence of structure, and the mechanisms of these reactions, such as substitution, addition, elimination, and rearrangement, have been used to classify the subject matter of the scientific research literature in which new reaction chemistry research results are reported.

1.3 CLASSIFICATION VS. SUBJECT ACCESS

When chemistry is broken down into subdisciplines or subdivisions as in the preceding section, certain aspects of the subject are grouped together based on pre-defined criteria: carbon-containing compounds, chemistry of living things, synthesis of compounds, etc. In every discipline this sort of **classifying** (division into logical subject categories) is done. It forms the basis of information storage and retrieval techniques, and it is apparent when you walk through the book stacks of a science library. Books on a certain topic tend to stand in one physical location, with books on related topics nearby. That is because the books were classified and assigned a call number (or shelf number).

1.3.1 Library of Congress Call Numbers

The call number (classification number) serves two functions. First, it allows the book to be located or accessed on the shelf. It serves as an **accession number**. With an exact call number, we have only to follow the proper alphabetical and/or numerical sequence to find a book on the shelf. The second thing which the call number does is sort the material on the shelves into some kind of logical subject order. In the United States, pure science books are classified in the Q's (Library of Congress System) or 500's (Dewey Decimal System). Table 1.2 shows the major subject divisions of the Library of Congress QD schedule for chemistry.

TABLE 1.2
Major Divisions of the Library of Congress Classification Schedule for Chemistry

Chemistry (General)	QD 1-69
Analytical Chemistry	QD 71-145
Inorganic Chemistry	QD 146-196
Organic Chemistry	QD 241-449
Physical and Theoretical Chemistry	QD 450-731
Crystallography	QD 911-999

Within the QD schedule are found more specific call numbers. For example, electrochemistry is classed as QD 273 or in the range QD 551-571. Other areas like biochemistry have one group of classification numbers in the QD range and others in completely different ranges, depending on the major emphasis of the work. QD 415-431.7 covers works on biochemistry, but animal biochemistry is dealt with primarily in the range QP 501-801. Appendix I is an index to many of the Library of Congress call numbers relevant for chemistry. A library in a specialized chemical company might use a much more detailed classification scheme than is available in either the Dewey or Library of Congress systems.

1.3.2 *Chemical Abstracts* Classification Scheme

The major library subject classification schemes can be used to assign call numbers to books and other library material. But there are other ways of classifying chemical knowledge in order to print references to it in a logical subject order. Table 1.3 lists the major divisions of the classification scheme used in *Chemical Abstracts* (*CA*), the world's premier abstracting journal for all fields of chemistry.

TABLE 1.3
Major Subject Divisions of *Chemical Abstracts*

Section Name	Section Numbers
Biochemistry	1-20
Organic Chemistry	21-34
Macromolecular Chemistry	35-46
Applied Chemistry and Chemical Engineering	47-64
Physical, Inorganic, and Analytical Chemistry	65-80

Books are found in a library by using the card catalog (or its online equivalent) to obtain the call number. References to particular subjects in *Chemical Abstracts* (and other abstracting services) are generally tied to an **accession number** which links the search keys in the subject and other indexes to the full entries (including the abstracts) in the main body of the work. An **abstract** is a paragraph describing the contents of the publication which has been indexed; it summarizes the most important parts of the work. (See Figure 1-1.)

93: 25540j **Formation kinetics of an amino carboxy type merostabilized free radical.** Bennett, Richard W.; Wharry, Donald L.; Koch, Tad H. (Dep. Chem., Univ. Colorado, Boulder, CO 80309 USA). *J. Am. Chem. Soc.* **1980**, *102*(7), 2345–9 (Eng). The redn. of isatin and *N*-methylisatin by the merostabilized free radical 3,5,5-trimethyl-2-morpholinon-3-yl (I) to isatide and *N*,*N*′-dimethylisatide is described. The reaction rates are 1st order in the meso and *dl* dimers (II and III) of I and zero order in *N*-methylisatin. The reaction rate is a measure of the rate of bond homolysis of II or III. Rate consts. and activation parameters are reported. The free energies of activation for homolysis of II and III in CHCl₃ are 24.6 ± 0.3 and 25.1 ± 0.2 kcal/mol, resp. The enthalpy of activation for combination of I is estd. at 4-5 kcal/mol. Measured rate consts. for bond homolysis were consistent within 1 std. deviation with the obsd. kinetics of isomerization of II to the equil. mixt. of II and III. The activation parameters in part are a measure of the effect of merostabilization on radical stability.

106: 205033s **The photochemistry of isatin.** Haucke, G.; Seidel, B.; Graness, A. (Dep. Chem., Friedrich Schiller Univ., DDR-6900 Jena, Ger. Dem. Rep.). *J. Photochem.* **1987**, *37*(1), 139–46 (Eng). Isatin does not show fluorescence or phosphorescence under any conditions. After excitation it undergoes rapid intersystem crossing to the triplet state (τ = 340 ns in deaerated benzene). This triplet state is capable of H abstraction, yielding a product analogous to pinacol, isatide. In the absence of a H-atom donor, isatoic acid anhydride is formed as the main product.

Figure 1-1
Typical Abstracts for Journal Articles in the Printed *Chemical Abstracts*

Now let's look at a reference to a book as it is classified with a Library of Congress call number, then as listed in *Chemical Abstracts* with its accession number.

QD 305 Gajewski, Joseph J.
.H5G27 Hydrocarbon Thermal Isomerizations. N.Y.: Academic Press, 1981. (Organic Chemistry; 44)

96:B67903x **Organic Chemistry, Vol 44: Hydrocarbon Thermal Isomerizations**. Gajewski, Joseph J. (Academic Press: New York, N.Y.). **1981**.

Chemical Abstracts entries are grouped in each weekly issue into one of eighty specific subject sections, yet are numbered consecutively during each six-month period of a volume. Hence, by itself the *CA* accession number tells nothing about the subject matter of the Gajewski book. It merely shows that the work can be found in volume 96 of *CA* as abstract number 67903x and that it is a book (indicated by the capital B, designating the **document type**). On the other hand, the LC number QD 305.H5 tells a person who knows the LC classification schedule that the book deals mainly with organic chemistry, more specifically, a general work on aliphatic hydrocarbons.

1.3.3 Subject Indexing

Browsing through particular areas of a classification scheme, either the shelves of a library or the pages of a classified abstracting journal or other source, is one way of locating material on a given subject. However, classifying a scientific work, whether it is a book, thesis, journal article, or patent, usually involves assigning a single place to that work in a subject scheme. There may be major aspects of the work which would tend to make a second or even third place in the classification scheme a logical place to assign it. In such cases a **cross reference** (SEE or SEE ALSO reference) may be found which leads to additional relevant information in another place in the classification scheme. Unfortunately, such cross references are not always made.

Another way to find material of interest is to use a purely subject approach. The use of subject words to describe the contents of a work is an attempt to overcome the problem of having to place only one call number on a book or assign only one accession number to an article or other work, thus placing it in only one section of a library or on only one page of an abstracting or indexing journal. Subject indexers or catalogers choose terms which create lexical paths to the work through the indexes, regardless of its location. This allows us to match our subject interests with published works. The subject words may lead to a call number for a book or to an abstract accession number and thence to an abstract. In an indexing journal, the subject words simply lead to a bibliographic description of the work. The original work or a copy of it must then be physically located in order to discover more about its information content.

Up to this point in the history of publishing, the usual method of finding information on a specific topic for which we have no previous information has been to do a search in a subject index. The process of assigning subject words to describe the information contained in a book, journal article, patent, etc. is known as **indexing**.

Indexers have several choices to make. They must decide whether all possible words are going to be allowed in the indexing vocabulary, or if only certain words (presumably the best or most commonly used or most descriptive terms) will be selected to describe the contents of the works. The first technique utilizes what we call **natural-language** or **uncontrolled vocabulary**. It is commonly found in **key-word indexes**, which are compiled by a computer. Such indexes make no attempt to show the relationships among synonyms, abbreviations, and related terms or to show a preference for one term over another. The second technique of indexing, which is restricted to the use of words selected from a preferred subject term authority list, is said to employ **controlled vocabulary**.

Indexers must also decide how far up or down the hierarchical classification scheme of a discipline to go in assigning subject words. For example, if an article is about the compound *flavan*, will the indexer also assign the name of the more general class of compounds *flavonoids*? This would allow a searcher to retrieve all works dealing with the general class of compounds without having to search for each individual flavonoid.

1.4 THE PUBLICATION PROCESS: PRIMARY LITERATURE

Let us now follow a scientific research project from its inception in order to see how people find out about the work. In the process, we will define some more of the terms which will be used in this book.

It is very rare in science today to find a solitary scientist working in isolation and carrying a research project to completion. More than likely, the scientist will be part of a research group working on a common problem. The members of the group may be in close contact with others in different geographical locations who are working on similar problems, using the same techniques, etc. The kind of informal information exchange group which these scientists form is sometimes referred to as the **invisible college**.

As the research progresses, there begin to be found reports of the results outside the confines of the research group itself. This might take the form of a seminar given in the scientist's own organization, a written progress report submitted to a funding agency, or a lecture given at a conference. When the research project is complete, there will be more formal presentations of the results, perhaps as an article or a series of articles in a conference proceedings volume or in scientific journals, through the publication of a dissertation or thesis, or the granting of a patent. A final, detailed report may also be filed with a government or private funding agency. All of these written forms of documentation are collectively referred to as the **primary literature**. This is generally the first place in the formal written communication system where people who have had no contact with the researchers during the research project can learn of the results. (Of course, there are a number of people in the invisible college who already know of the results through contacts with the researchers in the informal, largely oral, communication process.) The primary literature is sometimes referred to as the archival record of science, a record which permits many people to read it.

1.4.1 Journals

The most common form of primary publication is the journal article. A journal article in a refereed scientific journal such as *Tetrahedron* has undergone the scrutiny of editors and reviewers who see copies of the authors' manuscript. Depending on their reactions, the authors may have to make major or minor revisions to the text or may have the manuscript rejected outright. If the latter happens, the article may be submitted for publication to a journal of lesser reputation or one having less stringent editorial controls. This process of peer review is very important in maintaining the quality of scientific journal articles. There is a very definite pecking order to scientific journals, and many, many scientists strive to have their work published in the most prestigious journals. It may take a year or more between the actual completion of the research project and the publication of the results in a journal article. Some journals have been established to speed up this process and are known as **rapid communication** or **letters journals**. Still others appear weekly and are **news journals**, carrying short reports of the most significant research results months before lengthier and more detailed versions appear in journals which are published less frequently. Appendix II is a list of primary journals recommended by the Committee on Professional Training of the American Chemical Society.

In terms of their frequency of publication, scientific journals may appear weekly, twice monthly, monthly, quarterly, or even irregularly. There are thousands of scientific journals, and no one library can possibly take all of them which are relevant for the field of chemistry.

Scientific journals are not referred to as **magazines**, a term which is reserved for less scholarly works. There is a more general term, **serial**, which encompasses both journals and magazines, as well as books which are published as part of a numbered series. The term **periodical** can also be used to refer to both journals and magazines which appear regularly, but not to books in series or other serials which appear irregularly. Thus, the broadest term for a publication of this type is **serial publication** (or simply, **serial**): a publication which has a uniform title common to all of the issues or volumes, which appears regularly or irregularly at intervals of time, and for which the editors anticipated no ending point when it began. Examples of serials are *Journal of the American Chemical Society*, *Advances in Chemistry Series*, *Progress in Inorganic Chemistry*, *Chemical & Engineering News*, and *Organic Chemistry* (a series of monographs).

1.4.2 Conference Proceedings

A **conference proceedings** volume is ordinarily published as a book, but sometimes produced as a special issue of a primary journal or even as a government publication.

Collected between its covers are the lectures which were presented at a given conference (or at least the most important papers). A single editor may be responsible for the compilation of such a book (which may be published for a single conference or may be published regularly each time the conference is held). An example is the *Contributed Papers* of the XVIIth International Conference on Phenomena in Ionized Gases, published in two volumes. Since only the editor reviews the articles before publication, some people tend to place less value on conference proceedings volumes as primary sources of information. The main attraction should be their speed of publication relative to the completion of a research project. Unfortunately, publication of conference proceedings is often delayed for various reasons. Such publication delays may stretch to a year or more after the new research was presented at the conference.

1.4.3 Government Reports, Dissertations, Patents

Books and journals are relatively easy to find in scientific libraries. Some of the other forms of primary literature are not widely held by libraries and may have to be ordered from special places which make them available. The local science librarian should know how to procure copies of **government reports**, **dissertations** or **theses**, and **patents**. Many abstracting and indexing services include such publications in their coverage, so references to them are likely to be found in a subject or other type of search. Note the examples from *Chemical Abstracts* which are shown in Figure 1-2.

104:**192809f Ozonization of humic and fulvic acid (isolated from a lowland river water) and upland water: organic-by-products.** Killops, S. D.; Watts, C. D.; Fielding, M. (Water Res. Cent., Bucks., UK SL7 2HD). *Tech. Rep. -Water Res. Cent. (Medmenham, U. K.)* **1985**, TR224, 26 pp. (Eng).

83:**179348h Isatin studies and models of emetine via Reissert compounds.** Piccirilli, Robert M. (Clarkson Coll. Technol., Potsdam, N. Y.). 1975. 151 pp. (Eng). Avail. Xerox Univ. Microfilms, Ann Arbor, Mich., Order No. 75-16,941. From *Diss. Abstr. Int. B* **1975**, 36(2), 724.

106:**156491d Preparation of phenylquinazolinecarboxylic acids, their antitumor activity, and pharmaceutical compositions containing them.** Hesson, David P. (du Pont de Nemours, E. I., and Co.) U.S. US **4,639,454** (Cl. 514-259; A61K31/505), 27 Jan 1987, Appl. 692,412, 17 Jan 1985; 10 pp.

FIGURE 1-2
Typical Report, Dissertation, and Patent Entries from *Chemical Abstracts*.

1.4.4 Formats: Document Types

All of the types of primary publications discussed in the preceding sections have very distinctive looks to them. We say that they differ in **format**, with each being a certain **document type**. Indexers often code the document type for online searching. Thus, in a computer search of the *Chemical Abstracts* and other databases, a searcher may even use the format of a publication as a search key. The first document cited in Figure 1-2 (a technical report) might be coded with a T, the second (a dissertation) with a D, and the third (a patent) with a P in the online *CA*.

1.5 THE SECONDARY LITERATURE

A characteristic common to all of the primary literature is that when it first appears, it is difficult to find out about the information contained in it. The primary literature is simply not well organized from the point of view of information retrieval. Unless there is a way to discover soon after it is published, that an article or other primary work has appeared, information which is vital to a research project may be missed. That is why most scientists have developed the habit of regularly reading or scanning a dozen or so primary scientific journals which frequently publish information of interest to them. These may include a news journal or two, one or more journals which publish articles from all areas of chemistry, as well as some specialized journals which have a high percentage of articles in which they are interested.

It is the **secondary literature**, such as *Chemical Abstracts*, on which most scientists depend in order to cope with the "information explosion" of primary literature. Secondary publications (many of which are serials) exist to better organize the information in the primary literature for effective retrieval. At one extreme, the information from the primary literature is repackaged and published in the form of encyclopedias, handbooks, dictionaries, data compilations, monographs, textbooks, and treatises. The information *itself* is available in these works, organized in some manner for easy retrieval by the intended audience. At the other extreme, quick pointers to the new original literature are provided by indexing and abstracting companies and other information providers through current awareness services.

1.5.1 Treatises

A **treatise** is a multi-volume work whose contents have been organized in a logical, classified manner and which covers an entire subject field (perhaps a whole subdiscipline of chemistry). The arrangement of the material by subject classification presents the material in a form understandable by a specialist in the field. Treatises frequently include the word "comprehensive" in the title. Examples are Bailar's *Comprehensive Inorganic Chemistry* in 5 volumes and *Rodd's Chemistry of Carbon Compounds*, the second edition of which began in 1964 and which, with its supplements, had grown to more than 40 physical volumes by the end of 1989. Treatises include detailed bibliographies which document the primary literature and other sources from which the information has been gathered.

1.5.2 "Multigraphs" (Multiply-Authored Books) and Monographs

There is a type of book which has more than one author, and for which the editors have enlisted expert authors to contribute chapters on particular aspects of a new topic in science. Let's call these publications **multigraphs**. Some call them "managed texts," "composite works," or "multi-authored books". They are books written and produced by a team of authors under very close editorial control of the publisher. An example is *Practice of High Performance Liquid Chromatography; Applications,*

Equipment, and Quantitative Analysis, edited by H. Engelhardt (SpringerVerlag: Berlin, etc., 1986). This book has 26 contributors who collectively wrote the 15 chapters.

A **monograph** is a book on a particular topic written by a single author and, like a multigraph or treatise, is intended to be read by those well versed in the field. Gajewski's *Hydrocarbon Thermal Isomerizations*, cited in section 1.3.2, is a monograph. Monographs are generally considered to be secondary literature; multigraphs should also be thought of as secondary literature. Both are works on a much smaller scale and of much narrower scope than a treatise. They are usually published in one volume which is fairly comprehensive in its coverage of the subject matter. Like treatises, they tend to have the information organized in a logical (classified) manner. Also, like treatises, they have full documentation of the references to the primary literature from which the information was taken.

1.5.3 Encyclopedias and Dictionaries

Encyclopedias and **dictionaries**, while they may be broad in scope, usually present less detailed information than do treatises and monographs. They include relatively few, if any, bibliographic references to the original primary literature whose results they summarize and repackage. Dictionaries and encyclopedias are usually arranged in alphabetical order and are thus easier to use by the non-specialist than are monographs and treatises. Examples are the *McGraw-Hill Encyclopedia of Science and Technology* and Bennett's *Concise Chemical and Technical Dictionary*. It should be noted that the compilers of encyclopedias, dictionaries, and even handbooks may at times call their works by somewhat misleading titles, such as the *Dictionary of Organic Compounds*, which is really a handbook.

1.5.4 Handbooks and Data Compilations

Handbooks and **data compilations** contain data which have been taken from the primary literature and have usually been evaluated for accuracy. Examples are the *CRC Handbook of Chemistry and Physics* and *Sadtler Infrared Absorption Spectra*. The larger multi-volume data compilations are well documented, with full bibliographic references to the original literature from which the data were sifted. It would be impossible to repackage into handbooks and other data compilations all of the data which appear in the primary literature. Other types of secondary sources (for example, abstracting or indexing journals or databases) must frequently be used to lead us back to the original primary literature when a search through handbooks and data compilations has failed to turn up the needed information.

1.5.5 Reviews and Review Serials

Another type of secondary source which gives less actual information than the secondary sources mentioned above is the **review** or **review serial**. An example is the

Annual Review of Biochemistry. Many books of this type have titles which start with the words *Advances in . . ., Progress in . . .*, etc. Such a work usually looks like a book, although some review serials are published as journals, for example, *Chemical Reviews*. Review articles may also appear in a conference proceedings volume or as special articles in primary journals.

    Just as a movie or play critic evaluates an artistic work and reports an opinion along with some sketchy details of the work, so a reviewer of primary scientific literature reads the original research works and makes qualitative judgments about their contents. A **review article** is then written, which briefly summarizes sometimes hundreds of primary research articles or other works and gives the bibliographic information needed to find the original works. A reviewer usually stakes out a period of time (perhaps all of the articles published on a given topic in a year) and tries to succinctly summarize in a few sentences for each work reviewed the most important points about the major advances in that period. The review may appear one to two years after the publication of the original research results. Nevertheless, reviews provide a compact source with which to quickly scan a large body of literature. They are very popular with scientists as a way of gathering information from the primary literature, especially when a new research project is begun.

1.5.6 Bibliographies

Another secondary tool which allows a large volume of primary literature to be covered by consulting one secondary work is a **bibliography**, a compilation of bibliographic citations on a particular topic. An example is Ohno and Morokuma's *Quantum Chemistry Literature Data Base; Bibliography of Ab Initio Calculations for 1978-80*, which has had supplements issued each year. Aside from the fact that the primary publications were considered good enough to be chosen for inclusion in the work, no evaluative information about their contents is usually given in a bibliography. Also, no general statements can be made about bibliographies with respect to the time period of their coverage. It is up to the compilers of such works to decide how nearly comprehensive they want the bibliography to be. While bibliographies do not exist on all topics, a bibliography can be a tremendous time saver if one can be found in an area of interest.

1.5.7 Current Awareness and Retrospective Searches: Indexing and Abstracting Services

Finding out about new research one to two years or more after it appears is not very good from the standpoint of currency of knowledge. Even keeping up with a dozen or so primary journals (a typical number of journals for most scientists to routinely read or scan) makes it very likely that relevant information will be missed. This information might be published in any of the thousands and thousands of other scientific journals, not to mention information contained in books, patents, reports, and dissertations. In order to gauge the enormity of this problem, consider that the abstracting journal *Chemical Abstracts* includes in its coverage each year approximately half a million new scientific research works! What is even more astounding is that only about one in six publications considered for inclusion in *CA* is actually selected. Thus, scientists often supplement their reading of primary journals by scanning various types of publication lists or services which are specifically designed to promote current awareness of new research results.

Most scientists depend on various **indexing or abstracting journals**, such as *Chemical Abstracts* or *Science Citation Index*, or specific **current awareness journals**, such as *Chemical Titles* or *Current Contents*, to help them. Since it is the goal of current awareness services to be relatively up-to-date in their coverage of the primary literature, they must appear much more frequently than the types of secondary sources previously discussed. Thus, they are published no less frequently than once a month, and in many cases, are published weekly.

There are hundreds of abstracting and indexing services in science, some of them covering a whole discipline such as *Chemical Abstracts*, some multidisciplinary, like *Science Citation Index*, and others covering a subdiscipline, like *Analytical Abstracts*. Spinoff products, for example, the current awareness series *CA Selects* from *CA*, are subsets of the abstracting and indexing journals. Such publications are called **standard interest profiles**. Even more sophisticated individualized computer methods of maintaining current awareness are available through the technique of **SDI, Selective Dissemination of Information**. SDI searches concentrate on the most recent additions to an abstracting or indexing database in order to produce frequent customized bibliographies according to a given researcher's or research group's interests.

Naturally, as the abstracting and indexing publications continue to be produced over the years, the earlier volumes can be used for a **retrospective search** of the older literature. For example, *Chemical Abstracts* goes all the way back to 1907, and *Science Citation Index* covers the literature to 1945. Cumulative subject and other types of indexes assist in retrospective searches. Computer-readable versions of the abstracting and indexing services have made retrospective searches even easier.

1.6 THE RELATIONSHIP OF THE SECONDARY LITERATURE TO NEW PRIMARY LITERATURE

This completes the types of printed secondary sources to be covered in this book. Figure 1-3 shows the relationship of the secondary literature to new knowledge.

Let us take as time zero in Figure 1-3 the date when new knowledge is published as formal primary literature, whether published as a journal article, patent, dissertation, report, or other form of primary literature. Then the bars show the normal ranges of time *after* the appearance of the primary publication when the various types of secondary publications either make reference to that new knowledge or actually repackage it for easier access. Thus, it might take two to five years or more for a major new discovery to show up in handbooks, dictionaries, encyclopedias, monographs, multigraphs, or treatises. A review which makes mention of the discovery will likely not appear for one to three years after its publication in the primary literature. Abstracting services will require three to nine months on the average to both index and abstract the publication before the item might appear in a journal like *Chemical Abstracts*.

14   CHEMICAL INFORMATION SOURCES

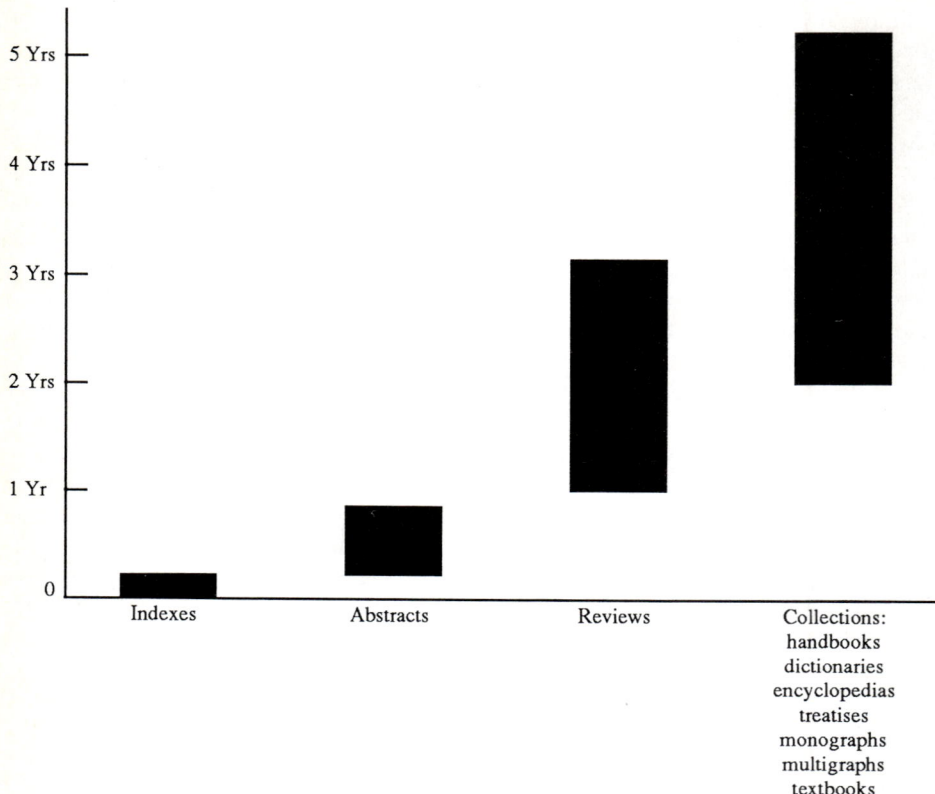

Figure 1-3
Time Lag in Secondary Literature Coverage of New Knowledge

For the quickest access, the most recent issues of indexes and current awareness services based on indexing databases are used, providing access at or near the time of publication up to three months after its appearance. Thus, an entry is likely to appear in *Science Citation Index* soon after the publication of the primary journal article, so current awareness from that source is quite up-to-date. Current awareness services based on *abstracting* databases will, in general, take longer to identify the new primary literature, appearing three to nine months after publication. This is true because the process of writing an abstract for a primary document requires considerably more time than the act of indexing that document.

1.7 COMPUTER-READABLE SOURCES

In the last few decades, it has been increasingly common to find secondary and even primary works produced with a computer. The resulting database has often been made available as a file which can be searched by various techniques to retrieve needed information. It is helpful to think of a database as being one of five types: bibliographic,

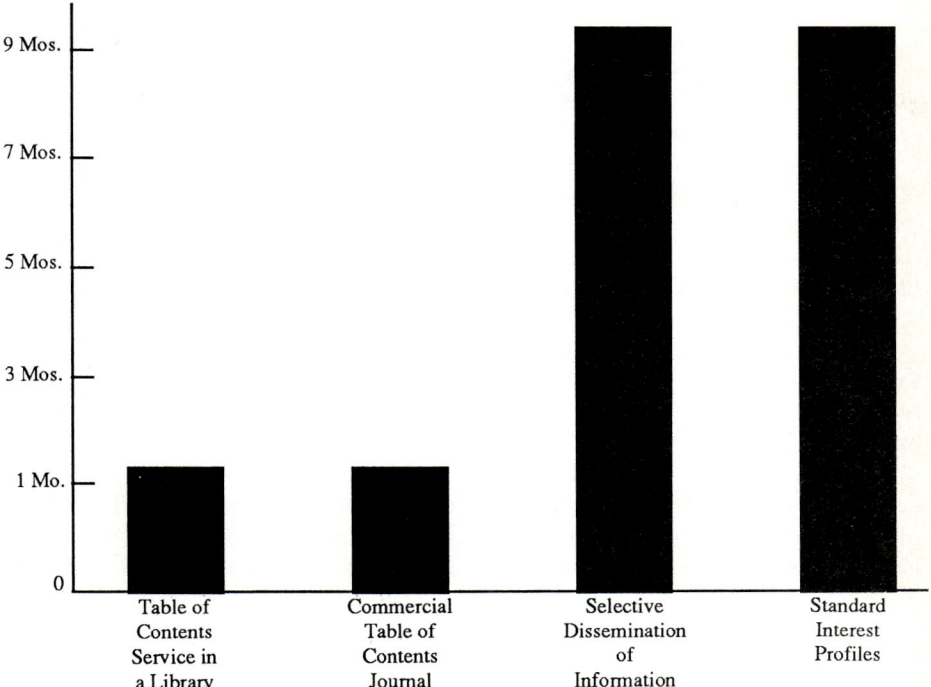

Figure 1-4
Time Lag in Current Awareness Services' Coverage of New Knowledge

numeric, text, dictionary, or directory.[2] (The last four are sometimes collectively referred to as **non-bibliographic databases**.) Let us see how these are related to the printed primary and secondary sources previously discussed.

1.7.1 Bibliographic Databases and Document Access

A **bibliographic database** is one that contains the information necessary to find a book, journal article, patent or other primary literature in a library or to obtain it from some other source. It gives the searcher a list of bibliographic references or citations. A bibliographic database can be searched by the same types of **search keys** which are found in the indexes of the printed abstracting or indexing journals. Thus, searches by author, subject term, molecular formula, or other search approaches are possible. There are hundreds of bibliographic databases available now, many of them in science and technology. They exist for both indexing and abstracting journals. Some even allow the texts of the abstracts themselves to be searched.

There are several advantages to searching a bibliographic database. First is that the number of index terms assigned to a given primary scientific work is frequently many more in the database than in the corresponding indexes of the printed product.

Therefore, there are potentially more paths to lead to the information. Second, a relatively large span of publication dates can be covered, often in a single search, in a very short period of search time compared to a manual search. Third, it is possible to combine concepts in ways which would be impossible or very difficult in a manual search of the printed literature. For example, the computer can be told to retrieve all works on HPLC (High Performance Liquid Chromatography) printed in English, but none of which were published as patents. Finally, there are some bibliographic databases which either do not have printed counterparts or which combine in one database the contents of several distinct printed abstracting or indexing services. Since bibliographic databases exist for all of the printed abstracting or indexing services mentioned thus far, a customized bibliography could be obtained by searching a given database. In addition, standard interest profiles or SDI services for current awareness are available from almost all bibliographic databases. The Chemistry Reference Sources Database includes most of the bibliographic, as well as other types of databases, which a chemist might find of interest.

It is important to realize that another step needs to be taken to actually get the desired information when citations have been obtained from a bibliographic database (or a printed bibliography or one which has been manually compiled from a printed abstracting or indexing service, for that matter). The bibliographic information from the search must be translated into the call numbers (or other types of location information) which allow the original primary source documents to be found.

It is common practice nowadays to refer collectively to all of the primary scientific works which are covered by abstracting or indexing services as **documents**, a term which previously was reserved for publications of official government bodies. If a library does not have a needed primary source, the material may be obtained through a **document delivery service**. One example of such a service is **interlibrary loan**, a resource-sharing library service which supplies either the original work or a photocopy from another library. Many other document delivery services are available through the vendors of online database services.

In summary, bibliographic searching results in a bibliography which leaves the desired information a step or two away. We say that it provides ''pointers'' to the original sources which themselves may or may not actually contain the answers to a search problem.

1.7.2 Numeric Databases

A numeric database, on the other hand, gives actual answers. In science, **numeric databases** are the computer-readable equivalents of handbooks and large data compilations. If a numeric database is searched for the physical constants of a particular compound, a match produces the actual values for the melting point, boiling point, etc., not pointers to articles or other primary documents where the answers *may* be found. Furthermore, some numeric databases allow the use of *numerical data* as the input values which are searched. If the melting point, boiling point, and perhaps an IR spectrum for an unknown have been measured in the laboratory, inputting those data may reveal the actual compound which has such attributes (or a list of candidate compounds which best fit the data). While more and more numeric databases are becoming available in science and technology, data are usually found only for the most common compounds.

1.7.3 Text Databases

A third type of database is a **text database**. Since many journal articles and even some encyclopedias are produced with the aid of a computer, the resulting database can be searched just as a bibliographic database would be. At this point in time, there is one major difficulty with text databases. Most of them do not handle very well such things as scientific symbols, graphs, tables, chemical structural formulas, etc. Thus, they are limited in the amount of information which can actually be retrieved.

Journals published by the American Chemical Society and other publishers are available for online searching in text databases. Text databases also exist for some encyclopedias, for example, the *Kirk-Othmer Encyclopedia of Chemical Technology* and the *Academic American Encyclopedia*. However, it is usually only the textual words and other index terms which can be searched and retrieved in a text database. Some work is being done to integrate graphics characters into text databases, but for the foreseeable future, not many databases will truly become full-text, fully searchable counterparts of printed sources. Nevertheless, more and more primary and secondary sources are becoming available in this format. Text databases make it possible to bypass even an online version of an abstracting or indexing journal and go directly into the text of a primary document to retrieve needed information.

1.7.4 Dictionary Databases

A fourth type of database has been called a **dictionary database**. In most cases, this is really a misnomer. The majority of the dictionary databases turn out to be authority lists of terms which are used in searching the controlled-vocabulary files on a given system. Thus, the MESH (Medical Subject Headings) file of the National Library of Medicine (NLM) really exists to help select the right terms for a medical subject search. Likewise, various chemical dictionary files allow the searcher to go from a known common name, trade name, semi-trivial name, or even a molecular structure for a chemical substance to the index name or registry number for that compound. These are not really dictionaries and in no sense are such dictionary databases equivalent to printed chemical dictionaries. On the other hand, the text databases of encyclopedias or journals, except for the graphics, duplicate the printed sources.

1.7.5 Directory Databases

The fifth and last type of database is a **directory database**. We have not mentioned printed directories up to now since they are not closely related to either the primary or secondary literature. A **directory** simply leads to a person, organization, company, etc. which you need to contact.

1.8 OPTIONS FOR DATABASE SEARCHING

When bibliographic databases first became available, quite a number of different organizations leased the tapes of interest to them and developed software to search the databases on their own computers. Most of the early usage was for SDI current awareness searching. However, a few organizations made a point of integrating the data from the new tapes with older data and provided searches for a fee to customers outside their organizations. These **vendors** of online services were instrumental in popularizing the use of databases. Today there are many vendors, such as STN International, DIALOG, ORBIT, and others, from which both current awareness and retrospective searches can be purchased. An advantage of searching different databases on the same vendor's system is that a uniform search software is usually used across all databases provided by that vendor. However, there may be widely varying costs for searching, depending on the vendor and the database selected.

An option which appeared in the latter part of the 1980s is to purchase a database in **CD-ROM** (compact disk—read-only-memory) format. The CD-ROMs can be searched with a microcomputer to which a player is connected. Many of the advantages of online computer-based searching on a vendor's remote computer are possible with CD-ROMs, including access to more data than is found in the printed counterparts and the capability to combine concepts in a search. However, the storage capacity of the CD-ROM (and other formats of laser disks), while impressive, is not high enough to hold a large database on a single CD. The CD-ROM version of the *Science Citation Index* for just a single year is provided on one compact disk. In order to search several years, the user would have to change the data disks a number of times. With an online search, the user can often opt to stay in a single file and go back in time perhaps ten to twenty years or more. Another drawback to CD-ROM databases is that they are generally not updated as frequently as their online counterparts. With CD-ROM, it is typical to receive the update on a quarterly basis, whereas an online database may be updated as often as weekly. Nevertheless, the CD-ROM option for acquiring databases is becoming increasingly popular, since it allows a known cost to be programmed into an annual budget. Fixed prices are set for the CD-ROM products, but charges for online searching through a vendor vary with the amount of usage. Another difference is that CD-ROM workstations allow only one user to access the database at a time, whereas many users can search a remote online database simultaneously. The software to access CD-ROM databases is typically quite "user-friendly," so searching proceeds smoothly with minimal instruction required. The number of databases available in CD-ROM format will undoubtedly increase rapidly. Some vendors are offering products which encourage consultation of the CD-ROM product for older references and a search of the online remote database for more current information.

A third option which, with the computing power and storage capacity of modern computers, has once again become attractive to individual organizations is to lease the database and mainframe computer software and to mount the files locally. Some of the major vendors are encouraging this. For example, the BRS On-Site option provides both the software and formatted databases tailored to an organization's computer system. This has the advantage of fixed cost for the searching, but requires a significant investment in computer equipment and local staff.

1.9 TERTIARY SOURCES: GUIDES TO THE LITERATURE

Some people call reference sources such as directories, lists of periodicals, buyers guides, biographical sources, etc. "tertiary" sources. However, we will simply consider them other reference works which cannot be related in a chronological sense to the primary and secondary literature and deal with them at appropriate places in this book.

There is a type of literature which explains, teaches you how to use, and leads you through the vast array of primary and secondary scientific literature. These are **guides to the literature**, books or other works written or compiled by scientists, librarians and others to help you use the literature. Since they cover both primary and especially secondary sources, they are often called **tertiary sources**. They are extremely important in mastering the scientific literature of a given field. This book is an example of a guide to the literature. Several others will be discussed in the next chapter.

1.10 SUMMARY

There is a continuum of information exchange from the inception of a research project culminating in reports of the research published in the primary literature (patents, conference proceedings, journal articles, reports, and dissertations). From the point where new scientific information joins the accumulated archival record of science in the primary literature, various secondary sources exist to make that scientific knowledge more accessible (indexes, abstracts, reviews, encyclopedias, dictionaries, handbooks, data compilations, treatises, monographs, and multigraphs). Current awareness services strive for rapid dissemination of the new research results, while indexing and abstracting services provide additional, fairly rapid access to the primary literature. Many of them use controlled-vocabulary subject indexing and publish abstracts which summarize the research. A broader, though less timely overview of new research is undertaken by those who write review articles. Finally, the most important research results are repackaged and made more easily accessible in various types of collected works such as encyclopedias, dictionaries, treatises, handbooks, and large data compilations. These present the most important data and research results from the primary literature several years after their original publication.

In the production of all of these secondary literature sources, the computer has been used to one degree or another, thus providing alternative ways of searching for information with computer databases. It is important to understand the function of each type of primary and secondary literature source and to be able to recognize the various formats in which they are published.

Since both the primary and secondary literatures are so voluminous, a third type of reference source has been developed to help master their use. Such tertiary sources are guides to the primary and secondary literatures. In the ensuing chapters, we will learn much more about the inter-relationships of the various types of scientific publications and databases and learn how to select the most appropriate ones for solving particular information problems in chemistry.

1.11 SELECTED READINGS

Rusch, Peter F. "Introduction to Chemical Information Storage and Retrieval." *J. Chem. Educ.* **1981**, *58*(4), 337-342.

Heller, Stephen R. "Online Chemical Information." *Chem. Int.* **1987**, *9*(4), 136-138.

Warr, Wendy A. "Online Access to Chemical Information: A Review." *Database* **1987**, *10*(3), 123-127.

Antony, Arthur A. "The Literature; Becoming Part of It and Using It." In *The ACS Style Guide: A Manual for Authors and Editors*; Dodd, Janet S., Ed.; American Chemical Society: Washington, DC, 1986; pp 159-184. [16]

REFERENCES

[1] Jones, David. "The Boundaries of Chemistry." *New Sci.* **1979**, *83*(1170), 661-663.
[2] Conger, Lucinda D. "Type of Databases—Some Definitions." *Database* **1984**, *7*(1), 94-95

CHAPTER 2

DATABASE DIRECTORIES, GUIDES TO THE LITERATURE, AND SEARCH STRATEGIES

2.1 INTRODUCTION

In this book, we will use the term **guide** in a very broad sense to refer to books and other printed, audio-visual, or computer-based products which help us to find or learn how to use chemical information sources. Thus, some directories are considered guides in this book. Since most guides deal with primary or secondary sources, they are sometimes called **tertiary sources**.

There are sound reasons to use guides to the scientific or technical literature. A guide may contain descriptions of primary or secondary sources or search tips which can lead to information on a specific topic. To that end, some guides include detailed instructions on how to use particular reference tools or computer-readable sources; others simply list many different sources. Guides are not restricted to the printed chemical literature. They may also lead us to computer-readable sources or even explain the techniques used in creating computer-based secondary chemical information sources.

A guide does not have to be published as a separate book. Some are journal articles or chapters in other books or encyclopedias. In fact, the instructions for use in most printed secondary reference sources can be thought of as "mini-guides".

Frequently, these instructions are found at the beginning of the work. For an abstracting or indexing journal, such detailed instructions on usage can often be found in the first issue published in a given calendar year. Scanning the introductory section of each reference work used is a good habit to form, even if the work is a familiar one. Over time, significant changes occur in the organization, indexing policies, and coverage of continuing reference tools.

Guides to the literature can be of tremendous help in selecting and learning to use appropriate sources to solve information problems. They should be part of an overall strategy for searching the printed and computer-readable reference sources. Other essential features of a plan of attack or **search strategy** for solving chemical information problems are presented in this chapter.

2.2 USER AIDS FOR COMPUTER-READABLE DATABASES

Unfortunately, good computer-readable databases are not always accompanied by good instructions for their use. Such documentation may come from two distinct sources.

TABLE 2.1
Major U.S. Vendors of Databases Relevant to Chemistry

Name	Address	Phone
BRS Information Technologies	8000 Westpark Drive McLean, VA 22102	800-468-0908 703-442-0900
Chemical Information Systems, Inc.	7215 York Road Baltimore, MD 21212	800-247-8737 301-321-8440
DIALOG Information Services, Inc.	3640 Hillview Avenue Palo Alto, CA 94304	800-334-2564
NASA Scientific & Technical Information Facility	P.O. Box 8757 Baltimore/Washington Airport, MD 21240	301-859-5300 301-621-0100
National Library of Medicine	8600 Rockville Pike Bethesda, MD 20894	800-638-8480
ORBIT Search Service (formerly SDC)	8000 Westpark Drive McLean, VA 22102	800-421-7229 703-442-0900
Numerica/TDS	10 Columbus Circle New York, NY 10019	212-245-0044
Questel, Inc.	5201 Leesburg Pike Suite 603 Falls Church, VA 22041	800-424-9600 703-845-1133
STN International	2540 Olentangy River Rd P.O. Box 02228 Columbus, OH 43202	800-848-6533 614-447-3698
Wilsonline H.W. Wilson Company	950 University Avenue Bronx, NY 10452	800-622-4002 212-588-8400

The database producer will sometimes publish detailed information on the use of the product, in some cases, even providing training manuals and learning files at low cost. **Database producers** are companies like Chemical Abstracts Service (CAS), the Institute for Scientific Information (ISI), Biosciences Information Service (BIOSIS) and many others which produce printed and/or computer-readable sources of information. In recent years Chemical Abstracts Service has developed some very good online user aids. Examples are *An Introduction to Searching CAS Online Using the Learning Files* [137] (1985) and *Getting Started in CAS ONLINE* [1886] (1988). These present basic instruction in online searching of the *Chemical Abstracts* database as it is available through the vendor STN International. Guides of this type which are specific to given reference works or databases will be noted in some of the following chapters as relevant.

Most of the manuals and user aids for online searching are published by the database vendors. **Database vendors** are companies which have leased databases from the producers and make them available for searching at a cost through their own large computer facilities. Table 2.1 lists the major database vendors for chemistry.

The vendors generally publish search manuals which contain details of the various features of their search systems. In addition, they routinely compile catalogs and lists containing descriptions of the databases. Examples of the database records and searching hints are included in some of these materials.

Many of the vendors in Table 2.1 offer the same databases. For example, several of them offer an online version of *Chemical Abstracts*, and NLM has a partial file of *CA*. It is a good idea to find out which database vendors are available in your organization and learn what user aids the vendors and producers have developed for databases of interest.

2.3 DATABASE GUIDES AND ONLINE AIDS FOR DATABASE SELECTION

With the range of databases available today, it would be easy to overlook a potentially very useful source. Thus, it is wise to survey the available databases before attempting a search. This is particularly true for searches on topics which are interdisciplinary in nature. There are a number of printed and online sources of help.

Oldest of the database directories is *Computer-Readable Databases: A Directory and Data Sourcebook* [150] (annual) which was published in its first edition in 1976. The latest edition covers all types of databases, including many databases in science, technology, and medicine. This is one of the most nearly comprehensive of all directories which cover publicly-available databases in computer-readable form. Over 4,200 entries in the directory give information on the size of the databases and the average number of items added each year and tell which database vendors make them accessible. All entries are annotated with a paragraph or two of descriptive material. Information on the availability of user aids and the types of indexing, coding, or classification used in the creation of the database is also found in the directory. There are several dozen broad controlled-vocabulary subject categories and hundreds of uncontrolled-vocabulary keywords which provide subject access to the database. The directory is available online on DIALOG.

Cuadra/Elsevier's *Directory of Online Databases* [144] (quarterly) is comparable to the previously-discussed directory in its coverage of databases and has the advantage of more frequent updating of the printed version. A subscription buys four issues per year, two complete main issues and two updates. In addition to the subject indexes, the *Directory of Online Databases* has a Telecommunications Index to match the networks which can be used for online searching to the over 400 online vendors around the world. The *Directory* is available online in the United States through Questel and ORBIT.

The *Data Base Directory* [141] (1984) is sponsored by the American Society for Information Science. Emphasizing databases available in the United States and Canada, the *Data Base Directory* describes approximately 2100 databases. This is considerably less than the number listed in either of the two directories discussed above, which give extensive coverage to databases outside North America. There are numerous other directories of databases, almost all of them in printed format only. Of those, Hall and Brown's *Online Bibliographic Databases* [145] (1986) is noteworthy for its coverage of the more important English-language online bibliographic databases, excluding only those which deal with patents. Less than 200 databases are covered in the work.

```
?ss flavonoid?
204: CA SEARCH 1967-1987 UD = 10616
    83 FLAVONOID?
205: ONTAP BIOSIS PREVIEWS
    32 FLAVONOID?
208: ONTAP COMPENDEX
    2 FLAVONOID?
210: AGRICOLA - 1979-87/MAR
    15 FLAVONOID?
213: ONTAP INSPEC
    0 FLAVONOID?
231: ONTAP CHEMNAME(tm) (26,696 Substances)
    0 FLAVONOID?
250: ONTAP CAB ABSTRACTS
    56 FLAVONOID?
251: FSTA - 69-87/MAY
    31 FLAVONOID?
254: ONTAP MEDLINE
    5 FLAVONOID?
272: ONTAP EMBASE
    19 FLAVONOID?
279: ONTAP/CLAIMS
    1 FLAVONOID?
280: ONTAP WORLD PATENTS INDEX
    1 FLAVONOID?
294: ONTAP SCISEARCH (Copr. ISI 1985)
    11 FLAVONOID?
```

FIGURE 2-1
Search on DIALINDEX for ONTAP Learning Files with Information on Flavonoids.

Some of the vendors have included a database selection feature in their search systems. This allows a global search of all databases or perhaps clusters of databases in certain broad subject areas which are available on the respective systems. The result is a list which shows the number of times a subject term or phrase (or combinations of search terms from the subject and other fields) has resulted in a **hit** or match in the databases listed. DIALOG's DIALINDEX [143] gives access to the documents indexed on a single term, a multi-word phrase, a prefix-coded field, or a full search statement of up to 240 characters. BRS's CROS file [140] does essentially the same thing. The user chooses a file category (for example, all databases, or just the life sciences or the physical sciences databases) before running the search. On ORBIT, the corresponding search is done in the Data Base Index [142] (file DBI).

Figure 2-1 shows the results of a search on DIALINDEX. Thirteen practice files were searched. Of the databases identified, the *Chemical Abstracts* database with 83 references has the most information on this class of chemical substances. At this stage, we do not have a bibliography of articles or other documents on flavonoids, simply a good indication of which of the many databases on DIALOG should be searched in order to compile such a bibliography. In that sense, the search has served the same purpose as consulting a printed guide to the literature: to narrow the range of possible sources to the best or to those most likely to have the answers.

Search **INDX = "DATABASES DIRECTORY"** or **DATABASES AND DIRECTORY** to find additional examples of database directories in the Chemistry Reference Sources Database (CRSD). There are over 150 databases of interest to chemists included in the CRSD. Each entry lists the vendor(s) which makes the database available for online searching.

2.4 COMPREHENSIVE CHEMISTRY GUIDES

The most common publication format for a guide to the literature in science and technology is a printed book such as this one. However, guides to the literature can also be found as journal articles, encyclopedia articles, audio-visual presentations, and even microfiche. Below are discussed some comprehensive guides to the chemical literature in book and other formats. Search the Chemistry Reference Sources Database (CRSD) for GUIDE AND [NAME of a discipline] to get a fuller listing of guides to the chemical and other scientific literatures. For example, a search with the strategy **GUIDE AND MEDICINE** produces at least six references. Many of the guides cover the same basic printed secondary reference tools. Only the unique aspects of each guide will be commented upon in this chapter. Guides which deal with a subdiscipline of chemistry or a particular reference tool or information problem will be noted in other chapters.

Communication, Storage and Retrieval of Chemical Information [24] (1985) is derived from a September 1982 British conference. The book of updated conference papers, although not constituting a guide to the literature in the traditional sense, really does an admirable job of surveying the area of computer-based chemical information retrieval. There is even a very good chapter on classical methods of communicating non-structural chemical information. Computer-based chemical information systems are the main subject of the book, however. Since the work is the product of British authors, there is considerable information about British and European online services. In the book the term **databank** is used to designate a machine-readable collection of *factual* information. This is a commonly used term. (Related terms are non-bibliographic database, source database, and numeric, text, dictionary, or directory database.)

The ability to search chemical information databases by using a chemical structure or substructure has become very important in chemistry. Hence, four of the eleven chapters in *Communication, Storage and Retrieval of Chemical Information* are concerned with various aspects of chemical structure searching. Computer science, theoretical chemistry, and chemical information science coalesce in the technique of **structure-activity relationships** which correlates physical, chemical, or biological properties with structural properties of a molecule. Sophisticated mathematical techniques such as multiple regression and discriminant analysis are used in structure-activity relationships. Computer-aided synthesis design receives considerable attention in the chapter "Techniques of Structure Manipulation". Many of the most important acronyms in chemical information science are listed in the appendix "Glossary of Acronyms, Trade and Product Names". The second appendix contains the addresses of a number of organizations which are involved in the storage and retrieval of chemical information via computers.

Mellon's *Chemical Publications; Their Nature and Use* [10] (1982) is a classic work now in its fifth edition. The book is divided into two major sections. Part 1 is entitled "Publications: Kinds and Nature". It is significant that this first part covers 256 of the 364 pages of descriptive material in the book. Part 2 deals with "Publications: Storage and Use". There is a brief chapter in the second part on libraries and information centers as well as longer chapters on both manual and computer searching. A final chapter contains 28 pages of library problems, ranging from searching for physical constants to preparation of a bibliography.

Mellon includes some historical information on the development of chemistry and chemical publications and follows a classical division of primary, secondary, and tertiary sources. In the category of tertiary sources, he lists compilations of information about chemists as well as works which help in the search for information on the production of and trade in chemicals and chemical equipment. One of the strengths of Mellon's book is that it provides extensive lists to illustrate the various formats discussed, such as reviews, bibliographies, treatises, etc.

Wolman's *Chemical Information; A Practical Guide to Utilization* [13] (1988) emphasizes real search problems and their solutions. Consequently, the chapter subheadings tend to reflect approaches to information problems: locating material in the library, locating melting points and/or boiling points, locating sources of fine chemicals, designing a synthesis, etc. A separate chapter on "Structural and Substructural Searches" covers formula indexes, chemical name indexes, and various notation systems. Over 60 pages are devoted to information on "Obtaining Numerical Data," the longest chapter in the book. Spectral, crystallographic, thermophysical and thermochemical, kinetic, and solubility data are included.

Skolnik's *The Literature Matrix of Chemistry* [12] (1982) is in some respects a rich historical work and in others a detailed bibliography from which to select potential reference tools. The bulk of Skolnik's book is organized by format (books, encyclopedias and treatises, numerical data compilations, patents, etc.). He does not include the dates of publications in the lengthy bibliography in Chapter 1 which lists reference works and books. Tables 5 and 6 are devoted to journals of the 17th-19th centuries. Table 7 includes currently-published chemical journals by subdiscipline of chemistry and even lists journals of chemical interest in industrial areas such as cosmetics, explosives, and the food industry. A large number of abstracting and/or indexing services are either discussed or listed in the work. The book concludes with two historical chapters emphasizing the major figures in the history of chemistry, with brief sketches of many American chemists.

Use of Chemical Literature [5] (1979), edited by R.T. Bottle, is a multigraph published in Great Britain. Hence, the British aspects of technical information are discussed more fully than are the American aspects. The book includes a useful chapter entitled "Translations and Their Sources, With Special Reference to Russian Literature". The literatures of subdisciplines of chemistry are treated in chapters on inorganic chemistry, nuclear chemistry, organic chemistry, and polymer science. A historical chapter outlines the most significant chemical achievements in successive chronological periods and includes an appendix which lists books on the history of chemistry and biographies of chemists. There are also appendices to the book itself which contain problems for library work and notes on their solutions.

Antony's *Guide to Basic Information Sources in Chemistry* [4] (1979) is essentially an annotated bibliography of sources. Earlier guides to the literature are discussed, followed by chapters organized by the formats of materials. "Guides to Techniques" lists works on analytical techniques, preparative techniques, data handling, formularies, and instrumentation. Antony makes a useful distinction by grouping reference works on biographies and directories of people in one chapter, while placing product, service, and company directories in another. The final chapter, "Chemical Information Search Strategy," contains 11 case studies of people with information needs and the sources used to solve their problems.

Maizell's *How to Find Chemical Information: A Guide for Practicing Chemists, Educators, and Students* [9] (1987) was written by a former industrial chemical information specialist. Therefore, the author pays close attention to some of the aspects of the chemical literature which are especially important in industry. Safety and related topics as well as chemical marketing and business information sources receive considerable coverage in this guide. There is even a brief chapter in Maizell's book on **process information**, the kind of information which tells how individual chemical products are best made in full-scale commercial operations. Document delivery services are of great interest to industrial libraries since their collections may be quite comprehensive in very narrow, specialized areas, but not very broad in scope. Consequently, Maizell has included some useful tips on obtaining material which might not be found in a given library. The three introductory chapters, while quite short, offer some valuable sketches of the flow of information and communication patterns in chemistry as well as tips on how to formulate an effective search strategy to solve chemical information problems. Finally, about one-sixth of the book is devoted to two chapters on the history, development, and use of *Chemical Abstracts*.

Woodburn's *Using the Chemical Literature; A Practical Guide* [14] (1974) is an older work which, nevertheless, has much useful material. In the chapter on nomenclature-structure correlation, the author describes the major systems used in this century for formula indexes, as well as the indexes of ring systems which have been developed.

Wilen's *Use of the Chemical Literature; An Introduction to Chemical Information Retrieval* [15] (1978) is a collection of audiotapes with a printed manual and exercise booklet. The 281-page manual has copious examples of the items presented in the tapes. Wilen's exercises are designed to expose the student to:

1.) retrospective searches, both for time-consuming exhaustive searches and for quick overviews or factual reference questions. (These are sometimes called "reconnaissance reading," searches for background information or facts using dictionaries, handbooks, treatises, reviews, monographs, and especially encyclopedias.)

2.) data searches

3.) the use of primary literature, current awareness

4.) the use of tertiary sources (including lists of periodicals and scientists, which Wilen's definition of tertiary sources encompasses).

Each of the 60 exercises is accompanied by a detailed answer and explanation of the search path recommended by the author.

A complete package of guides and other user aids is available in Atkinson's *Chemistry: A Learning Package on Sources of Information* [133] (1981). Included are tape/slide programs on "Finding Information in Chemistry" and "How to Use Chemical Abstracts," as well as audiocassette programs on other chemistry reference works, posters, and a student handbook.

Additional practice in the use of the chemical literature can be had by working through Gorin and Dermer's *Library Exercises in Chemical Literature* [17] (1986). Written as a manual for an actual course in chemical information science, the book advocates extensive practice as the most effective way to acquire skill in literature searching. Literally hundreds of different problems are included to test the mastery of the tools and services which are briefly presented in the manual.

Peck's *Chemical Industries Information Sources* [11] (1979) covers both industrial chemistry and chemical engineering, as well as the engineering aspects of related areas like agriculture, biosciences, food sciences, and **materials sciences** (plastics, metals, ceramics, rubbers, and textiles). Nuclear engineering, paper and pulp engineering, and petroleum engineering comprise the final three chapters, all of which contain extensive lists of both primary and secondary sources.

Dorman's *Chemical Industries: An Information Sourcebook* [347] (1988) is a guide to over 550 English-language handbooks, indexes, abstracts, encyclopedias, directories, etc. in many areas of industrial chemistry. The author has selected a "core library collection" of 158 most important research sources.

2.5 SEARCH STRATEGY FORMULATION

When there is a need to find information, the natural tendency is to try first the sources which are close at hand. It has been shown that most scientists will search their own personal files or private libraries before trying other routes to information. Frequently, the second source they turn to is an **information gatekeeper**, a trusted colleague who is known to possess a tremendous store of facts and references.

Libraries and information centers rate farther down on the scale of potential sources. One reason might be the perception that they are just too difficult to use. The sheer volume of material in a chemistry or science library can be overwhelming. Thus, it is especially important for the user of a library and its services to have a clearly defined search strategy and to consult the guides to the literature in developing that strategy.

A correct **search strategy** is an appropriate path which results in the completion of a library research project, in short, leads to the answer. The searcher must always bear in mind the kind of information sought and the sequence in which the sources of information should be consulted. To that end, the guides to the literature discussed in section 2.4 can be of invaluable help.

For any search some basic questions must be answered. These include:
- What is the nature and type of information needed?
- How deeply into the topic is it necessary or desirable to go?
- How large a time span should be covered?
- Are there certain languages of publication or formats of the primary or secondary sources which can be eliminated or in which the search should be concentrated?

Obviously, the search strategy will vary depending on the type of information sought. Quick searches for a particular physical or chemical property, for a method of testing or analysis, or for a way to make a particular substance will take different routes than will a detailed search which seeks to comprehensively review the past literature on a topic. In general, a search for information on a particular compound is less demanding than a search for a technique of measurement, theoretical concept, etc. The tools which have been developed to deal with compounds are now well defined and very sophisticated.

These are the steps to be taken in a search for information:
1. Precisely state the information need and define the main concepts, including a statement of why the information is needed and what time span should be covered.
2. Select the search terms or other search keys.
3. Categorize the information need, select the kinds of library sources most likely to have the answers, and list the appropriate secondary sources in order of the likelihood of satisfying the search.
4. Locate the items and search them until the answer is found or you are satisfied that no answer is available.

2.5.1 Statement of the Information Need

If necessary, the topic should be clearly defined using an encyclopedia article or chapter from a textbook or other secondary source. This type of background reading helps to delineate the major concepts. A number of the later chapters in this book cover sources useful for background reading, including reviews. It is important at this stage to concisely state what information is needed and to sort out the knowns from the unknowns. A list of the major concepts should be formulated and any restrictions on time period, languages, or formats of publication should be clearly stated.

2.5.2 Preliminary Selection of Search Terms and Other Search Keys

A detailed list of subject terms or other search keys (author or organization names, molecular formulas, etc.) should be developed. Many search topics involve terminology which ultimately calls for vocabulary choices by the searcher. In step 4 it will be necessary to match the search terms to those in the available indexes.

A card catalog can be thought of as an index to books and other formats of material. However, it is an index in most libraries only at the broadest level of access, for example, the names of the authors or editors, the titles of books and serials, and a few subject terms to broadly describe their content. (We must depend on abstracting or indexing services or other types of indexes for access at the level of chapters in books or at the level of articles in journals.) In order to use a card catalog effectively, even one which is available as an online catalog, it is often necessary to crack the code of the controlled subject vocabulary. A tool which helps in this task is a subject term authority list like the *Library of Congress Subject Headings* catalog.

2.5.3 Selection of Appropriate Secondary Sources

Much of the floundering which can occur in a library could be avoided if the searcher has a clear idea of the various formats of primary and secondary information sources, the types of information they are likely to contain, and the various access routes to the publications. Figure 1-1 shows that it does no good to look in dictionaries, encyclopedias, handbooks and the like for the most recent information. Defining when the information is likely to have appeared in printed or computer-readable secondary sources is one of the key steps in developing the search strategy. It is difficult to become familiar with the kinds of information contained in different types of sources. Nevertheless, it is a necessary skill to acquire. That is one of the aims of this book, and with practice, it will become easier to select appropriate sources for any kind of search. Listed below are the places in this guide which can help to identify productive sources for particular information needs.

FOR HELP CONCERNING:	CONSULT CHAPTER OR APPENDIX
A. Need for specific fact(s)	
1. Meaning (definition)	13
2. Structural data, numerical data	8-10
3. Historical or biographical information	15
4. Nomenclature, terminology, or symbols	5, 7, 14
5. Identification of unknown compounds	9, 10, 12
6. Safety or hazard information	12
7. Bibliographic or writing-style information	14
B. Methods ("How-to" types of searches) and techniques of measurement	9, 11-12
C. Synthesis or preparation of compounds	11
D. Comprehensive subject searches	5-9, 12, I, III
E. Publications by an author or corporate body or which have cited a known publication	4
F. Keeping up-to-date (current awareness)	13
G. Patent information	6
H. Sources of chemicals or laboratory equipment	15
I. Miscellaneous Information Needs	15

Chemical Literature Guides and Aids: Bibliography and Index [19] (1984) and the Chemistry References Sources Database will help identify other guides or sources which are specific to the relevant fields or help find other general guides to the literature.

Always list first the sources which you think are most likely to have the information. If a source has collective indexes, those should be used.

2.5.4 Search of the Appropriate Sources

For subject searches (those involving the names or abbreviations of chemical compounds, processes, techniques, equipment, etc.), the sources which have been selected may employ subject term authority lists. For any access tool which uses words as the entry points, the user must discover how the index terms are handled. Remember that a key-word index which uses uncontrolled vocabulary requires the use of synonyms, abbreviations, etc. for the same subject concept. On the other hand, a controlled-vocabulary index allows the searcher to find the unique term actually used in the index for a given concept by consulting the subject term authority list.

It is a good practice to keep a list of the sources which have been searched for a given research project. Without such a list, time can be wasted during a comprehensive search by consulting the same reference work on two different occasions without fully realizing it. For large sets which are arranged chronologically, be sure to write down the volume numbers and dates which have been consulted. Search until the answer is found or you are satisfied that no answer is available.

Knowing when to stop or admitting that the answer cannot be found is sometimes the hardest step of all. It has to be realized that despite the best search plans, some questions have no answers in the resources available. This may not always be bad. If a company intends to invest a lot of time, energy, and resources in the development of a new product, it certainly does not want a search of the patent literature to turn up the identical product. Likewise, a graduate student who has spent much time developing a research proposal does not really want to find something very similar in a search of *Dissertation Abstracts International*. A searcher simply has to be honest about the effort which has been put into a search. If the effort has been sufficient, there is nothing wrong with concluding that the answer could not be found or that the information is just not there.

2.6 SUMMARY

Even the tertiary literature offers a wide range of choices to the user of chemical information. Chemists have always prided themselves on their command of the chemical literature. However, as that literature grows larger and larger, it becomes increasingly difficult to cope with. Fortunately, a large number of database directories, guides to the literature, and other user aids have been developed over the years. The chemist today has over ten major database vendors and hundreds of potentially relevant databases to contend with. Directories of databases and computer techniques which assist in selecting the most promising databases should be used by the chemist. However, before embarking on any search for information, it is a good idea to formulate a search strategy. Four basic steps leading to a successful search are described in this chapter.

2.7 SELECTED READINGS

Houck, Michael. "A Look at the Database Directories Online." *Database* **1987**, *10*(3), 67-75.

Tenopir, Carol. "Database Directories: In Print & Now Online." *Libr. J.* **1985**, *110*(13), 64-65.

Tenopir, Carol. "Database Directories: The Rest." *Libr. J.* **1985**, *110*(15), 56-57.

Tenopir, Carol. "Database Selection Tools." *Libr. J.* **1988**, *113*(18), 52-53.

CHAPTER 3

GENERAL TECHNIQUES AND BENEFITS OF COMPUTER-BASED SEARCHING

3.1 INTRODUCTION

Online searching of computer-readable databases began to be readily accessible in the early 1970s from DIALOG and ORBIT. Since that time the online searching industry has grown into a multi-million dollar business which is extremely competitive. Such competition has given rise to very sophisticated search systems from the vendors and ever-increasing numbers of databases from the producers. Whereas the lines between these two groups and the other major players in the online search industry were originally well defined, they have recently become somewhat blurred. Thus, we find producers starting to vend search services for their own and other companies' databases, vendors acquiring their own telecommunications networks, and even vendors providing gateways to the databases on other vendors' systems. In this chapter we will learn more about the capabilities of online bibliographic search systems, their limitations, and the requirements for using them. For the most part, non-bibliographic databases are dealt with in other chapters.

3.2 COMMON FEATURES OF ONLINE RETRIEVAL SYSTEMS

Table 2.1 lists vendors of database searching in the area of science and technology. While there are obviously many differences among such a large number of companies, there are also some common features. Let us look at some of those.

3.2.1 Time Periods and Subjects Covered by the Databases

Even the oldest of the bibliographic databases cover only a few decades of the primary literature, with few exceptions. The database corresponding to *Dissertation Abstracts International* includes the entire file of records extending all the way back to 1861. Chemical Abstracts Service has begun to provide limited online searching access to pre-1967 records. But these are rare exceptions. The normal case is for a database producer to concentrate on building a database forward in time and to rely on printed products for searches of the older literature. Nevertheless, the number of scientific and technical databases has grown dramatically, and those which might hold information of interest to chemists number far in excess of 100 by now.

There are databases which are devoted to a particular subject like chemistry or a subfield of a subject like analytical chemistry. There are also databases which cut across many disciplines and deal with mission-oriented problems like energy or environmental questions or problems of space exploration. Finally, there are some databases that are created for particular document types like patents or dissertations, but which are multidisciplinary in their subject scope. Each vendor chooses certain databases to market to particular segments of the user population. Some even allow **cross-database searching**, performing a single search simultaneously in a number of related databases. Certain databases are available on only one search system, but no one vendor has all of the relevant databases. The Chemistry Reference Sources Database contains information about the main databases which are important for chemistry. Search DATABASE AND NAME of the subject you are interested in, for example, **DATABASE AND INDX = "MASS SPECTROMETRY"**. For broader coverage, consult the database guides discussed in section 2.3.

3.2.2 Virtues and Benefits of Online Searching

It has been said of online searching that it is "exhaustive, but not exhausting." Searching a file of several million records can yield a bibliography numbering several thousand items in a few minutes of computer connect time. The speed of online search systems is so phenomenal that any researcher should factor in the time saved when weighing the cost-effectiveness of an online search versus a manual search of the printed literature. In addition, the wide selection of primary literature abstracted and indexed today virtually assures that a number of items will be retrieved which are not in the local library. This expansion of resources has often been cited as a major benefit of online searching.

Perhaps the most important feature of online searching is the ability to search for multiple concepts. Documents which deal with synonymous or related concepts can be gathered into one set by the use of the OR logical operator. The OR operator generally broadens a search. Likewise, dissimilar concepts can be intersected with AND logic to find items containing both concepts in the same document. The AND operator thus narrows the scope of a search. Finally, there is the possibility of further narrowing the search by *excluding* concepts through the use of the NOT operator. All database vendors have built these basic **Boolean search operations** into their system search software. Most have added even more sophisticated techniques such as those discussed in section 3.6.6.

It is not uncommon for the database producers to structure their files in ways that allow more retrieval possibilities than can be found in the printed counterparts. For example, searches in printed subject indexes sometimes turn up irrelevant citations because a term or concept in one subject area is used in a totally different context in another. Take the case of ICR (ion cyclotron resonance). The same acronym is used in the biological sciences to designate a particular type of mouse sometimes used in biochemistry experiments. Thus, a search in a comprehensive chemistry database using ICR as a search term is likely to produce irrelevant references. Such irrelevant references or **false drops** might be unavoidable in a manual search of a printed multi-disciplinary subject index. But an online search permits several options for solving the problem. One could combine the subject term AND the relevant classified section numbers of the abstracting journal, thus helping to eliminate the false drops. Another alternative is the addition of a phrase like "NOT MICE" to the online search strategy. This would probably eliminate most or all of the false drops. Another way in which producers enhance their online products is by adding more index terms to the computer-readable version. This keeps down the cost of production of the printed product, but allows more in-depth indexing to be placed in cheaper electronic storage media.

Finally, since the databases are usually updated on a regular basis (normally every two to four weeks) it is easy to have a custom bibliography printed out for just that information which is new to the database. This is the technique known as **SDI (Selective Dissemination of Information)**. The new records are loaded into the database in **LIFO** (Last-In, First Out) order, thus assuring the retrieval of the latest information when the vendors update their files. The LIFO order also permits the searcher to cut off a retrospective search at a certain point in time in a database covering a number of years.

3.2.3 Costs of Online Searching

There has been a trend in the online industry to base charges more on the information retrieved than on other expense categories. In general, the costs of an online search are based on three factors: telecommunications network charges, computer connect time, and royalties for the information extracted from the database (usually charges per citation, known as **hit charges**).

An online searcher works from a computer terminal or microcomputer which can be many miles distant from the host computer containing the database. To access that computer, the searcher uses a telephone line which the vendor leases from a **telecommunications company** (See Table 3.2). Such companies typically charge around $10 - $15 per hour for using the lines, and those costs are part of the bill which the vendor issues. There are obviously costs of operating a large mainframe computer which the vendor must recover, and those too are factored into the bill.

The most expensive portion of the total online search bill is likely to be the retrieval charge (hit charge). This charge may vary with the amount of information extracted, ranging from a minimal charge for a list of abstract numbers to a maximum charge for a printout containing all of the data associated with the records, including perhaps the abstracts. Records can be printed online at the time the search is done or printed offline at the remote computer site and sent to the user.

Recently there has been a movement away from connect time as the major cost component of the search. Database producers are more and more attempting to incorporate into their charges the value to the searcher of making use of the database. They factor in the value of the time saved and the benefits from computerized search techniques which are impossible in printed products, as well as the value of the specific records retrieved. Thus, we are likely to see higher costs for the hit charges and even fees based on the number of terms input to the search query.

The user must pay for the retrieved information regardless of whether it is actually relevant to the search question. Therefore, it is wise to plan an online search carefully before connecting to the computer. One can save money by printing only the accession numbers and looking up the actual bibliographic references and abstracts in a printed source. However, a lot of time is required to do that for a list containing many accession numbers. Each user must decide how valuable is the time expended in a search.

Patent databases and chemical dictionary files tend to be the most expensive types of databases used in chemistry. Databases produced by government agencies like the National Library of Medicine and the National Technical Information Service tend to be the cheapest to search.

3.2.4 Problems With Online Searching

In addition to the cost of online searching (which can be a problem for some), there are a number of other factors which make online searching somewhat problematical. There are equipment costs, software costs, training costs, and perhaps phone charges which must be reckoned with. But a novice user has barriers other than monetary to cross on the path to mastery of online searching. The search software packages of all vendors were originally written as command-driven systems, and all of those were written by different people. Thus, the software used by the vendors is by no means uniform. For example, on the STN International system, a searcher gives the command **DISPLAY** to print the references on a local printer or video monitor (sometimes called a VDT, video display terminal, or a CRT, cathode-ray tube). If an offline print is desired from the location of STN's host computer, the command **PRINT** is used. On ORBIT, however, **PRINT** yields a local printout and **PRINT OFFLINE** produces an offline print from ORBIT's host computer. Table 3.1 shows the variety of commands used by several vendors to perform the same tasks. The lack of uniformity among command languages makes the initial choice of a vendor a critical one, for it is unlikely that a searcher will become very proficient in using more than one or two systems.

TABLE 3.1
Comparison of Commands Used on Various Online Searching Systems

Messenger® Commands Comparison Chart

COMMAND FUNCTION	STN	DIALOG2	ORBIT	BRS	ESA/IRS	QUESTEL+
System prompt	=>	?	USER:	:	?	?
Change files	FILE	BEGIN	FILE	..CHANGE/	BEGIN	..BA
List files	HELP FILE NAMES	?FILES	FILES	DOC=ALL (in NEWS file)	?FILES	..FI
File content	HELP CONTENT (in current file)	?FILE no.	EXPLAIN filename	Enter FILE file Search filename	?FILEDATA	..DF filename
File charges	HELP COST (in current file)	?RATES	EXPLAIN PRICES	DOC= no. (in NEWS file)	?CHARGES	N/A
Execute a search - in steps	SEARCH SEARCH STEPS	SELECT SELECT STEPS	none (default) N/A	..SEARCH N/A	SELECT N/A	none (default) N/A
Restrict a search	RANGE	LIMIT	date ranging	..LIMIT/	LIMIT	..LIMIT
Precedence of Boolean operators	1. AND, NOT 2. OR	1. NOT 2. AND 3. OR	1. AND 2. NOT 3. OR	1. AND 2. NOT 3. OR	1. NOT 2. AND 3. OR	nesting required
View inverted index	EXPAND	EXPAND	NEIGHBOR	ROOT	EXPAND	..DI
Save information - temporarily	SAVE SAVE TEMP	SAVE SAVE TEMP	STORE SAVE	..SAVE PS ..SAVE	END/SAVE N/A	..SV N/A
Execute saved item	ACTIVATE	RECALL	RECALL	..EXEC PS ..EXEC	EXECUTE	..EX
Remove saved item from storage	DELETE	RELEASE	PURGE	..PURGE PS	RELEASE	..ER QU
View answers online	DISPLAY	TYPE	PRINT	..PRINT	TYPE	..LI
Print answers offline	PRINT	PRINT	PRINT OFFLINE	..PRINTOFF	PRINT	..PR
Order original document	ORDER	ORDER ORDERITEM	ORDER	Use MSGS file	ORDER ISSUE	..OR
Automatic current awareness search	SDI	SAVE SDI	SDIPROFILE	..SDI	BEGIN...END/SDI	..PF
View session history	DISPLAY HISTORY	DISPLAY SETS	HISTORY	..DISPLAY	DISPLAY SETS	..HI
End online session - with temporary save of entire session	LOGOFF LOGOFF HOLD	LOGOFF LOGOFF HOLD	STOP As warm restart	..OFF ..OFF CONT	LOGOFF LOGOFF HOLD	..ST N/A

Learning the vendor's software command language is only half of becoming a good searcher. The other half comes from a thorough knowledge of the databases themselves: how they are structured, the time periods covered, the type of indexing used, etc. Inconsistent indexing in a given database and the use of quite different terminology when different databases are searched for the same concept are sometimes encountered. The database producers' and vendors' user aids are quite important in overcoming such problems. It is an axiom of online searching that the truly successful searcher of an online database has first mastered its printed counterpart. This can be a large task, given the number of databases available, including relevant general interest or multi-disciplinary databases.

The trend now is toward teaching the **end-user** (the person who is actually going to use the information) to personally do the searching. Nevertheless, the complexities of online searching have led many to conclude that it is best to rely on a librarian or information specialist for online searches. Thus, it is still quite common for scientists to work with these so-called **search intermediaries** who actually perform the searches. A skilled online searcher who is familiar with both the database and the system search software has the best chance of obtaining a proper balance between precision and recall.

Precision (sometimes known as relevance) is simply the proportion of documents retrieved in a search which are judged by the user to be relevant. A low number of **false drops** (irrelevant citations) is obviously desirable when one is paying for each reference retrieved. However, it is necessary to balance precision with **recall**, the proportion of relevant items retrieved from the total store of relevant items *in* the database. If a tight and narrow search strategy is written, the chances of high precision are good, but there are always nagging questions in such cases. Did I retrieve a high percentage of the good references actually in the database? Was my search really comprehensive? In some cases, this is not really a problem. For instance, a "quick and dirty" search is one where only a few good relevant references are desired by the user. But in a comprehensive search, recall becomes more critical. In general, precision and recall are inversely related: the higher the precision, the lower the recall and vice versa.

Not every item found in an online search is going to be held by the local library. Uncovering exciting bibliographic references to items which are not immediately available can lead to frustration. Document delivery in this age of electronic communication is still a seriously weak link in the total communication process. Some ways to deal with this problem are discussed in chapter 13.

3.2.5 Other Factors

As mentioned earlier, the online search business is a competitive one. This has led some database producers to place certain restrictions on their products. A prime example is the decision by Chemical Abstracts Service to make available online the text of the *CA* abstracts from January 1967 to the present and the indexing of the pre-1967 *CA* bibliographic file only through STN International (in which CAS is a major shareholder). Some database producers like Derwent, a major producer of patent databases, have restrictions on who can gain access to their files or to portions of them.

Technology marches on, and as new tools arise, the vendors and producers are quick to react. Thus, the widespread use of the microcomputer, coupled with a device to search at faster speeds, made it possible to rapidly download the results onto a disk without waiting for a slower printer. The connect-hour charges were reduced by printing the references after disconnecting from the host computer. This prompted a shift from connect-time charges toward more emphasis on hit charges. Downloading was originally viewed with great suspicion by the database producers since anyone with a microcomputer and enough storage capacity could conceivably pirate the entire database! Most producers have now worked out reasonable licensing agreements which, for a fee, permit portions of the database to be downloaded in a given time period, and retained and reused indefinitely. Even without such a license, there is usually no additional charge for references which are downloaded, edited, and printed out, as long as the downloaded file is destroyed.

As the new media of **CD-ROM** (Compact-Disk Read-Only Memory) and other optical laser disks begin to penetrate the market, some database producers have created stand-alone microcomputer searching systems with their own databases. The laser systems get around the unpredictability of the cost of an online search since there are no connect charges or costs tied to the information retrieved.

Online searchers have frequently made suggestions for improvements in the software. Minor improvements are usually quickly adopted by the vendors. Less frequently, the vendors make expensive major revisions to the software. Producers also make changes in their products, including changes in indexing policy, revisions of the classification scheme, or even major restructuring of the database. There has been increasing emphasis on the production of text databases in recent years. This has fostered the development of new search techniques to accommodate the different types of data structures in those files. As non-bibliographic databases are produced, more innovative retrieval methods are required, particularly where numeric and chemical structure databases are concerned. Keeping abreast of such developments is very time consuming and requires periodic re-training of all online searchers.

3.3 EQUIPMENT AND SOFTWARE NEEDED FOR ONLINE SEARCHING

There is a wide range of hardware (computer equipment) and software (programs) available for online searching. A simple printing computer terminal with an acoustic coupler for linking the terminal to a telephone is the minimal equipment needed for online searching. However, when personal computers became popular, software for telecommunications was soon developed which allows the microcomputer to emulate a terminal. In most situations, the microcomputer has an asynchronous serial port (RS-232) to which a modem or acoustic coupler is attached. These devices transmit data across telephone lines in a serial fashion, that is, one data signal following another. Alphanumeric characters are coded into series of 1s and 0s (**bits** or *bi*nary digi*ts*), according to the American Standard Code for Information Interchange (ASCII). Some microcomputers are equipped to handle the transmission and receipt of graphics characters which can be displayed on the video monitor or printed on a dot matrix or laser printer. If the host computer has graphics databases, the microcomputer may also have a light pen or mouse for input of graphics characters. These devices make it relatively easy to draw a chemical structure, for example.

The key to successful use of a microcomputer for online searching is a good telecommunications software package. There are dozens of such packages available, some of them free or nearly free. Most incorporate features for downloading and uploading data. A *de facto* standard in the industry is the set of protocols developed for the Hayes line of modems. Therefore, it is important to acquire software which can accommodate the Hayes protocols (rules for transmitting data). Many software packages allow automatic dialing and re-dialing from a telephone directory which is stored on a menu in the microcomputer and which contains the phone numbers of the telecommunications networks. Another required software feature is the capability to change the data transmission speed. Some software, for example, STN Express, includes features to log onto the telecommunications network, transmit the proper address of the remote host computer, and enter the appropriate password to the remote computer. The software may even be programmed to automatically select the most frequently-used database. All of this is done with little or no effort on the part of the end-user. After the search has been run, the results can easily be downloaded with such telecommunications software. (See section 3.6.3 for more details). Search **INDX = "TELECOMMUNICATIONS SOFTWARE"** in the Chemistry Reference Sources Database for some representative telecommunications packages.

3.4 TELECOMMUNICATIONS WITH THE HOST COMPUTER

There are a number of telecommunications networks used by the major vendors of online searching. Table 3.2 lists those which access most online searching systems in the United States.

The telecommunications companies maintain local phone numbers to access their systems in most locations in the United States, thereby saving the full cost of a long-distance phone call dialed direct to the remote computer. However, if it is necessary to dial direct, all of the vendors have numbers which can be used for that purpose. The address code of the host computer and a terminal code for the terminal or microcomputer must be given when making the connection through the telecommunications company. That is not the case for searches performed through the gateway networks like Easynet and InfoMaster. These systems do all of the logon work, and, if desired, select the appropriate vendor and database to be searched.

3.5 PROTOCOLS FOR SIGNING ON

In computer jargon, **protocols** are the rules and controls followed by the software on both the user's end and the host computer's end to establish and maintain communications links. A part of the protocol has to do with how fast the data are being transmitted. This is measured in terms of the **baud rate** (the number of bits or binary digits transmitted per second). The baud rate for online searching over telephone lines at the present time is either 300, 1200, or 2400. The modem or acoustic coupler and the software on both ends of the data link must be set at the same baud rate. Each basic piece of data transmitted (bit) is a part of an ASCII character. These characters are coded series of 1s and 0s assembled into 7- or 8-bit segments called **bytes**. They

TABLE 3.2
Telecommunications Companies for Online Searching in the United States

Name	Address	Customer Assistance Telephone Number
CompuServe	5000 Arlington Centre Blvd. P.O. Box 20212 Columbus, Ohio 43220	800-848-8199
DIALNET	DIALOG Information Services 3460 Hillview Avenue Palo Alto, CA 94304	800-334-2564
EasyNet	Telebase Systems, Inc. 763 Lancaster Avenue Bryn Mawr, PA	800-841-9553 215-526-2800
InfoMaster	Western Union/InfoMaster Dept. 503 9229 LBJ Freeway, Suite 234 Dallas, TX 75243	800-247-1373
INFONET	Computer Sciences Corporation 2100 East Grand Avenue El Segundo, CA 90245	213-615-0311 703-538-7200
Telenet	GTE Telenet Communications Corporation 8229 Boone Blvd. Vienna, VA 22180	800-336-0437
Tymnet	2070 Cam Bridge Road Vienna, VA 22180	800-336-0149

Not all systems can be used for all vendors. For example, Questel is accessible through Telenet, Tymnet, and INFONET for direct searching. EasyNet and InfoMaster are gateway systems which serve a number of different vendors. The IQuest option on CompuServe does the same thing.

designate letters, numbers, punctuation marks, and graphic symbols. The ASCII code for the letter "R" is 1010010.

The basic unit of transmission varies from system to system, depending on how the control information around the ASCII code for a byte is designated. The computer needs a way to distinguish the start of one character from the end of another, so a start bit and a stop bit are appended to each character coded in ASCII. In addition, a check on the validity of the transmitted data is sometimes employed by placing a parity check bit before the stop bit. The parity bit can either be odd or even. If an error is detected in the parity bit, the character is re-transmitted. Thus, designations such as 8N1 (8 data bits, no parity, 1 stop bit), and 7E1 (7 data bits, even parity, 1 stop bit) are used. STN International uses 7E1, whereas ORBIT will take either odd or even parity with a 7-bit character. Each character transmitted has around ten bits in its code. Therefore, the baud rates mentioned earlier (300, 1200, and 2400), correspond roughly to transmission rates of 30, 120, and 240 characters per second respectively.

The final item of protocol which must be considered is the duplex setting, which for online searching is usually full duplex. This means that a character sent to a remote computer is echoed back to your video monitor or printer as soon as it is received. With half duplex, characters are sent from the keyboard to the monitor and to the host computer simultaneously.

The searcher must keep in mind that telephone lines are used for online searching. Consequently, there is always a chance that noise may intrude on the data being transmitted. It might even happen that the communications link will be completely broken at times. If that occurs, most vendors have built a fail-safe feature into the system search software. This puts the search back to the point which was reached before the interruption, provided the link is re-established within a short period of time. Table 3.3 lists some common problems encountered when searching online and some tips to solve them.

TABLE 3.3
Telecommunications Problems and Potential Solutions

	SYMPTOM	POSSIBLE CAUSE	POSSIBLE SOLUTION
1.	GENERAL PROBLEMS		
	Occasional gibberish	Noise on phone line NB: Lower baud rates are less susceptible than higher rates.	Log off and redial.
	Disconnected	1. Call Waiting or other signals	Eliminate feature or possibility of interruption.
		2. Phone extension use (may just cause gibberish with some modems)	Eliminate feature or possibility of interruption.
		3. If using acoustic coupler:	Try another phone instrument.
2.	PHONE NETWORK PROBLEMS		
	Fast or slow busy signal	Overload of phone company circuits or those of the network	Try another network or redial in a few minutes.
	Network phone rings, but no answer	Wrong number	Re-dial.
	Get the high pitched tone, but no connect	May have caled a 300 baud number at 1200 baud, etc.	Try another number.
3.	SOFTWARE PROBLEMS		
	Doubling of characters	Communication program set for half-duplex instead of full duplex	Reset duplex.
	Nothing typed appears, but everything by other computer appears ok	You are in full duplex; other is in half duplex.	Reset duplex.
	No response or gibberish at the start	Baud rate settings of your computer and the other may be different.	Reset baud rate.
	Losing data on the printer	Not enough delay between line transmissions	Change terminal identifier: Telenet: Try A9 for D1; Tymnet: Try I for A

3.6 FEATURES OF THE SYSTEM SEARCH SOFTWARE

The vendor's system search software takes control of the terminal or microcomputer once the link has been established through the telecommunications network. In this section some of the major features of that software are discussed.

3.6.1 Logging On and Logging Off

There is a certain amount of security required in the process of signing onto the vendor's host computer. Whether called a "username," "user id," "account number," or something else, the user will first be asked to give a code to identify the place to be billed for the search. As a check on unauthorized use of the account, a password is then requested. Figure 3-1 shows a crib sheet to lead end-users through the logon and logoff sequences for STN International. The Telenet telecommunications service is used, as is an external Hayes-compatible modem and an IBM PC XT microcomputer running the telecommunications software PROCOMM [247] from a hard disk. User responses are in boldface type in this example and others throughout the book. The designation (carriage return) means to strike the key labeled "Return" or "Enter". (On some keyboards, the key is simply indicated by an arrow coming down and to the left.) This must be done after each command or search statement is typed on all vendors' systems. Consequently, it will not be repeated in future examples.

3.6.2 Modes of Searching: Levels of Proficiency and Defaults

The totally command-driven systems of the early years of online searching have now been complemented with "user-friendly" (some would say, "user-tolerant") aspects. Certain of the vendors have even produced versions of their software which are tailored for the novice or infrequent user. DIALOG's Knowledge Index and BRS's After Dark and Colleague series are priced lower than normal searching, but are available only at times other than the normal workday. Even during hours of peak use, BRS offers two options to the user. Upon logging on, the choice is given to search with the traditional BRS system commands or through a menu-driven format. STN International's software assumes a novice user if the full form of the command is given instead of an abbreviation, for example, **DISPLAY** instead of **D** or **DIS** to view the retrieved items. The user may then be prompted for the number of items to display and in what format (abstract numbers only, bibliographic information, include the abstracts, etc.).

All systems have certain safeguards. It is easy to run up a large bill by inadvertently searching a huge topic and/or requesting a printout of many items. In order to protect against such actions, the systems have included default options which take over in the absence of user input. (A **default** is what the computer assumes you want to do if no explicit information to the contrary is given.) Thus, a print command by itself might print only one reference, but not in a format which includes all of the data for the record. Likewise, upon entering a search statement which requires a lot of

44 CHEMICAL INFORMATION SOURCES

computer connect-time, the searcher might be asked at frequent intervals if the search should be continued. It is important to realize that defaults are for the user's protection, but can be over-ridden with the proper commands.

3.6.3 Retrieval Software

The databases available through the online database vendors are normally searched through a **command-driven software** system. This means that the computer expects the searcher to know what command to give in order to perform a certain function. (In contrast, **menu-driven software** systems present a list from which the searcher may choose one of several options.) Thus, a considerable amount of training and practice is needed to master a command-driven system and learn the various search techniques. For one person to master all of the vendors' search systems would be very

LOGON
1. Turn on the microcomputer, video display, and attached printer.
2. Turn on the external modem.

After a few minutes a list of choices to select will appear on the video display screen. You will be selecting **Telecommunications**.
YOU RESPOND: 3 (CARRIAGE RETURN)
A list of telecommunications packages will appear on the screen. You will be selecting **ProComm**.
YOU RESPOND: 1 (CARRIAGE RETURN)
When the ProComm copyright information appears on the screen:
YOU RESPOND: (CARRIAGE RETURN)
When the status line is displayed across the bottom of the screen:
YOU RESPOND: HOLD DOWN THE Alt KEY AND TYPE d
DISPLAYED ON SCREEN: DIALING DIRECTORY
YOU RESPOND: 2 (or 3 or 4) (CARRIAGE RETURN)
The modem now dials the Telenet telephone number you selected.
DISPLAYED ON SCREEN (IF YOU CHOSE 2):
             Dialing Telenet: Bloomington
             DIAL TONE
             9,,3321344
             CONNECT 1200
YOU RESPOND: (CARRIAGE RETURN) (CARRIAGE RETURN)
DISPLAYED ON SCREEN: TELENET
                  812 13B (or some similar designation)
                  TERMINAL=
YOU RESPOND: d1 (CARRIAGE RETURN)
DISPLAYED ON SCREEN: @
YOU RESPOND: c 614 21i (CARRIAGE RETURN)
DISPLAYED ON SCREEN: 614 211 CONNECTED
YOU RESPOND: Type in your login ID (CARRIAGE RETURN)
DISPLAYED ON SCREEN: PASSWORD:
YOU RESPOND: Type in your password (CARRIAGE RETURN)
DISPLAYED ON SCREEN: TERMINAL (ENTER 1, 2, 3, OR ?):
YOU RESPOND: 3 (CARRIAGE RETURN)

FIGURE 3-1
Logging On and Off STN International's CAS ONLINE Academic Program via Telenet.

difficult. Instead, it is the usual case that an online searcher chooses a system or two on which to specialize. The choice might be influenced by the range of databases available on the systems or the availability of some specialized databases which no other vendor provides. Alternatively, a special price deal which is offered by one vendor to a particular class of users (for example, lower rates for academic users) might sway the user toward that system.

An online searcher does not have to learn all of the command languages for all of the online systems. **Front-end** and **gateway software** systems have been developed to assist in both formulating and running an online search. The front-end software is placed on the user's microcomputer, whereas gateway software resides on a larger remote computer located between the user's PC and the vendor's host computer. Both front-end and gateway software assist the user in constructing a search and often provide access to more than one online vendor. Since at least some of the cost of

---

The welcome message for STN appears on the screen along with news messages. After the news, the time and date of login appears. Record the time as shown on the screen on the usage slip. You are now in the LCA file. You will choose **FILE CA** (for bibliographic or subject searching) or **FILE REG** (for chemical searching, including structure searches) or **FILE CAOLD** (to search pre-1967 literature by CAS Registry Numbers of chemical compounds).

YOU RESPOND: file ca cost = your last name (CARRIAGE RETURN)
or **file reg cost = your last name (CARRIAGE RETURN)**
or **file caold cost = your last name (CARRIAGE RETURN)**
(type only the first 8 characters of your last name)
You can now enter your search strategy.

PRINTER
You may toggle the printer on or off at any time during your search. To toggle the printer on or off:
   HOLD DOWN THE Alt KEY AND TYPE L
The status line across the bottom of the screen will say **PRT ON** or **PRT OFF** on the right side.

TO STOP THE SCREEN FROM SCROLLING, HOLD DOWN THE Ctrl KEY AND TYPE s.
TO RESUME SCROLLING, HOLD DOWN THE Ctrl KEY AND TYPE q.
(N.B. Software packages differ in the way they turn the printer on and off. Check your manual for the telecommunications software to see how your software performs this task if you are not using PROCOMM.)

LOGOFF
When you are finished searching, you must disconnect from STN and enter the time on the usage slip.

YOU RESPOND: logoff (CARRIAGE RETURN)
DISPLAYED ON SCREEN: ALL L-NUMBERED QUERIES AND ANSWER SETS ARE
                     DELETED AT LOGOFF.
                     LOGOFF? (Y)/N:
YOU RESPOND: y (CARRIAGE RETURN)
                     To logoff in one step, type: **log y (carriage return)**.
YOU RESPOND: **HOLD DOWN THE Alt KEY AND TYPE h**
This disconnects the telephone line. The status line across the bottom of the screen will say **DISCONNECTING** in the far left corner.

FIGURE 3-1
Logging On and Off STN International's CAS ONLINE Academic Program via Telenet.

online searching is tied to the amount of time the searcher is connected to the host computer, these software packages strive to minimize the connect time. This is done in part by allowing the searcher to formulate the search strategy *before* connecting to the online vendor's computer. The search is then *uploaded* at the appropriate time. Once the references have been found in the database and an answer set has been formed, the answers can be *downloaded* into a microcomputer system for later browsing, editing, and printing. There are several front-end software packages which assist with online searching, some specific to a particular vendor (like Grateful Med [195] for the National Library of Medicine) and others with the capability to search multiple vendors (like Pro-Search [1887] or STN Express [534]). Gateway systems like InfoMaster or EasyNet are available through large remote computers and allow online searching with a credit card. Table 3.4 lists a number of these packages. Some of these packages have complementary software components to work with downloaded data.

3.6.4 Records and Fields in the CA File; Displaying an Answer Set

It is helpful to think of a **record** as a short description of the material which leads back to the original document or piece of information. Organizing a database into records allows the user to sort the records into a desired order, retrieve a group of references to primary documents according to a given search strategy, utilize the format codes for the documents, access information about the content of the documents through abstracts or indexing terms, and perhaps to identify the locations of the documents.[1] These are essentially the purposes of any document storage and retrieval system based on records which represent the original documents. Records are further subdivided into **fields**, the parts of the record which from one document to another are generically similar. Thus, we speak of an author field, a title field, a subject-term (or **descriptor**) field, a publisher field, etc. Figure 3-2 shows the full records for two articles from the CA File, the database which provides bibliographic information and abstracts corresponding to the printed *Chemical Abstracts* on STN International. Compare them with the same abstracts from the printed *CA*, as reproduced in Figure 1-1.

The **CAS ONLINE CA FILE** on STN International contains over 9,000,000 bibliographic records, abstracts, and associated indexing terms which correspond to the entries in the printed *Chemical Abstracts* from January 1, 1967 to the present. Companion files are the **CAOLD File**, which leads to pre-1967 references in the printed *CA*, and the Registry File. The **CAS ONLINE Registry File** is a dictionary database containing information about a chemical substance, such as its molecular formula, various types of names used for the substance, a structure drawing, etc. Figure 3-3 shows the relationships among these CAS ONLINE files and the types of searching which can be done with each. In that figure, **Registry Number** refers to the unique number assigned to each chemical substance in the CAS Registry File. Examples of registry numbers are found in the controlled-vocabulary fields of the CA File records reproduced in Figure 3-2.

TABLE 3.4
Selected Front-End or Getway Systems for Online Searching

Name	Company	Vendors Searched
Pro-Search	Personal Bibliographic Software, Inc. P.O. Box 4250 Ann Arbor, MI 48106-4250 313-996-1580	BRS DIALOG
Grateful Med	National Library of Medicine 8600 Rockville Pike Bethesda, MD 20894 800-638-8480	NLM
InfoMaster*	Western Union/InfoMaster Dept. 503 9229 LBJ Freeway, Suite 234 Dallas, TX 75243 800-247-1373 Access Number: 800-422-4664	BRS DIALOG ORBIT Others
IQuest* SmartSCAN	CompuServe Suite 380 175 South Third Street One Capitol South Columbus, Ohio 43215 800-848-8199 614-457-0802	DIALOG Others
STN Express	STN International 2540 Olentangy River Road P.O. Box 3012 Columbus, Ohio 43210 614-447-3600/800-848-6533	STN BRS ESA Dialog
EasyNet*	Telebase Systems, Inc. 763 Lancaster Avenue Bryn Mawr, PA 800-841-9553 215-526-2800 Access Number: 800-EASYNET	DIALOG Others
microCAMBRIDGE	Cambridge Scientific Abstracts 5161 River Road Bethesda, MD 20816 301-951-1400	DIALOG (Cambridge files only)
Medical Connection	DIALOG Information Services, Inc. 3640 Hillview Avenue Palo Alto, CA 94304 800-227-1927	DIALOG
WILSEARCH	The H.W.WilsonCompany 950 University Avenue Bronx, N.Y. 10452 800-367-6770 800-462-6060, N.Y. only	Wilsonline

* = gateway systems. Telebase has licensed its product to various companies including Western Union and CompuServe.

48 CHEMICAL INFORMATION SOURCES

BASIC INDEX	Other Searchable fields			SAMPLE FORMAT	Other Display Formats	
		AN	CA93(3):25540j		CAN	
	/TI	TI	Formation kinetics of an amino carboxy type merostabilized free radical			
	/AU	AU	Bennett, Richard W.; Wharry, Donald L.; Koch, Tad H.			
	/CS	CS	Dep. Chem., Univ. Colorado			
	/LO	LO	Boulder, CO 80309, USA			
	/SO	SO	J. Am. Chem. Soc., 102(7), 2345-9			
	/SC	SC	22-4 (Physical Organic Chemistry)		BIB	
	/SX					
	/DT	DT	J			
	/CO	CO	JACSAT			
	/IS	IS	0002-7863			
	/PY	PY	1980			
	/LA	LA	Eng			
		GI	Diagram(s) available in offline prints and/or printed CA Issue.			
	/AB	AB	The redn. of isatin and N-methylisatin by the merostabilized free radical 3,5,5-trimethyl-2-morpholinon-3-yl (I) to isatide and N,N'-dimethylisatide is described. The reaction rates are 1st order in the meso and dl dimers (II and III) of I and zero order in N-methylisatin. The reaction rate is a measure of the rate of bond homolysis of II or III. Rate consts. and activation parameters are reported. The free energies of activation for homolysis of II and III in CHCl3 are 24.6 .+-. 0.3 and 25.1 .+-. 0.2 kcal/mol, resp. The enthalpy of activation for combination of I is estd. at 4-5 kcal/mol. Measured rate consts. for bond homolysis were consistent within 1 std. deviation with the obsd. kinetics of isomerization of II to the equil. mixt. of II and III. The activation parameters in part are a measure of the effect of merostabilization on radical stability.		ABS*	ALL
		KW	isatin reaction merostabilized radical; morpholinonyl dimer cleavage			
	/CV	IT	Bond cleavage (in trimethylmorpholinolyl dimers)			
	/CV	IT	Kinetics, reaction (of isatin or N-methylisatin, with merostabilized free radical)		IND*	
		IT	57765-60-3 74043-30-4 (reaction of, with isatin or N-methylisatin, kinetics of)			
		IT	***91-56-5*** 2058-74-4 (reaction of, with merostabilized free radical, kinetics of)			

**Basic Index contains single terms only.

*ABS and IND formats include the AN field.

FIGURE 3-2
Sample Records for Journal Articles from the CA File on STN International

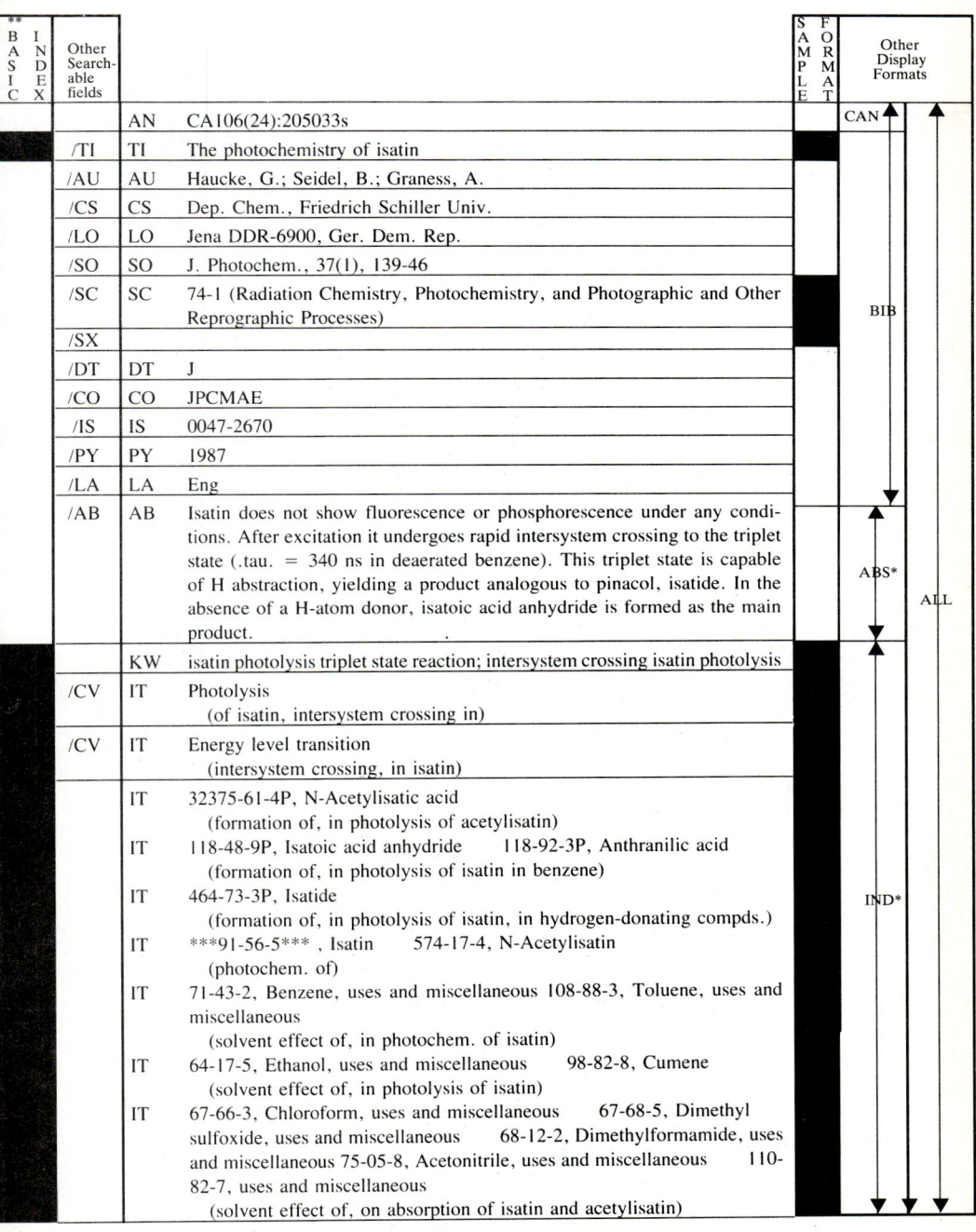

FIGURE 3-2
Sample Records for Journal Articles from the CA File on STN International

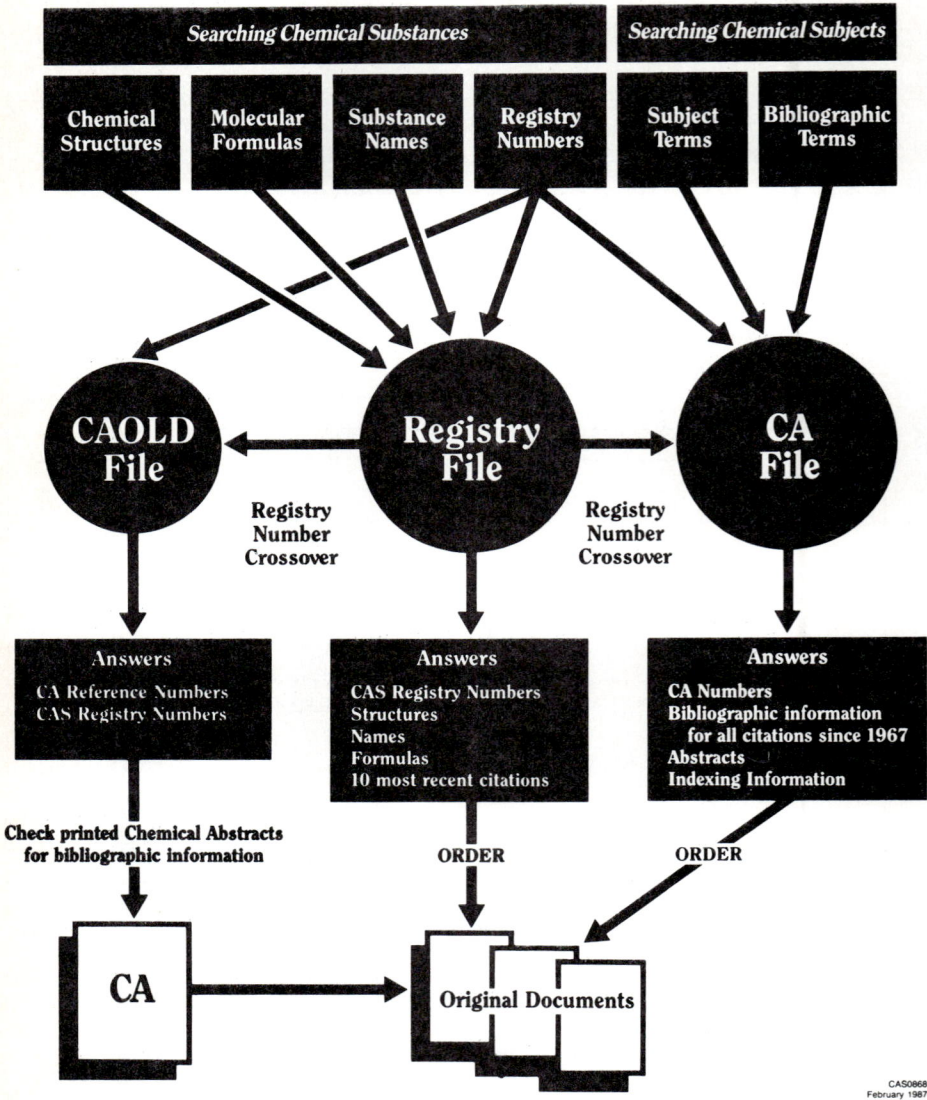

Figure 3-3
Relationships Among the CAS ONLINE CA, CAOLD, and Registry Files

Referring again to Figure 3-2, the records show that the **DISPLAY** (or **PRINT**) command gives various options for obtaining part or all of the data from the records in the CA File. The **SAMPLE** option is a free format which allows the inspection of the title for the document, plus the *CA* section code and the indexing terms. This permits the searcher to judge the relevance of some or all of the retrieved documents and to select other indexing terms if a modification of the search strategy is necessary.

The **BIB** format gives the *CA* abstract number and the relevant bibliographic information. (**CBIB** is a condensed format which displays the same fields as BIB without field labels.) In order to see both the bibliographic information and the abstracts, enter: => **DISPLAY BIB ABS** where the "=>" is the system prompt on the STN International system. Entering => **DISPLAY ALL** would retrieve every piece of information in every field of a record. Note that if we do not specify an answer set number (designated on STN with a capital L), nor a range of answer numbers to be displayed from the answer set, the system will default to the first answer in the most recent answer set which has been formed.

It is also possible to display just the information in a particular field or fields by combining the field code(s) with the display command. For example, => **DISPLAY L5 1-50 TI** results in a list of the first fifty titles and the associated answer numbers within answer set L5. After scanning the titles, the searcher might then choose the answers considered to be the most relevant for display of fuller information, for example,

=> **DISPLAY L5 3, 17, 28-30, 47 BIB ABS**

3.6.5 Assistance in Search Formulation and Retrieval

One of the most powerful features of online bibliographic search systems is their capability to help the user construct and refine an online search. Already noted in section 2.3 were the online database indexes which assist in the selection of the proper database. Other features of the system search software which are especially useful are counts of the documents retrieved and online dictionaries of searchable terms.

3.6.5.1 Inverted Dictionary Files (Indexes).

There are two basic ways to search files online. One way is to enter the database at the top of the file and search the various records sequentially (serially), one after the other. For a small file such a sequential search may not be a problem. But for the kind of searching done in enormous online databases, the speed of the search would be impossibly slow if sequential searching were done. Therefore, it is the usual case that the records are searched indirectly through **inverted dictionary files**. This means that each allowed search term is placed in an alphabetical file, and each term is linked to the accession numbers for the records containing that term. Thus, it is really the group of accession numbers which are retrieved, and those numbers provide the links to the actual **records**, the bibliographic data, abstracts, and indexing terms which correspond to documents from the primary literature.

It might be helpful to visualize a sequential search in a printed book for a particular fact. In order to find the information, we would start with page one and read or scan each page until we found what we were looking for. With an inverted dictionary file approach, we would use the subject index in the back of the book where the appropriate terms are listed alphabetically and linked to the proper pages in the book. (Think of the page numbers as accession numbers in this case; the index would be equivalent to an inverted file with the page numbers posted on the index terms.) In a real inverted dictionary file, it is the accession numbers of the records which are posted on the terms in the alphabetical file, the online index.

At its simplest level then, an online search of an inverted file database for "gas-liquid chromatography" might use the strategy:

gas AND liquid AND chromatography

The computer would then proceed to search an inverted dictionary file (alphabetic index) of all search terms and compare the document accession numbers posted on each of the three search terms. If there are documents containing all three terms, those accession numbers would be gathered into the answer set. A display or print command would then match the accession numbers to the records and print out the requested data. In Figure 3-4, document number 101453 is the only one which satisfies the search for gas AND liquid AND chromatography. The result will be a statement like "1 posting". A print command then has to be given in order to see the record.

ACID	CHROMA-TOGRAPHY	GAS	LIQUID	ZEOLITES
101017	101091	101091	101216	101453
101587	101216	101395	**101453**	101972
101659	**101453**	101421	101852	
	101972	**101453**		

FIGURE 3-4
Posting of Accession Numbers in a Hypothetical Inverted Dictionary File.

3.6.5.2 The Basic Index.

The concepts of an inverted dictionary file and a basic index are crucial to understanding what takes place in an online search. An alphabetized, inverted dictionary file of subject terms (on most systems, *single* terms, not phrases) from all SUBJECT fields of all records in a bibliographic database is called a **basic index**. In Figure 3-2, such words would be taken from the title, key-word, and index-term fields of the CA File records.

It is possible to search more precisely by limiting a search to subsets of the subject terms listed in the basic index, for example, just those subject terms and phrases used in the controlled-vocabulary, index-term fields. Thus, one could enter the phrase => **SEARCH ENERGY LEVEL TRANSITION/CV** for a complete search of all documents dealing with this subject. If the searcher does not specify that the search should be limited by field, it will automatically be performed in the basic index by default. Thus, entering => **SEARCH ENERGY LEVEL TRANSITION** on STN International without the /CV field code would probably result in more answers since the computer would default to the basic index of single-word terms from titles and keywords, as well as controlled-vocabulary index terms. Each of the words in the phrase would be in the dictionary. The results of this search are likely to be less precise than when the controlled-vocabulary phrase is used.

3.6.5.3 Other Inverted Dictionary Files.

It is not just the subject fields which have inverted files in online databases. All terms, phrases, or other search keys which are available for searching in all fields are linked to the records by means of inverted files. For example, there are inverted files for authors' names or for journal CODENS (standard six-character abbreviations for journal titles). The inverted files show the number of documents (or records) posted on each term. Looking at a snapshot of part of the basic index or other inverted files before a search strategy is actually executed is called **neighboring** or **expanding** the search term. Figure 3-5 shows how the /AU inverted file is used in a search. In later chapters we will see how this feature greatly aids in term selection for other types of searches.

```
=>      expand wiggins, g/au
E1                  2       WIGGINS, EDWIN W/AU
E2                  4       WIGGINS, F N/AU
E3                  1       WIGGINS, G/AU
E4                  2       WIGGINS, GAIL/AU
E5                  1       WIGGINS, GARY/AU
E6                  4       WIGGINS, GERALDINE L/AU
E7                  6       WIGGINS, GLENN C/AU
E8                  1       WIGGINS, H G/AU
E9                  9       WIGGINS, H S/AU
E10                 1       WIGGINS, HUGH/AU
E11                 3       WIGGINS, HUGH S/AU
E12                16       WIGGINS, J/AU
=>      search e5
L1                  1       "WIGGINS, GARY"/AU
=>      display L1 1 bib abs

L1      ANSWER 1 OF 1

AN      CA106(11):83611f
TI      Chemical information science coverage in Chemical Abstracts
AU      Wiggins, Gary
CS      Chem. Libr., Indiana Univ.
LO      Bloomington, IN 47405, USA
SO      J. Chem. Inf. Comput. Sci., 27(1), 1-3
SC      20-5 (History, Education, and Documentation)
DT      J
CO      JCISD8
IS      0095-2338
PY      1987
LA      Eng
AN      CA106(11):83611f
AB      Coverage of publications on chem. documentation or chem. information science in Chem. Abstrs.
        is examd. Most of these publications are found in section 20 of Chem. Abstrs. although many
        relevant articles are scattered throughout several sections of Chem. Abstrs. Many articles were
        scattered among 39 sections other than section 20 during 1984-1985. Data are provided on journals
        contributing the most refs. on chem. information science and on the languages of publication.
```

FIGURE 3-5
Inverted File Author Search in the CA File on STN International.

3.6.6 Search Commands and Other Software Features

Depending on what system is used, the command to perform a search could be "FIND," "SEARCH," "SELECT," "QUERY," some other command, or nothing! In this section we will look at some of the most important searching features of the system search software.

Obviously, the capability to combine sets with the Boolean Search operators (AND, OR, NOT) is very important in online searching, and all system search software performs these operations. However, they may not all work in the same way. What if we were searching for either gas chromatography or liquid chromatography and formulated a strategy like this:

gas OR liquid AND chromatography

The system may have been programmed to look at OR statements first. If that is the case, this strategy should work. But what if the reverse were true? If the system works on the AND statement first, then the results will definitely include works on liquid chromatography. However, the search strategy will also pull a lot of material dealing with gas which has nothing to do with chromatography. Referring again to Figure 3-4, documents 101395 and 101421 would be pulled with our strategy even though they have nothing to do with chromatography. Documents 101216 and 101453 would also be pulled. The precision of the search will be significantly lower than if the strategy

gas AND chromatography OR liquid AND chromatography

were used.

Table 3.1 shows the order of precedence of Boolean operators on various systems. One of the safeguards which have been developed to prevent mistakes with the order of the search operators is a technique known as nesting. **Nesting** is simply the use of parentheses or brackets to group together in a logical manner the components of a search statement. For example:

(gas OR liquid) AND chromatography

produces the correct result regardless of the order of precedence on most systems. When nesting is used, the parts of the search query inside the parentheses are acted on first, then the remaining steps are performed. It is good practice for beginning searchers to make frequent use of nesting.

Another way around the problem of the sequence of a search is to perform each logical step separately. When the answer to each search statement is obtained, it is assigned a set identifier (usually a number). The set number can be used to stand for the documents in that set when later statements are entered. For example,

SS1. **Search gas OR liquid**
Result: 1 (4793 postings)
SS2. **Search 1 AND chromatography**
Result: 2 (397 postings)

The use of the AND operator implies nothing about the sequence of the terms in a subject phrase or title or abstract. If searched in the basic index, it does not even imply that the words all had to occur in the same field. Suppose we were searching for information on the use of automated Venetian blinds to control light in an ophthalmic hospital. Suppose also that we found a database which covers the broad field of vision, with both medical and related optical engineering documents included in the database. Our search strategy is:

Venetian AND blind

The search produces thirteen documents, but much to our surprise, when the first hit is printed, it is titled, "A Study of the Blind Venetian: The Impact of Pizza on Loss of Sight in a Floating Society." The second is "A Plan for Illuminating Blind Alleys with Venetian Gondola Lights." False drops!

This particular problem and many others like it could be solved by the use of a **positional** or **proximity operator** which tells the computer to make the match only if a certain word order is found. The usual proximity operators indicate:

adjacency   specifies that the search terms must be side by side, but no particular word order is required, for example,

Venetian(ADJ)blind

sequence   specifies that the search terms must occur in the order in which they are entered, for example,

Venetian(WITH)blind

In terms of our hypothetical search, the last statement is the best one to use, the one which is most likely to eliminate false drops. Nevertheless, the second false drop would have been eliminated even if we had used the adjacency operator.

The proximity operators also handle the problem of double-word phrases such as "protein-protein interactions". Using a strategy like "protein AND interactions" in an inverted file which just posted on the term "protein" may not be successful for such a topic. However, proximity operators solve this problem by using a search statement like:

protein(WITH)protein(WITH)interactions

There is a third special variant of the AND operator in common use now. That is the **field-specific operator** or **linking operator**. In this case the terms which are joined are required to be from the same field. For example, a match occurs only if all words in the search statement are from the title field, or only from the key-word field, or only from the controlled-vocabulary subject terms or phrases. The use of such an operator helps prevent the kind of false drops where one term comes from one field and the other from a second field. A work with a title such as "Venetian Case Studies of a Retinitis Pigmentosa Support Group," which had subject indexing "Blind, support groups" would be avoided by the strategy **Venetian(LINK)blind**.

These special operators are particularly valuable in searching full-text databases. Each of the special variants of the AND operator has another feature in many systems. That is the ability to define a variable number of intervening words which can stand between the two terms. Thus,

gas (2WITH) chromatography

would pick up cases where phrases such as "gas-liquid chromatography," "gas-phase chromatography," and the like were used in the title or indexing phrases. On STN International, it is assumed that a phrase entered without the /CV (or other) field code is to be searched as if (1W) were between each term. That explains why the strategy => **SEARCH ENERGY LEVEL TRANSITION** works on that system. If this were not the case, the result would be zero, since the computer would be searching in the basic index of single-word terms for a multi-word phrase.

There is also a technique known as **string searching**. When a relatively small set of documents has been isolated into an answer set, it is possible on some systems to use a command which will do a sequential search on all of the words in that set of document records, looking for a particular string of characters. This technique could be used to find double-word phrases like "protein-protein interactions" or to find words on a stop list which would normally not be searchable. It is the type of search which wordprocessing software performs when a particular character string is sought.

Finally, there is sometimes a need to limit a search to a particular time period. A search might be limited to a certain range of dates of publication of the primary documents or to a range of dates when they were entered in the database. Each system has developed ways of specifying the ranges of time which a search should encompass.

3.7 SUMMARY

Success in online searching depends on many factors, including the quality of the database searched and the use of a correct search strategy. Mastery of the command-driven search language is important, but equally important is a thorough understanding of the database. The online searcher today has a number of choices to make, ranging from the vendor to the database and from the type of equipment selected to the type of software. Software packages are available which lessen the requirement to learn the vendor's system search software. While the trend in recent years has been for the user to perform the online search, there are still many times when a librarian or information specialist should be consulted.

There are numerous benefits to be had with online searching, but trade-offs must be made. Time saved with an online search must be paid for in real money. Computer-based searching is of relatively recent origin. Therefore, one must always be aware that not all information found in printed sources in libraries has counterparts in online or other computer-readable formats.

3.8 SELECTED READINGS

Badgett, Tom. "Online Databases: Dialing for Data." *PC Magazine* **1987**, *6*(9), 238-240, 242, 244-246, 248, 257-258.

Buntrock, Robert. "Chemcorner," a regular column in *Database* from March 1979 to the present. From July 1978 until it moved to *Database*, the column appeared in *Online*.

Coons, Bill. "Frontiers in Front-Ends and Gateways." In *Online '86 Conference Proceedings*; Online, Inc.: Weston, CT, 1986; pp 30-35.

Hawkins, Donald T.; Levy, Louise R. "Front-End Software for Online Database Searching. Part 1: Definitions, System Features, and Evaluation." *Online* **1985**, *9*(6), 30-37.

Levy, Louise R.; Hawkins, Donald T. "Front-End Software for Online Database Searching. Part 2: The Marketplace." *Online* **1986**, *10*(1), 33-41.

Hawkins, Donald T.; Levy, Louise R. "Front-End Software for Online Databases. Part 3. Product Selection Chart and Bibliography." *Online* **1986**, *10*(3), 49-58.

Saffady, William. "Communications Software Packages for the IBM Personal Computer and Compatibles." *Libr. Tech. Rep.* **1985**, *21*(4), 355-456.

Williams, Brian. "Macintosh Communication Software." *Libr. Software Rev.* **1986**, *5*(5), 319-323.

REFERENCES

[1] Stibic, Vladimir. *Personal Documentation for Professionals*; Elsevier North-Holland: Amsterdam, New York, 1980; p 39.

CHAPTER 4

SEARCHING BY AUTHOR OR ORGANIZATION NAMES AND BY KNOWN CITATIONS

4.1 INTRODUCTION

It is often the case that a searcher knows the name of a person who is associated with a particular topic on which information is sought. The name may have been found through a previous search or might have turned up through background reading. Subject searches in a source like *Science Citation Index* (*SCI*) frequently reveal the names of key individuals who have been very active in a particular research area. A bibliography in an encyclopedia article or review article is another good source for identifying relevant names. It might also be found that certain organizations or companies are very productive, so the searcher could decide to identify the major publications which have been written by employees of that organization.

    Whatever the reason for doing an author search, there are numerous benefits to using this technique. In the first place, author searching is generally much easier than subject searching. Even in the very old literature of chemistry, author indexes are likely to be found. Furthermore, the powerful technique of citation searching and the use of organizational indexes prepared by the Institute for Scientific Information considerably broaden the approaches which a searcher can use.

4.2 SCIENCE CITATION INDEX AND OTHER ISI PRODUCTS

Since the early 1960s, the Institute for Scientific Information (ISI) has made a major contribution to indexing the world's scientific and technical literature by publishing the *Science Citation Index* [549]. *Science Citation Index* is a multidisciplinary index covering over 3300 of the world's most important science and technology journals. Its subtitle is "An International Interdisciplinary Index to the Literature of Science, Medicine, Agriculture, Technology, and the Behavioral Sciences." ISI's approach is unique. By using sophisticated statistical analyses of citation patterns, the world's most important scientific and technical journals are identified. These form the core of literature from which the many ISI indexes are produced. The relative importance of a journal can be judged quite well by the number of times authors put into their own bibliographies citations to earlier articles published in that journal. The more citations a journal receives in a given time period, the more important it is assumed to be. By keeping track of these relationships, the compilers of *Science Citation Index* are able to concentrate on a relatively small number of journals.

The SciSearch database [266] on DIALOG or ORBIT contains all of the records which appeared in the bi-monthly or quarterly issues of *Science Citation Index* (*SCI*) from 1974 to the present. In addition, the database is supplemented with records taken from about 1000 journals indexed in the five *Current Contents* versions for science and technology which are not source journals in the printed *SCI*. The records from these sources do not have the references from the bibliographies processed for citation searching, but they do give the database a substantial edge over the printed *SCI* in total number of source records. Thus, about 4400 major scientific and technical journals are covered in SciSearch. Those 4400 journals are estimated to contain 90 percent of the world's significant scientific and technical literature. *Chemical Abstracts*, on the other hand, includes over 12,000 journals in its coverage. While arguments can be raised against the methods for selecting the journals covered in *SCI*, there can be no arguing with the success of the *Science Citation Index* since it was founded in the early 1960s.

The *Science Citation Index* is published bi-monthly, and the issues are cumulated each year into annual volumes. There are three major parts to *SCI*: the Source Index [545] (which contains the full bibliographic references to the source journal items, including article titles), the Permuterm Subject Index [546] to the source journal articles, and the Citation Index [547] to the source journal articles. (Smaller *SCI* indexes

TABLE 4.1
Time Periods Covered in *Science Citation Index* Cumulative Volumes

	Source Index	Permuterm Subject Index	Citation Index
1945-54	x		x
1955-64	x		x
1965-69	x	x	x
1970-74	x	x	x
1975-79	x	x	x
1980-84	x	x	x

allow searching by patent numbers and by corporation names.) Multi-year cumulations can now be purchased to replace the annual volumes. Table 4.1 shows the time periods for which cumulations are now available. Annual volumes of the printed *SCI* began only in 1961. The retrospective volumes covering the years prior to 1965 were produced long after that date.

4.2.1 Citation and Author Searching in the *Science Citation Index*

Whenever a new article is written, the author of the scientific or technical journal article invariably draws upon the past work in that area of research. It is the usual case that each new article includes a number of older references in the bibliography. Those items are referred to as **cited references**. The new article can then be termed the **citing reference**. (Bibliographic **citation** and bibliographic **reference** refer to the same thing: the bibliographic information which completely identifies a document.) In *Science Citation Index*'s Source Index each year are listed the new articles (citing references) which have appeared in the more than 3300 primary source journals included in *SCI*'s coverage. The Source Index arranges those entries alphabetically by the first author (primary author) of the article. A maximum of nine secondary authors (co-authors) of the same article are cross-referenced to the primary author. Thus, each year the *Science Citation Index* provides an author index to the hundreds of thousands of new primary source articles which are published that year.

The *SCI* Citation Index is an index which uses the cited references in the bibliographies of the source journal articles as a way of finding the newer articles. The reason for constructing a citation index is simple. There is usually a direct subject relationship between the previously published articles, books, patents, etc. which are cited in a new article and the subject content of that new work. Consequently, if we know an older article, book, patent, etc., the Citation Index for *SCI* lets us find *newer* articles which have cited that work.

Figure 4-1 shows how the *SCI* Citation Index is used to find relevant references. In the example, it is assumed that the 1975 article by Davidson, "The Iterative Calculation of a Few of the Lowest Eigenvalues and Corresponding Eigenvectors of Large Real-Symmetric Matrices," published in volume 17 (1975) of the *Journal of Computational Physics*, was already known to the searcher.

The unique feature of citation indexing is that it leads to a bibliography of newer works on a subject without having to use subject terms. In other words, the *Science Citation Index* lets the searcher work forward in time to compile a bibliography by using an earlier relevant work. Since the earlier work (the cited reference) might be cited over a time period of a large number of years, the *SCI* leads to relevant newer citing references published right up to the present. This means that even an article published in the 19th century, if cited after World War II in any of the *SCI* source journals, could be used as an entry point in the *SCI* Citation Index. Furthermore, since *Science Citation Index* is multi-disciplinary in its coverage, citing references may be found from related disciplines which might be overlooked in a search of a traditional subject index.

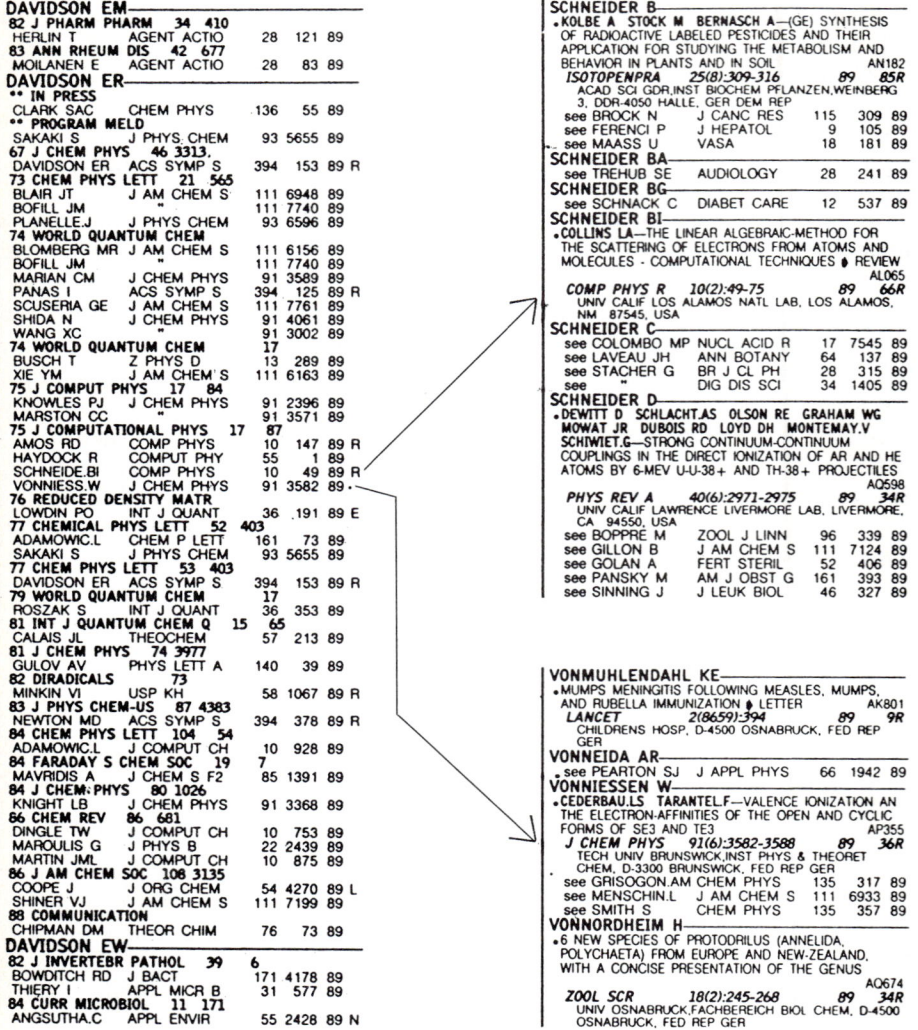

Figure 4-1
Use of the Citation Index to *Science Citation Index*

In 1980-84 the average article in *SCI*'s Source Index cited over 14 references, thus providing a depth of indexing which is usually not found in a traditional subject index. Important scientific works continue to be cited decades after their publication. Even though subject terms may evolve and change over the years, a correct bibliographic citation should be a constant. Unfortunately, scientific authors are not always skilled in constructing a correct bibliographic citation, so errors can be found in the

Citation Index. If the compilers of *Science Citation Index* are made aware of such errors in time, they correct them in the printed annual and cumulated editions. ISI has also developed computer routines to find variant references to the same cited article and to unify them in the Citation Index.

There are certain conventions followed by ISI in listing names in the Source Index and the Citation Index. In the Citation Index, a maximum of 15 characters is allowed for the last name of a cited author. If the last name is longer than that, a period is used to truncate the name after the fourteenth character. For the citing authors, a maximum of 9 characters is allowed, with truncation again being indicated by a period after the eighth character if necessary. Compound names are fused, and hyphens are dropped in such names. Since 1981, articles like van, Von, de La, Di, etc. which precede the name are also fused to the last name. (The exception to this rule is the Spanish or Portuguese y and e, which are treated as initials.) Prior to 1981, the *SCI* volumes list the name as the individual author preferred it. All names in the 1980-84 cumulation were standardized, however. Russian names are transliterated according to the conventions laid down by the British Standards Institute in cases where they have not already been transliterated. If the source journal has transliterated a Russian name, then the standard used in the journal is accepted for *SCI*'s entry. In the case of an anonymous citing work, the journal's abbreviated title appears in place of the author with a black square block indicating that it was anonymously written. Anonymous cited items follow the letter Z in the Citation Index and are arranged by journal title abbreviation.

4.2.2 Citation and Author Searching in SciSearch

At the present time, the only U.S. vendors which have the online *Science Citation Index* database SciSearch are Dialog and ORBIT. The appropriate Dialog files, which take the online coverage back to 1974, are files 34, 434, 433, and 432. Table 4.2 shows both the DIALOG and ORBIT files for SciSearch. By using the OneSearch feature on Dialog, it is possible to search all of the SciSearch files at once or segments covering 1980 to the present or 1987 to the present. (See Figure 5-3.)

The online version of *Science Citation Index* is not edited, and, therefore, is not as accurate as the printed annual or cumulative editions of *SCI*. Even though the database could long be searched retrospectively to 1974, no attempt was originally made to correct errors using ISI's computer routines for SciSearch. Thus, the SciSearch database contained the records in the form in which they were originally produced for

TABLE 4.2
SciSearch on DIALOG and ORBIT

DIALOG File no.	Time Period	ORBIT Filename	Time Period
34	current: 6 mos.-1 year	SCINEW	Latest 4-7 wks
434	1987-beginning of File 34	SCISEARCH	1985 to date
433	1980-86	SCISEARCH74	1974-84
432	1974-79		

the weekly tapes sent to the vendors. The recently reloaded file of SciSearch now incorporates the corrections which went into the annual *SCI* issues and contains the many additional references which went into the *SCI* cumulations. Over 38,000 new source articles were added to the 1980-84 *SCI* cumulation alone. Those references were added because the new source journals had emerged as more important to their fields than was originally thought. Thus, the SciSearch database contains many more records than are found in the annual or cumulative printed editions of *Science Citation Index* due to the inclusion of additional records from *Current Contents*. The online database is updated weekly, making it extremely valuable for current awareness purposes.

Author searches in SciSearch can retrieve publications by any individual author's name, whether primary or secondary. The strategy on DIALOG would be something like:

?S AU = LASTNAME IIIII

(The "?" at the beginning of the statement is the DIALOG prompt, and the "S," the command to search.) Up to five initials are allowed. All of the conventions for names in the printed *SCI* are, of course, applicable for the database. That is, compounds names are fused, and no punctuation is used.

For a cited reference search, the format is:

?S CR = LASTNAME II, YEAR, V#, P#

Thus, only the primary author's name, year of publication, the volume number of the journal, and the beginning page number of the article are necessary for an online citation search. The search in Figure 4-2 was run in file 294, DIALOG's ONTAP SciSearch practice file, using a 1979 paper by C. H. Seager. Note that the first step was to use the EXPAND command (abbreviated "e") in order to examine the inverted file of cited references by authors with last names similar to Seager.

The limit for last names in a cited reference search is 15 characters. To find all of an author's works which have been cited in SciSearch, use the strategy:

?S CR = LASTNAME I,?

(The second ? is a truncation symbol, indicating anything is acceptable at this point.) It must be remembered that only the primary (first-listed) author can be searched as a cited reference in both the printed *SCI* and the SciSearch database. For the prolific author Jay K. Kochi, many, if not most of the entries in his long list of publications do not show him as the primary author using ISI's definition. Nevertheless, papers with Dr. Kochi as co-author are some of the most highly-cited works in chemistry. He is listed as the co-author of 27 works in the Source Index of the 1980-84 *Science Citation Index* and of 26 works in the 1982-86 Collective Author Index to *Chemical Abstracts*. Obviously, looking up all of those first-named authors would be very time-consuming, both in a manual and an online search. Nevertheless, it could be done! A quick check of the original source journal (the original primary article) will usually reveal who is the senior author. An asterisk (*) generally signifies the person to whom correspondence should be addressed.

File 294:ONTAP SCISEARCH (Copr. ISI 1985)
Set Items Description
?e cr = seager ch
Ref   Items   Index-term
E1   1   CR = SEAGARD JL, 1983, V58, P432
E2   1   CR = SEAGER AF, 1953, V30, P1
E3   0   *CR = SEAGER CH
E4   3   CR = SEAGER CH, 1979, V34, P337
E5   1   CR = SEAGER CH, 1981, V52, P1050
E6   2   CR = SEAH MP, 1972, V32, P703
E7   1   CR = SEAH MP, 1973, V3, P1538
E8   1   CR = SEAH MP, 1973, V335, P191
E9   1   CR = SEAH MP, 1973, V40, P595
E10   1   CR = SEAH MP, 1975, V31, P627
E11   1   CR = SEAH MP, 1975, V53, P168
E12   1   CR = SEAH MP, 1977, V25, P345
            Enter P or E for more?
?s e4
    S1   3   CR = "SEAGER CH, 1979, V34, P337"

?t 1/5/1-3

1/5/1
1806810 DOC TYPE: ARTICLE   GENUINE ARTICLE#: RV933 11 REFS
 POLYCRYSTALLINE ZN3P2/INDIUM-TIN OXIDE SOLARCELLS (ENGLISH)
 SUDA T; SUZUKI M; KURITA S
 INST VOCAT TRAINING,DEPT ELECTR,1960
 AIHARA/SAGAMIHARA/KANAGAWA
229/JAPAN/; KEIO UNIV,DEPT ELECT ENGN/YOKOHAMA/KANAGAWA 223/JAPAN/
 JAPANESE JOURNAL OF APPLIED PHYSICS PART 2-LETTERS , V22, N10,
PL656, 1983

1/5/2
1804212 DOC TYPE: ARTICLE   GENUINE ARTICLE#: RV489   9 REFS
 PASSIVATION OF DRY-ETCHING DAMAGE USING LOW-ENERGY HYDROGEN
IMPLANTS (ENGLISH)
 WANG JS; FONASH SJ; ASHOK S
 PENN STATE UNIV,ENGN SCI PROGRAM/UNIVERSITY PK//PA/16802;
CHUNGNAM NATL UNIV/CHUNGNAM//SOUTH KOREA/
 IEEE ELECTRON DEVICE LETTERS , V4, N12, P432-435, 1983

1/5/3
1794863 DOC TYPE: ARTICLE   GENUINE ARTICLE#: RU897   18 REFS
 DEEP LEVEL EFFECTS IN SILICON AND GERMANIUM AFTER PLASMA
HYDROGENATION (ENGLISH)
 PEARTON SJ; KAHN JM; HALLER EE
UNIV CALIF BERKELEY LAWRENCE BERKELEY LAB/BERKELEY//CA/94720
 JOURNAL OF ELECTRONIC MATERIALS , V12, N6, P1003-1014, 1983

FIGURE 4-2
Cited Reference Search in the ONTAP SciSearch File.

One very powerful technique in searching SciSearch for cited references is **co-citation**. Finding a new paper which has cited two or more key papers increases the chances that the retrieved item will be a good hit. To do a co-citation search, simply truncate after the first initial of each primary article as in:

?S CR = AUTHOR1 I? AND CR = AUTHOR2 I?

For assistance with SciSearch searching, call the ISI toll-free number 800-523-1857.

4.2.3 Using the Corporate Source Approach in SciSearch and the *Science Citation Index*

One of the fields available in the SciSearch database is the Corporate Source (CS) Field. This can be thought of as an address field for the authors who publish in the SciSearch source journals. The CS field can be of great help when searching online for source publications of authors with common names. The authors' work locations are generally included in scientific journal articles. If the searcher knows the location, that information can be combined in an online search, for example:

?S CS = (INDIANA(S)CHEM) AND AU = WILLIAMS D?

It is necessary to use the parentheses and the (S) subfield operator in the Corporate name portion of the strategy on DIALOG.

```
? ss cs=(univ?(w)calif? (s) los(w)angeles) or cs=ucla
      S2        13079    CS=UNIV?
      S3          757    CS=CALIF?
      S4          671    CS=UNIV?(W)CS=CALIF?
      S5          254    CS=LOS
      S6          217    CS=ANGELES
      S7          217    CS=LOS(W)CS=ANGELES
      S8            0    CS=UCLA
      S9          134    CS=(UNIV?(W)CALIF? (S) LOS(W)ANGELES) OR
                         CS=UCLA?
?ss s9 and gene?
                  134    S9
     S12          762    GENE
     S13           18    S9 AND GENE?
?t 13/2/1
```

1814188 DOC TYPE: ARTICLE   GENUINE ARTICLE#: RW850   52 REFS
  A MULTIVARIATE GENETIC ANALYSIS OF RIDGE COUNT DATA FROM THE OFFSPRING OF MONOZYGOTIC TWINS (ENGLISH)
   CANTOR RM
   UNIV CALIF LOS ANGELES,LOS ANGELES CTY HARBOR MED CTR,SCH MED,DEPT PEDIAT,DIV MED GENET/TORRANCE//CA/90509
     ACTA GENETICAE MEDICAE ET GEMELLOLOGIAE , V32, N3-4, P161-207, 1983

FIGURE 4-3
Corporate Name Search in the ONTAP SciSearch File.

In Figure 4-3, the search statements are input with the DIALOG command SS, "Select Steps," which causes a set number to be assigned to each individual step in the search process. This technique is useful when a search may have to be modified, since it removes the necessity to key in previous elements of the search statement. The "t 13/2/1" is the manner in which an answer is displayed on DIALOG, where 13 refers to a set number, 2 refers to a pre-defined format for the output, and 1 is the answer number to be displayed.

The printed *Science Citation Index* includes a Corporate Index before the Source Index. The Corporate Index is actually divided into two sections. The Geographic Section is arranged alphabetically by location, that is, by country and city for all locations except the U.S.A. In the U.S.A. section, the arrangement is by state and city. Under each city's name is an alphabetical list of the organizations which the authors list as their home institutions. Standard abbreviations are sometimes used, such as CUNY for the City University of New York. The listing under the corporation contains the primary author, abbreviated journal, volume, page and year. Thus, the Corporate Index provides a means of finding all publications by the personnel of a given organization which have appeared in the source journal publications for that year. Figure 4-4 shows the *SCI* Corporate Index listing and a reference from the Source Index for an Indiana University Department of Chemistry author.

The second part of the Corporate Index is an Organization Section. A searcher would use this when not sure of the location of an organization. For example, looking under "Eli Lilly & Co." in the Organization section of any *SCI* annual Corporate Index shows two locations in Indiana: Greenfield and Indianapolis. It is then a simple matter to find publications written by Lilly personnel by using the Geographic Section as before.

One fairly common mistake which novice searchers make when doing a subject, author, or citation search is to enter the Corporate Index section, thinking that they are in the Source Index. Those who fail to recognize that the Corporate Index is published in the first volume of the annual *SCI* and precedes the Source Index entries can be perplexed if an author's name is sought in the Corporate Index section.

4.2.4 Author and Corporate Searches in Other ISI Products

Both *Current Contents* and other ISI products, such as the *Index to Scientific and Technical Proceedings*, include author indexes. Each weekly issue of *Current Contents* has such an index which lists the first author's name and address. Since most libraries discard *Current Contents* at frequent intervals, the author indexes in these works are only useful for the most recent literature. In fact, they are not cumulated.

The *Index to Scientific and Technical Proceedings* not only has an index for the authors of papers in conference proceedings volumes and the editors of the volumes, it also contains a Sponsor Index. There one can find up to ten corporate sponsors of a given conference tied to the main listing in the Contents of Proceedings section. There is also a Meeting Location Index and a Corporate Index. The Corporate Index to *ISTP* is structured just like the Corporate Index to *Science Citation Index*. There is

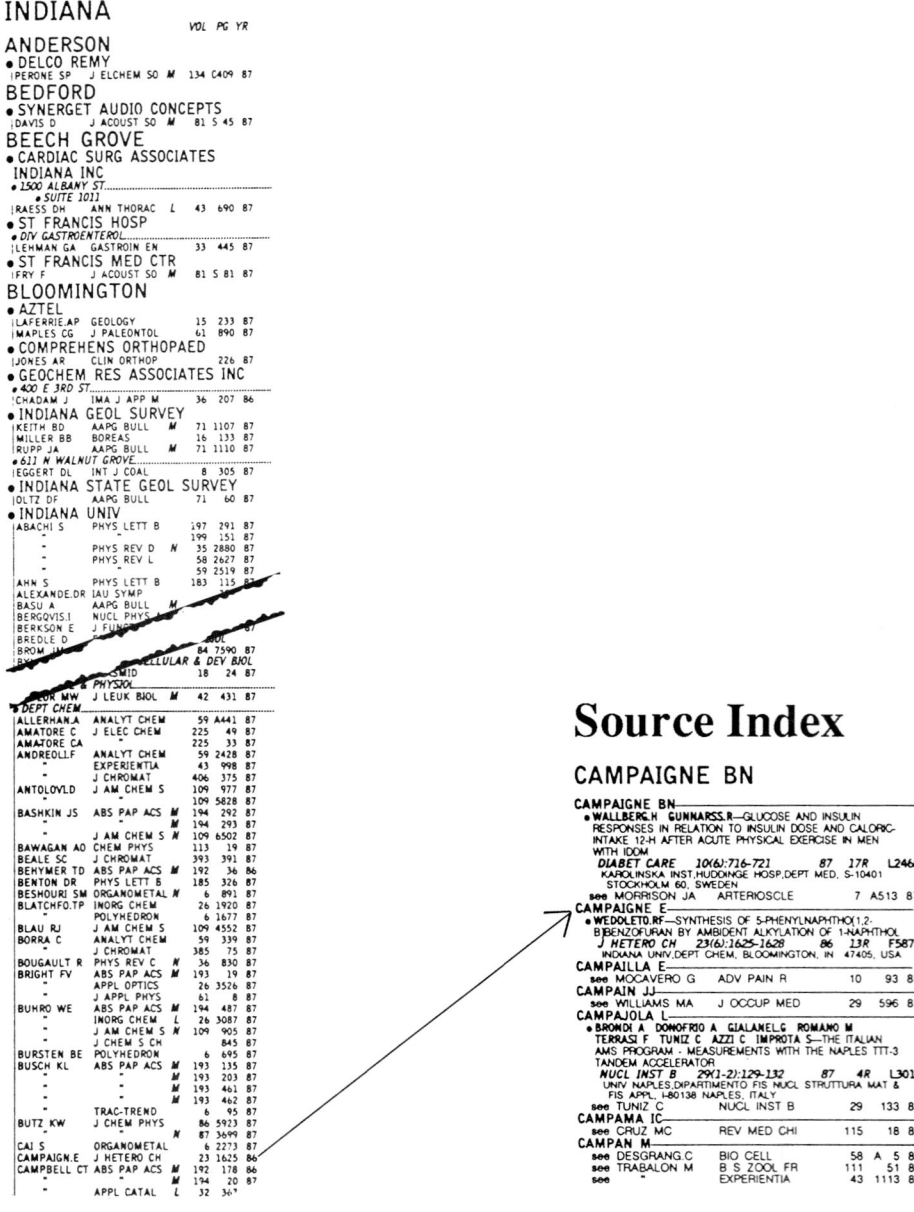

Figure 4-4
Search by Corporate Name in the *Science Citation Index* Corporate Index

a Geographic Section which leads to the papers in the Contents of Proceedings section and an Organization Section to help those who do not know the geographic location of an organization.

4.3 AUTHOR AND CORPORATE SEARCHES IN *CHEMICAL ABSTRACTS*

Chemical Abstracts has always included an author index. The index covers not only authors of journal articles, dissertations, books, reports, and other primary literature, but also patentees and patent assignees. A **patentee** is the person who has applied for and received a patent. In the United States and some other countries, a patent can be issued only to a person, not to a corporation. The patentee may then assign the rights to use an invention to the **patent assignee**, normally the company the inventor works for or sells the invention to. Since patents do not constitute a part of the primary source literature for *Science Citation Index*, there are no patentees or patent assignees found there. However, about one-sixth of the half million documents added to the *Chemical Abstracts* database each year are patents. Hence, the patentees and patent assignees make up a substantial portion of the entries in the printed *CA* Author Index. In that index, references to patents can easily be recognized by the "P" inserted between the volume number and the abstract number, as in 103:P160286w, found in the 11th Collective Author Index under Grieco, Paul A.

In the *Chemical Abstracts* Author Indexes, co-authors are included at the name of the first author. Cross-references are made from their names to the first author's name, and a long dash is used in place of the first author's name at the point where co-authors are listed under the first author. Unlike *Science Citation Index*, *Chemical Abstracts* inverts corporate names. In *CA*, the proper place for "Eli Lilly & Co." is: "Lilly, Eli, and Co.", interfiled with the personal names "Lilly".

The printed Author Indexes to *Chemical Abstracts* are easy to use if a few rules are known. Entries for authors with the same last name are distinguished first by the *initials* of the given names. This is true even though the full given names may be spelled out in the original publication! While it may seem strange to do so, this keeps all entries for an author together, regardless of the form of the name in the original. Thus, the entry for "Davidson, Ernest Roy" comes *after* the entry for "Davidson, Eugene Abraham" and before the entry for "Davidson, Elizabeth West" in the *CA* 11th Author Index!

In the alphabetizing rules for the *Chemical Abstracts* Author Indexes, personal and other names which have the least information for the given names come first. The characters in the given names which follow the initial letters come into play in cases of conflict, with something preceding nothing. Thus, "Brown, Donald R." precedes "Brown, D. Robertson" since *D*onald *R*. comes before *DR*. Like *SCI*, the printed *CA* alphabetizes hyphenated names as a unit, but retains the hyphen in the printed product. However, in the online CA File, hyphens are dropped in compound names, as shown in Figure 4-5 in a search for all publications of Aksel A. Bothner-By. The second component of the name is simply ignored in this case. Note that it must be input *without* the hyphen. If necessary for alphabetizing, the second part of a hyphenated last name does come into play. For example: OLIVER RUIZ, MANUEL/AU precedes OLIVER SMITH, DAVID/AU, which in turn precedes OLIVER YANEZ, S/AU in the inverted author file. Note also in Figure 4-5 that the EXPAND is very useful for spotting misspellings or data input errors, as with Bothner, Mich*a*el H. vs. Bothner, Mich*ea*l H.

Chemical Abstracts continues to follow the traditional library practice of filing names with prefixes "Mc" or "M" as though they were spelled "Mac". Likewise, "St." is alphabetized like "Saint". All other abbreviations and acronyms are alphabetized as they are spelled. Foreign prefixes of names like De, Des, Della, etc. are spelled as written by the authors, but alphabetized as if the names were fused. German names with an umlaut are spelled "ae," "ue," or "oe" in place of the respective umlauted letter.

In the online versions of *Chemical Abstracts* (and most other databases, for that matter) it is very advisable to use an EXPAND or NEIGHBOR command in conjunction with the author entry to discover variant forms of the author's name in the database. Although Chemical Abstracts Service has been encouraging authors in recent years to publish with their full first names, there are still many authors who only use initials in place of given names. This practice was much more common in earlier years. Consequently at least two versions of an author's name are likely to be in the database

```
=>   expand bothner by/au
     E1             1      BOTHMANN, PETER/AU
     E2             1      BOTHMIEDEMA, R/AU
     E3             0      BOTHNER BY/AU
     E4             3      BOTHNER BY, A/AU
     E5            26      BOTHNER BY, A A/AU
     E6            38      BOTHNER BY, AKSEL A/AU
     E7             1      BOTHNER BY, AKSEL A EDITOR/AU
     E8             1      BOTHNER BY, AKSEL ARNOLD/AU
     E9             3      BOTHNER, BRIAN/AU
     E10            8      BOTHNER, M H/AU
     E11            5      BOTHNER, MICHAEL H/AU
     E12            1      BOTHNER, MICHEAL H/AU
=>   expand bothner-by, a/au
     E1             5      BOTHNER, MICHAEL H/AU
     E2             1      BOTHNER, MICHEAL H/AU
     E3             0      BOTHNER-BY, A/AU
     E4             1      BOTHO, BOEHNKE/AU
     E5             1      BOTHOF, TH/AU
     E6            12      BOTHOR, C/AU
     E7             1      BOTHOR, CARLDIETER/AU
     E8             1      BOTHOR, W/AU
     E9             1      BOTHOREL RAZAZI, M P/AU
     E10            1      BOTHOREL, ALAIN/AU
     E11            1      BOTHOREL, MARIE PIERRE/AU
     E12            1      BOTHOREL, MELLE M P/AU
```

FIGURE 4-5
Use of the EXPAND Command in an Author Search for a Hyphenated Name in the STN CA File.

for older, established authors. Thus, in an online search, two (or possibly more) variants of the name may have to be used depending on how common the last name is, for example, => **EXPAND Davidson, E R/AU** as well as => **EXPAND Davidson, Ernest/AU**.

Although it is not possible to be absolutely sure that two variants of a name represent a single author, it is better to err on the safe side and include all forms which appear to satisfy the search. Alternatively, the name can be combined with the work location (Corporate Source field) or even the zip code of the author's home organization, if desired. However, the CA File includes the Corporate Source only for the

```
FILE 'CA' ENTERED AT 18:10:46 ON 23 MAR 86
COPYRIGHT 1986 BY THE AMERICAN CHEMICAL SOCIETY
 =>      expand parmenter, c/au
         E1              1       PARMENOVA, K V/AU
         E2              4       PARMENOVA, V A/AU
         E3              0       PARMENTER, C/AU
         E4              2       PARMENTER, C M/AU
         E5             44       PARMENTER, C S/AU
         E6              1       PARMENTER, CAROL M/AU
         E7             25       PARMENTER, CHARLES S/AU
         E8              1       PARMENTER, D V/AU
         E9              1       PARMENTER, DOUGLASS V/AU
         E10             1       PARMENTER, KEITH/AU
         E11             1       PARMENTER, M/AU
         E12             1       PARMENTER, MARK/AU
 => search e5 or e7
                        44       "PARMENTER, C S"/AU
                        25       "PARMENTER, CHARLES S"/AU
         L12            69       "PARMENTER, C S"/AU OR "PARMENTER, CHARLES S"/AU
 => s L12 and jacsat/co range=(1975,)
                     20998       JACSAT/CO
         L13             2       L12 AND JACSAT/CO
 => display 1

L13 ANSWER 1 OF 2

AN    CA83(2):18204t
TI    Extended view of the benzene 260-nm transition via single vibronic level fluorescence. II. Single
      vibronic level fluorescence as a probe in the assignment of the absorption spectrum
AU    Knight, A. E. W.; Parmenter, C. S.; Schuyler, M. W.
CS    Dep. Chem., Indiana Univ.
LO    Bloomington, Indiana, USA
SO    J. Am. Chem. Soc., 97(8), 2005-13
SC    73-3 (Spectra by Absorption, Emission, Reflection, or Magnetic Resonance, and Other Optical
      Properties)
DT    J
CO    JACSAT
PY    1975
LA    Eng
```

FIGURE 4-6
Use of the EXPAND Command in the STN CA File for an Author Search Limited to a Particular Journal.

first author listed. Figure 4-6 shows another expanded author entry in the CA File on STN, in this case, combining the results with articles appearing in the *Journal of the American Chemical Society* and ranging the date of coverage in *CA* from 1975 to the present. Note that the CODEN "JACSAT" is used in place of the journal title. CODENs can be found in the *Chemical Abstracts Service Source Index (CASSI)* [229] and are often published in the journals. Also, note that the answer is displayed in the BIB format, the default option in the STN CA File.

Another alternative which can be used in an online search is truncation (see 5.2.1.) to gather into one set all variants of an author name. This technique is generally less precise than using the EXPAND option, so most author searches should be performed with the EXPAND option. However, if the complete name of an author is not known or the searcher is unsure of the spelling, truncation can be an effective solution. Thus, using:

=> **S PARMENTER, C?/Au**

as a search strategy would pull 3 false drops in Figure 4-6.

The Corporate Source field (/CS) is where patent assignees are listed in the CA File. Thus, a search using a strategy with an adjacency operator:

=> **S ELI/CS(A)LILLY/CS AND P/DT**

should find all patents assigned to the Eli Lilly Company.

Available from Chemical Abstracts Service is a good computer-assisted instruction program *STN Mentor: Spotlight on Authors* [242]. Search **CAI** in the Chemistry Reference Sources Database to find other computer-assisted instruction programs for other facets of online searching.

4.4 SUMMARY

Searches which use the name of a personal or corporate author can be quite effective. When a very productive author is known to be working in a field of interest, relevant references can often be found which might be overlooked in a traditional subject search. This is especially true if the technique of citation searching is employed, using either the printed *Science Citation Index* or the SciSearch database.

While author searches are generally easier than subject searches, the conventions for spelling and alphabetizing names must be learned before an author search is done. Many of those conventions for *Chemical Abstracts* and *Science Citation Index* are discussed in this chapter.

4.5 SELECTED READINGS

DIALOG Basics: A Brief Introductory Guide to Searching; DIALOG Information Services, Inc.: Palo Alto, CA, 1986. [926]

SciSearch User Guide: Dialog; Institute for Scientific Information: Philadelphia, PA, 1989. [267]

Chronister, Diane. "ISI Databases: Cited Reference Searching." *Database* **1989**, *12*(5), 101-103.

CHAPTER 5

SEARCHING BY SUBJECT

5.1 INTRODUCTION

A commonly-used technique for gathering information is subject searching. The amount of information which can ultimately be found through subject searches in bibliographic and text databases (or in printed abstracting or indexing services and other secondary reference works) is almost limitless. In such works, one can find the results of research on chemical compounds, including their reaction chemistry and properties, as well as information on many other topics. Reference tools such as *Science Citation Index*, *Chemical Abstracts*, and others are well suited to subject searches.

5.2 SUBJECT SEARCHES

Subject searches are word searches. The key to their success is the use of the right word or groups of words in a properly formulated search strategy. Many of the information needs of chemists are met by subject searches. In fact, using a name of a chemical compound to search an index is really a type of subject search. The difficult part of any subject search is the selection of the appropriate word or words for the technique, process, type of reaction, name of a compound, or other concept in which the searcher is interested. In order for a subject search to be successful, the words chosen by the searcher must match the words in the subject index as prepared by a human indexer or compiled by a computer program.

Some problems which may arise in subject searches are lack of uniformity in the assignment of index terms, variant spellings, variant treatments of punctuation, conventions which have been developed for abbreviations or segmentation of words, and the problem of synonyms and related terms. It is beyond the scope of this work to deal with all of these situations in depth. However, a few hints will be given in this section to lessen the impact of such problems. There are a number of user aids and online search techniques which can be of assistance when searches involving subject words are performed. Among them are truncation, expanding, and the use of subject term authority lists.

5.2.1 Truncation and Masking vs. Neighboring or Expanding

Many computer search systems offer the option of truncating search terms or masking characters in order to broaden an online search. **Truncation** is the technique which signals the computer to gather into one set all of the records containing a given character string, including all search terms in the inverted dictionary file which contain those characters *or more*. (Some call this the use of "wild cards".) In contrast, **neighboring** or **expanding** allows the searcher to look at an alphabetically arranged section of the inverted dictionary file and select for searching only the relevant terms.

Truncation can occur on the left side of the character string to retrieve terms which have the string embedded in them. This is called **left-hand truncation**. Very few systems offer left-hand truncation. Much more common is **right-hand truncation**, which occurs on the other side. An example of a search on a system in which an asterisk (*) is used as a truncation symbol would be:

Search *phenyl*

Not only terms which start with the character string "phenyl," like "phenylanaline" and "phenylation," but also words like "biphenyl" would be retrieved.

Look at the truncation and masking options which STN International offers, as shown in Table 5.1. The # and ? must be used at the end of a term, whereas the ! may be used either at the end or internally. "Cataly!e" would retrieve only the variant spellings "catalyze" or "catalyse," but not "catalyzed". "Alcohol#" would select "alcohol" or "alcohols," but not "alcoholic". "Cataly?" would select "catalyze," "catalysis," "catalyst," "catalysts," and any variant spellings based on "catalyse". Other vendors such as Dialog, Questel, and ORBIT use different truncation symbols to perform the same tasks.

Caution must be exercised in using truncation. One inexperienced searcher tried to pull all references to "ions" or "ionization" in an SDI search system which allows both left-hand and right-hand truncation. The search was input as "*ion*" to retrieve cation, anion, etc., as well as ion, ions, ionized, ionization, etc. An enormous number of false drops resulted because of the many words which end in English with the suffix "ion," such as trunca*tion*! It is far better to be on the safe side in using truncation and not truncate too near the beginning of a word.

Despite its unpredictable results, truncation is sometimes a very valuable technique. Truncation can be used in searching both the Registry and CA Files on STN

TABLE 5.1
Truncation and Masking Symbols on STN International

Symbol	Function	Example
exclamation point (!)	Exactly one character	Search cataly!e
hash mark (#)	Zero or one character	Search alcohol#
question mark (?)	Any number of characters, including zero	Search cataly?

International. Thus, molecular formulas, compound names, etc. found in the Registry File, as well as subject words, authors' names, and the like in the CA File can be truncated. However, there are limits in all systems to the number of unique terms which can be handled through truncation. For instance, in both the CAS ONLINE Registry and CA Files, the limit of search terms assembled in one search statement by truncation is 2000. Inputting => **SEARCH ORGANO?** in the CA File yields the response "TRUNCATION LIMITS EXCEEDED—SEARCH ENDED."

Truncation can be a very powerful retrieval technique. However, it can also result in false drops. Consider a search for an English-language review article on organomagnesium complexes. A searcher might intuitively assume that truncating the input after the second "o" in "organo" would not be a reasonable choice, even in a small database. In large databases such as those corresponding to *Chemical Abstracts*, a quick check of the inverted dictionary file with an EXPAND or NEIGHBOR command is sometimes a better option than truncation. For example, inspection of the Basic Index of the CA File on STN International with the command => **EXPAND ORGANO** revealed at the time of this search that 130 terms start with the character string organo*a* alone! Using the EXPAND option would be a better strategy, as Figure 5-1 shows.

Ten reviews on organomagnesium complexes which were published in English were found in the search in Figure 5-1. Note that the E#'s for the entries in the snapshot of the Basic Index are combined with an OR statement. The expanded list even reveals one document in set E9 which has the term misspelled as "organomagnesium". Without the EXPAND, we might have missed this document.

5.2.2 Key-Word Searches

The terms in a subject search comprise one "**key**" (search term or phrase) which unlocks the treasure chest of knowledge in chemistry. Some people, therefore, refer to such words as **key words**. Let us restrict the phrase "key-word search" in this book to the kind of uncontrolled-vocabulary subject search that is possible with some indexes, generally, those subject indexes which are computer-produced (See Figure 13-2). Before a key-word subject index is compiled, someone has programmed a computer to look at a body of text in a given work, and the computer selects the terms to be used for indexing. The computer program then produces the alphabetized key-word subject index.

It is important to know two things about key-word, computer-produced subject indexes (generally known as "KWIC," Key-Word-in-Context or "KWOC," Key-

```
=>   expand organomagnesium
ENTER FIELD CODE (BI):.
E1                    1    ORGANOMAGNESIATE/BI
E2                    1    ORGANOMAGNESIATES/BI
E3                  823    ORGANOMAGNESIUM/BI
E4                    1    ORGANOMAGNESIUMALUMINUM/BI
E5                    2    ORGANOMAGNESIUMOXANE/BI
E6                   59    ORGANOMAGNESIUMS/BI
E7                    1    ORGANOMAGNETIC/BI
E8                    1    ORGANOMAGNETISM/BI
E9                    1    ORGANOMAGNSIUM/BI
E10                   1    ORGANOMANGANATE/BI
E11                   1    ORGANOMANGANATES/BI
E12                  74    ORGANOMANGANESE/BI

=>   s e3 or e6 or e9
                    823    ORGANOMAGNESIUM/BI
                     59    ORGANOMAGNESIUMS/BI
                      1    ORGANOMAGNSIUM/BI
L5                  861    ORGANOMAGNESIUM/BI OR ORGANOMAGNESIUMS/BI OR
                           ORGANOMAGNSIUM/BI
=>   s L5 and review
                 751781    REVIEW
L6                   26    L5 AND REVIEW
=>   s L6 and eng/la
                5156444    ENG/LA
L7                   10    L6 AND ENG/LA
=>   d bib abs

L7   ANSWER 1 OF 10


AN   CA111(11):97301v
TI   Bridging the fields of organometallic and classical coordination chemistry: localized and delocalized
     bonding in polynuclear complexes of (C5R5)(CO)2Mn
AU   Kaim, Wolfgang; Gross, Renate
CS   Inst. Anorg. Chem., Univ. Stuttgart
LO   Stuttgart D7000/80, Fed. Rep. Ger.
SO   Comments Inorg. Chem., 7(5), 269-85
SC   29-0 (Organometallic and Organometalloidal Compounds)
DT   J
CO   COICD2
IS   0260-3594
PY   1988
LA   Eng
AN   CA111(11):97301v
AB   A review contg. 37 refs. The capability of the title fragment to undergo efficient .pi. back bonding,
     to exist in two kinetically stable neighboring oxidn. states, and to exhibit a relatively small ligand
     field spitting allow the formation of unusual coordination compds. Several mono- and polynuclear
     examples with an odd or even electron count and different degrees and direction of electron delo-
     calization between ligand and metal fragment(s) are discussed.
```

FIGURE 5-1
Use of the EXPAND Command During a Subject Search for a Review in the CA File.

Word-Out-of Context indexes). First, the subject terms selected are those used by the authors of the primary works being indexed, generally taken from the titles of the works. This can cause problems. Suppose we were searching for information on X-ray photoelectron spectrometry (XPS) using a key-word subject index. This particular analytical technique is also known as electron spectrometry for chemical analysis (ESCA). Where should we look in a key-word subject index? If we search only in the E's, but some authors used "X-ray photoelectron spectrometry" or "XPS" in the titles of their articles or other primary works on this topic, the search is in trouble. Why? Because the computer will construct the key-word index with some of the primary works listed in the "E" section of the alphabetical index and others in the "X" section *with no cross references or see references* to tie the two together. Therefore, before beginning a search in a key-word subject index, likely synonyms and acronyms for names of the compounds, processes, analytical techniques, etc. must be identified. *Each* of those terms must then be found in the appropriate place in the alphabetical index. Reference works with uncontrolled-vocabulary subject indexes include *Current Contents*, *Science Citation Index*, *Chemical Titles*, and the weekly subject indexes to *Chemical Abstracts*.

The second important point about key-word subject indexes is that they frequently employ a **stop list** to cut down on irrelevant posting of references on terms such as "the," "if," "for," "toward," etc. These words and others like them have little or no retrieval significance. The computer simply checks each word in the text against the stop list and does not index on that term if a match is found. Since key-word subject indexes are usually produced from the titles of the works being indexed, it is extremely important for authors of scientific works to compose very descriptive titles and to be aware of the role of stop words in key-word indexes. Each significant (non-stop) word becomes a subject index term in the key-word subject index.

5.2.3 The Permuterm Subject Index to *Science Citation Index*

In chapter 4 is a full discussion of citation searching. Here we will concentrate on the use of the Permuterm Subject Index to *Science Citation Index*. Every article, letter to the editor, review, or correction notice, in short, everything that is not an advertisement, news notice, or other ephemeral item in the journals which are covered by *SCI* becomes a source item in *SCI* and, as a consequence, gets Permuterm Subject Index entries. Each source item is coded, as shown in Table 5.2.

All articles from review journals are coded R or REVIEW as are most papers with more than 100 references. To search online in SciSearch by document type on DIALOG, use the strategy ?**S DT = CODE AND SUBJECT**. "Review or Bibliography" must be truncated as **REVIEW?** for the period through 1988. Beginning in 1989, reviews are separately indexed from bibliographies, so the strategy ?**S DT = REVIEW** retrieves reviews and ?**S DT = BIBLIOGRAPHY** retrieves bibliographies. To find only journal articles on DIALOG, an answer set must be limited, as in: ?**L S2/ART**, where S2 is the designation for the answer set formed in a previous search and L is the DIALOG **Limit** command.

TABLE 5.2
Codes for Source Items in *Science Citation Index* and SciSearch

SCI Code	SciSearch Code	DIALOG or ORBIT	Meaning
blank	ARTICLE	U	journal article
B	BOOK REVIEW	B	book review (only from *Science* or *Nature*)
C	CORRECTION, ADDITION	C	correction
D	DISCUSSION	D	discussion (from a conference)
E	EDITORIAL	E	editorial or interview
I	ITEM ABOUT AN INDIVIDUAL	I	biographical item
K	CHRONOLOGY	K	chronology
L	LETTER	L	letter to the editor (concerning previously published material)
M	MEETING ABSTRACT	M	meeting abstract (from a conference)
N	NOTE	N	note (brief article)
R	REVIEW, BIBLIOGRAPHY	R	reviews plus bibliographies, 1974-88
R	REVIEW	R	reviews, 1989-
BI	BIBLIOGRAPHY	7	bibliographies, 1989-
			computer reviews for:
W	HARDWARE REVIEW	8	hardware, 1987-
	SOFTWARE REVIEW	9	software, 1987-
	DATABASE REVIEW	0	databases, 1987-
	PRESS DIGEST	1	press digest

ORBIT allows the use of both words and one-character codes when performing searches by document type.

In the Permuterm Subject Index to *SCI*, every significant word in the title of a source item is paired with every other significant title word which is not on the stop list. Full stop words include words such as "of," "the," and the like. Other terms are given key-word primary entries (the main alphabetical entries) and paired with all other key-words in the title. However, certain terms are considered semi-stop words (words like "methods," "analysis," "preparation," etc) and are only used as **co-terms**, the words paired with primary key-word terms. Subtitles are permuted separately from the main titles. The Permuterm Subject Index can be thought of as a key-word subject index to the source items which have appeared in a given time period in the source journals covered annually by *SCI*.

Full bibliographic citations with titles of the articles are found only in the *Science Citation Index* Source Index. Thus, the Source Index is the only place to find the titles of the source items published each year in the *SCI* core journals. The Source Index is arranged alphabetically by author, and the Permuterm Subject Index lists just enough information to locate the full reference to the article in the Source Index. Figure 5-2 illustrates the use of the SCI Permuterm Subject Index to find articles in the Source Index.

The task of selecting terms in both the printed *SCI* and SciSearch has been made easier by *SCI*'s editorial policy. At the input stage, spelling dictionaries are used to cut down on the number of places which must be checked in the alphabetic listing of

SEARCHING BY SUBJECT

PERMUTERM® SUBJECT INDEX   SOURCE INDEX

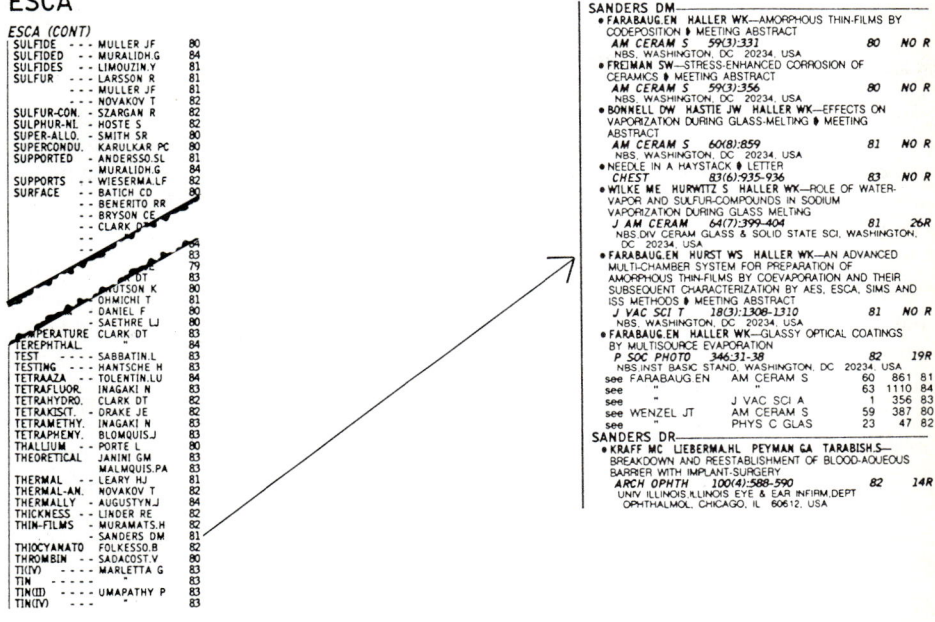

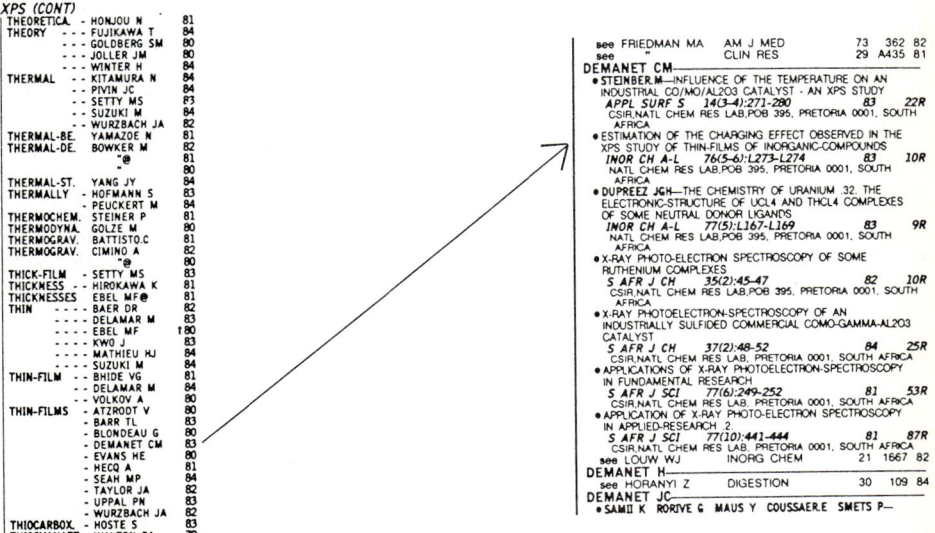

Figure 5-2
Typical Subject Search in the *Science Citation Index*

subject terms. This standardizes British to American spellings by using the suffixes -ize, -or, -er, etc. when words like generalise, colour, or fibre are encountered. Appendix A of the *SciSearch User Guide* [267] lists the preferred form of spelling for many words, like "inquiry" for "enquiry". Some cross references are now found in the printed work, for example, EPINEPHRINE see also ADRENALINE. However, it is a good idea to look up both singular and plural forms, as well as synonyms, acronyms, etc. Remember that the Permuterm Subject Index is an uncontrolled-vocabulary key-word index, so new words which appear in scientific titles will be picked up quickly by the PSI. Foreign language titles are translated into English, with a language field indicating the original language of publication.

In an online search on DIALOG or ORBIT, the SciSearch Basic Index contains words from the titles of the source items and, from 1983 onward, **research front names**. A research front is identified by a citation analysis, and a code number is assigned to each research front. Think of them as the really hot areas of research in science.

In addition to the **EXPAND** option in the Basic Index, truncation may be used effectively in subject searches in SciSearch. DIALOG uses only a question mark to indicate truncation. ALCOHOL? pulls any terms like "alcohol," "alcohols," "alcoholism," etc. ALCOHOL? ? with one space between the question marks would allow only one variable character to follow the stem, whereas ALCOHOL?? with no intervening space between the question marks retrieves the stem, or the stem plus one or two characters. Thus, in the latter case, ALCOHOL?? pulls "alcohol," "alcohols," or "alcoholic," but not "alcoholism". Internal masking is also indicated on DIALOG by a question mark, as in WOM?N, which would retrieve "woman" and "women". Figure 5-3 illustrates the use of truncation in a subject search in DIALOG. The OneSearch option on DIALOG is used to search all SciSearch files at once. A total of 661 references were found, and the latest in the database at the time of the search is printed in anabbreviated format.

Punctuation marks and other symbols generally cannot be searched in SciSearch. To compensate for this, proximity operators can be used as in:

?S non(w)linear OR nonlinear

However, chemical isotopes appear as hyphenated words, for example, C-13 or CARBON-13. The only stop words applied to the titles in the database are the following: or, on, not, and, of, the. When molecular formulas are encountered, they are broken apart and should be searched with the proximity operator:

?S Zn(w)OH(w)2

Super or subscripts are written on the line. Either a double *or* a triple bond is indicated by an equal sign: = .

The SciSearch database is updated weekly. Since the publisher, the Institute for Scientific Information (ISI), is so prompt about entering the currently received journals, SciSearch is one of the most up-to-date avenues to information in the current journal literature. Chemical Abstracts Service usually requires on the average two to three months to input data into the *CA* database for a record of a newly received journal

System:OS - DIALOG OneSearch

  File 34:SCISEARCH - 1989 WK 1-39
        (COPR. ISI INC. 1989)
* See also files 434 (1987-88), 433 (1980-86) & 432 (1974-79)
* Use 'BEGIN SCISEARC' to search all of SciSearch
* 1988 material has "rolled off" into file 434
  File 434:SCISEARCH - 1987-88
        (COPR. ISI INC. 1989)
* See also file 34 (1989-), 433 (1980-86) & 432 (1974-79)
  File 433:SCISEARCH - 1980-1986
        (COPR. ISI INC. 1988)
* See also file 34 (1989-), 434 (1987-88) & 432 (1974-79)
  File 432:SCISEARCH - 1974-1979
        (COPR. ISI INC. 1988)
* See also file 34 (1989-), 434 (1987-88) & 433 (1980-86)

Set Items Description
?**set detail on**
DETAIL SET ON
?**select superconduct? and ceramic?**

34: SCISEARCH - 1989 WK 1-39
        4977   SUPERCONDUCT?
        1610   CERAMIC?
         323   SUPERCONDUCT? AND CERAMIC?

434: SCISEARCH - 1987-88
        9321   SUPERCONDUCT?
        2726   CERAMIC?
         280   SUPERCONDUCT? AND CERAMIC?

433: SCISEARCH - 1980-1986
       11433   SUPERCONDUCT?
        6371   CERAMIC?
          57   SUPERCONDUCT? AND CERAMIC?

432: SCISEARCH - 1974-1979
        5089   SUPERCONDUCT?
        2135   CERAMIC?
           1   SUPERCONDUCT? AND CERAMIC?

TOTAL: FILES 34,434,433 and ...
       30820   SUPERCONDUCT?
       12842   CERAMIC?
    S2   661   SUPERCONDUCT? AND CERAMIC?

?**type 2/ti,so/1**
SUPERCONDUCTING AND ANELASTIC EFFECTS IN PB-DOPED BISRCACUO CERAMICS
PHYSICA C, 1989, V160, N1, P25-29

FIGURE 5-3
Use of Truncation in a Subject Search of *SciSearch*.

article or other primary work. On the other hand, the printed *SCI* is not published as frequently as the printed *Chemical Abstracts*. In contrast to the weekly publication schedule of *CA*, the printed *SCI* appears only every other month. For those who need more current tracking of the journal literature in a printed product, CAS publishes *Chemical Titles* bi-weekly, and ISI publishes *Current Contents*. In its five weekly science or technology series, *Current Contents* covers essentially all of science. The individual series titles are: *Life Sciences*; *Clinical Medicine*; *Agriculture, Biology, and Environmental Sciences*; *Physical, Chemical, and Earth Sciences*; and *Engineering, Technology, and Applied Sciences*. Each bi-weekly issue of *Chemical Titles* and each weekly *Current Contents* issue has a key-word subject index based on words from the titles. Further information on these and related current awareness products is found in chapter 13.

One of the virtues of ISI's Permuterm Subject Index volumes is that the index terms reflect the current living vocabulary of science. When a new name for a compound or concept is used in the title of a journal article, it becomes an indexing term in ISI's products shortly thereafter. The burden is then on the searcher to identify all of the possible relevant terms in order to perform a thorough search with high recall.

5.2.4 Subject Term Authority Lists

There is no such thing as the *correct* term to use in a key-word subject index (for example, the *Science Citation Index* Permuterm Subject Index) since potentially any word in a title could be an index term if it is not a stop word. If the indexing terms are not chosen from a list of allowed terms, we say that the work is indexed using **natural language** or **uncontrolled vocabulary**. If, on the other hand, the indexer consulted a **subject term authority list** of permissible indexing terms and made a conscious choice of one term over another, the indexer has used **controlled vocabulary**. Some authority lists are arranged in a manner which shows a great deal about the hierarchical subject relationships of the various terms. A list of this type is known as a **thesaurus**.

Examples of subject term authority lists are the *Library of Congress Subject Headings* catalog [213], the *Thesaurus of Engineering and Scientific Terms* [208], and the *Chemical Abstracts* Index Guide [176]. When searching a controlled-vocabulary subject index, it is critical that the correct indexing term be identified *before* the search is begun. The proper term to use may be found by consulting the authority list used by the indexer. A **See reference** is usually included in the subject term authority list or sometimes in the actual subject index. For example, the Index Guide to *Chemical Abstracts* has the following entry:

> **ESCA (electron spectroscopy for chemical analysis)**
> See Photoelectric emission
> x-ray
> See Photoelectron spectroscopy
> x-ray

The same see references are found at **XPS (x-ray photoelectron spectroscopy)**.

In general, subject searching in a subject index with a tightly controlled indexing vocabulary results in a precise and thorough search, assuming the indexers have done a good job. Mastering the use of controlled-vocabulary indexes gives the searcher a locksmith's tools with which to open the treasure chest of chemical information. Far fewer places in the alphabetical subject index must be consulted when controlled-vocabulary indexing is used, because the one *right* key is selected before attempting to enter. On the other hand, the controlled-vocabulary indexes do not usually adapt quickly to changes in scientific terminology. For many years into this half of the 20th century, the Library of Congress used the subject term "aeroplane" instead of the much more common American spelling "airplane." Most science subject indexes which use controlled vocabularies are still produced by human indexers, although computers may be used to assist them in the selection of the terms. Consequently, the number and choice of indexing terms are not limited to the words in the title of the document when controlled-vocabulary indexing is employed.

One should always determine if a subject term authority list is available for the reference work or database selected for use before beginning a subject search. Further examples include *SHE* (Subject Headings for Engineering) [209] for *Engineering Index* and the corresponding database COMPENDEX, *Guide to Index Terms in Analytical Abstracts* [355] (1986), and *Agricultural Terms: As Used in the Bibliography of Agriculture* [183] (1978).

5.2.5 *Chemical Abstracts* Subject Term Authority Lists: The Index Guide and Others

Chemical Abstracts Service decided to split the subject indexes to *CA* into two parts beginning in 1972. At that time the Chemical Substance Index and the General Subject Index to *Chemical Abstracts* made their debut. The Index Guide had been introduced three years earlier. There is no true online counterpart to the *CA* Index Guide. Although the **EXPAND** command can be used in the controlled-vocabulary (CV) field to see if a term is used in indexing, the scope notes and see also references are not available online, nor is the hierarchical list of General Subject Headings. The *Chemical Abstracts* Index Guide controls the vocabulary used in both parts of the subject indexes to *CA*.

Each five-year *CA* collective index period has its own Index Guide volume, and the controlled-vocabulary subject terms may change when the boundaries of those periods are crossed. At the beginning of each collective index period, Chemical Abstracts Service introduces changes in the controlled vocabulary for subjects. A booklet describing those changes for the twelfth collective index period (1987-91) has been issued. It is called *New and Revised Chemical Abstracts Indexing Terms—1987* [351]. An overview of the changes to subject terms is presented, followed by the new subject headings, cross-referenced to earlier terminology when appropriate. Thus, references can be found in one section such as:

Weed (11CI)
use *Plant*, Weed (12CI)

and in another:

Polyamines (12CI)
use *Amines*, polymers (11CI).

Changes in the registration, indexing, and naming of chemical substances are also indicated in a general fashion in this publication.

CAS has also issued the *CA Headings List: General Subjects* [352] and *CA Headings List: Plants and Animals* [353] (1985). These works list index headings and cross references for the 9th (1972-76), 10th (1977-81), and 11th (1982-86) collective index periods. Thus, collected in one place are the changes in terminology which have occurred for general subjects, plants, and animals over a 15-year period in *Chemical Abstracts*.

Another helpful work is "Standard Abbreviations, Acronyms, Special Characters, and Symbols in CAS Computer-Readable Files and Publications" [354] (1982). Abbreviations are routinely used in *CA* in both the indexes and abstract text. See in Figure 3-2 the abbreviations "redn" and "consts" in the abstract field of 93:25540j and "photochem" in the text modifications of certain of the controlled-vocabulary

CHEMICAL ABSTRACTS®

1989   **Index Guide**

INDEX GUIDE

CONTENTS

References given in this table of contents are to paragraph (¶) numbers.

Introduction	
General	1
Valid General Subject Index Headings	2
Cross-references for Chemical Substances	3
Cross-references for General Subjects	4
"See also" Cross-references	5
Indexing Notes	6
Homograph Definitions	7
Illustrative Structural Diagrams	8

INDEX GUIDE: An alphabetic sequence of cross-references, indexing notes, and valid general subject headings

Appendix I: Hierarchies of General Subject Headings
  Introduction
  Hierarchies
  Hierarchy Index

Appendix II: Indexes to *Chemical Abstracts*: Organization and Use	
Introduction	9
Chemical Substance and General Subject Indexes	10
Illustrative Keys	10A
Index Heading Subdivisions	10B
Ordering of Entries	10C
Illustrative Structural Diagrams	10D
Formula Index	11
Index of Ring Systems	12

Figure 5-4
Contents of the *Chemical Abstracts* Index Guide for 1987-91

index terms in 106:205033s. It is very important to include relevant abbreviations and correct forms of other symbols in a computer-based search.

It is best to consult the printed subject term authority lists for *Chemical Abstracts* before beginning a subject search. Nevertheless, the online search option permits the searcher to view the indexing of a relevant document on STN International using the **DISPLAY IND** command. Additional controlled-vocabulary subject terms or phrases might be found by inspecting the IT display field. They could then be input to broaden or narrow a search. In order to be searched in the controlled vocabulary field, a phrase must be labeled with the field indicator ''/CV'' in the CA File, for example, => **SEARCH BOND CLEAVAGE/CV**. If the ''/CV'' is omitted, the phrase will be searched in the **CA File Basic Index** which by definition is restricted to *single* words from the titles, key-word phrases, controlled-vocabulary index terms or phrases, and text modifications, plus compound registry numbers. Multi-word subject terms which

CAS Registry System	13
Selection of Index Entries	14
Specific Chemical Substances	14A
Chemical Reaction Studies	14B
General Subjects	14C
Data Collection and Analysis Centers	15
Abbreviations and Symbols Used in CAS Publications	15A
Appendix III: Selection of General Subject Headings	
Introduction	16
Types of General Subject Headings	17
General Guidelines	18
Hierarchical Relationship of General Subject Headings	19
Analytical Chemistry	20
Chemical Engineering	21
Drugs, Hormones, and Toxic Substances	22
Energy	23
Environmental Chemistry	24
Food and Agricultural Chemistry	25
General Biochemistry and Enzymes	26
History, Education, and Documentation	27
Industrial Organic Substances	28
Inorganic Chemistry	29
Macromolecular Chemistry	30
Metallurgy, Minerals, and Rocks	31
Natural Products	32
Nuclear Chemistry and Energy	33
Organic Chemistry	34
Physical Chemistry and Properties	35
Physiology, Metabolism, and Taxonomy	36
Reactions	37
Appendix IV: Chemical Substance Index Names	
Introduction	101–102
A. Nomenclature Systems and General Principles	103–139
B. Molecular Skeletons	140–163
C. Principal Chemical Groups (Suffixes)	164–177
D. Compound Classes	178–201
E. Stereochemistry and Stereoparents	202–212
F. Specialized Substances	213–224
G. Chemical Substance Names for Retrospective Searches	225–293
H. Illustrative List of Substituent Prefixes (Radicals)	294
J. Selected Bibliography of Nomenclature of Chemical Substances	295–308
K. Chemical Prefixes	309–311
L. Chemical Structural Diagrams from *CA* Index Names	312–318
M. Index	

Figure 5-4
Contents of the *Chemical Abstracts* Index Guide for 1987-91

are not labeled with field codes are searched in the Basic Index by default, and the WITH proximity operator is assumed between the terms. Thus, searching a multi-word phrase in the Basic Index of the CA File generally yields more hits than the same search in the CV field. Nevertheless, the answer set from such a search would probably contain a lower percentage of relevant references than would the set retrieved from a controlled-vocabulary search on the topic.

5.3 THE STRUCTURE OF *CHEMICAL ABSTRACTS*, THE CA FILE, AND THEIR SUBJECT INDEXES

Chemical Abstracts is the premier abstracting service for the field of chemistry and related areas. Each year nearly a half million documents are cited in *CA*. Each of them is classified into one of eighty subject divisions and assigned to either an odd-numbered or even-numbered issue. The weekly issues with odd numbers cover biochemistry in sections 1-20 and organic chemistry in sections 21-34. The issues with even numbers (those which appear every other week) cover macromolecular chemistry in sections 35-46, applied chemistry and chemical engineering in sections 47-64, and physical, inorganic, and analytical chemistry in sections 65-80. (See Figure 5-5.)

The numbers of the abstract sections can be effectively used to refine an online search. *Subject Coverage and Arrangements of Abstracts by Sections in Chemical Abstracts* [344] in its various editions describes any changes which might have occurred over time. In addition, appendix E of *Using CAS ONLINE: The CA File* [235] is "A Guide to Use of CA Sections for Retrospective Searching." It is advisable to consult these works because there have been changes in the classification scheme over the years. For example, a search which extended to 1967 in the CA File and utilized the section codes for "History, Education, and Documentation" would use 1/SC for the period 1967-71 and 20/SC for the period beginning with 1972, when documents in these subject areas were given a new section number. One search strategy would be:

=> SEARCH subject AND 1/SC RANGE=(1967,71)
=> SEARCH subject AND 20/SC RANGE=(1972,)
=> SEARCH L1 OR L2.

The **RANGE** command is used to limit a search statement to a particular time period. In a broad sense, the section codes are the *CA* counterparts to library call numbers. They classify the material into subject groupings. Documents assigned to the major subject categories can be isolated in an online search of the CA File by using codes for the sections, as shown in Table 5.3. Thus, the strategy

=> SEARCH (ICR OR ION(W)CYCLOTRON(W)RESONANCE) NOT BIO/SC

should eliminate practically all instances where ICR is used in the literature to refer to the type of mice used in biochemistry research, since documents with ICR used in that sense are likely to be found in sections 1-20 of *CA*.

SEARCHING BY SUBJECT 87

ABSTRACT SECTIONS

Biochemistry Sections

1. Pharmacology 224639
2. Mammalian Hormones 225339
3. Biochemical Genetics 226039
4. Toxicology 226649
5. Agrochemical Bioregulators 227115
6. General Biochemistry 227258
7. Enzymes 227612
8. Radiation Biochemistry 228018
9. Biochemical Methods 228132
10. Microbial Biochemistry 228618
11. Plant Biochemistry 228908
12. Nonmammalian Biochemistry 229282
13. Mammalian Biochemistry 229414
14. Mammalian Pathological Biochemistry 229823
15. Immunochemistry 230148
16. Fermentation and Bioindustrial Chemistry . 230580
17. Food and Feed Chemistry 230727
18. Animal Nutrition 231081
19. Fertilizers, Soils, and Plant Nutrition .. 231197
20. History, Education, and Documentation 231330

Organic Chemistry Sections

21. General Organic Chemistry 231400
22. Physical Organic Chemistry 231453
23. Aliphatic Compounds 232006
24. Alicyclic Compounds 232109
25. Benzene, Its Derivatives, and Condensed
    Benzenoid Compounds 232166
26. Biomolecules and Their Synthetic Analogs . 232353
27. Heterocyclic Compounds (One Hetero Atom).. 232465
28. Heterocyclic Compounds (More Than One Hetero
    Atom) 232605
29. Organometallic and Organometalloidal Com-
    pounds 232877
30. Terpenes and Terpenoids 233209
31. Alkaloids 233292
32. Steroids 233334
33. Carbohydrates 233365
34. Amino Acids, Peptides, and Proteins 233482

ISSUE INDEXES

    Keyword Index
    Patent Index
    Author Index

Sections 35–80 covering Macromolecular Chemistry, Applied Chem=
istry and Chemical Engineering, and Physical, Inorganic, and Ana=
lytical Chemistry appear in alternate issues of *Chemical Abstracts*.

ABSTRACT SECTIONS

Macromolecular Chemistry Sections

35. Chemistry of Synthetic High Polymers 233689
36. Physical Properties of Synthetic High
    Polymers 233967
37. Plastics Manufacture and Processing 234171
38. Plastics Fabrication and Uses 234509
39. Synthetic Elastomers and Natural Rubber .. 234780
40. Textiles and Fibers 234905
41. Dyes, Organic Pigments, Fluorescent Brighteners,
    and Photographic Sensitizers 235033
42. Coatings, Inks, and Related Products 235062
43. Cellulose, Lignin, Paper, and Other Wood
    Products 235312
44. Industrial Carbohydrates 235454
45. Industrial Organic Chemicals, Leather,
    Fats, and Waxes 235480
46. Surface-Active Agents and Detergents 235615

Applied Chemistry and Chemical Engineering Sections

47. Apparatus and Plant Equipment 235680
48. Unit Operations and Processes 235761
49. Industrial Inorganic Chemicals 236028
50. Propellants and Explosives 236125
51. Fossil Fuels, Derivatives, and Related Products .. 236164
52. Electrochemical, Radiational, and Thermal
    Energy Technology 236538
53. Mineralogical and Geological Chemistry ... 236680
54. Extractive Metallurgy 236922
55. Ferrous Metals and Alloys 237103
56. Nonferrous Metals and Alloys 237612
57. Ceramics 238114
58. Cement, Concrete, and Related Building
    Materials 238510
59. Air Pollution and Industrial Hygiene 238685
60. Waste Treatment and Disposal 238875
61. Water 239033
62. Essential Oils and Cosmetics 239288
63. Pharmaceuticals 239343
64. Pharmaceutical Analysis 239614

Physical, Inorganic, and Analytical Chemistry Sections

65. General Physical Chemistry 239645
66. Surface Chemistry and Colloids 240494
67. Catalysis, Reaction Kinetics, and
    Inorganic Reaction Mechanisms 241000
68. Phase Equilibriums, Chemical Equilibriums,
    and Solutions 241185
69. Thermodynamics, Thermochemistry, and
    Thermal Properties 241449
70. Nuclear Phenomena 241525
71. Nuclear Technology 241949
72. Electrochemistry 242342
73. Optical, Electron, and Mass Spectroscopy
    and Other Related Properties 242655
74. Radiation Chemistry, Photochemistry, and
    Photographic and Other Reprographic Processes .. 243905
75. Crystallography and Liquid Crystals 244498
76. Electric Phenomena 245033
77. Magnetic Phenomena 246289
78. Inorganic Chemicals and Reactions 246629
79. Inorganic Analytical Chemistry 246788
80. Organic Analytical Chemistry 247188

ISSUE INDEXES

    Keyword Index
    Patent Index
    Author Index

CA Abstracted Publications: Additions and Changes

Sections 1–34 covering Biochemistry and Organic Chemistry appear
in alternate issues of *Chemical Abstracts*.

Figure 5-5
Sample Tables of Contents for Two Successive Weekly Issues of *Chemical Abstracts*

    Each weekly issue of *Chemical Abstracts* includes a **Keyword Index** for subjects. This index, since it uses the uncontrolled terminology as written by the author(s), has no relationship to the Index Guide. That is, one should not consult the Index Guide seeking the "proper" term to use in the Keyword Index since it is based on a concept which has no unique "proper" or "correct" search term. A thorough search of the

TABLE 5.3
Codes for Major Subject Sections of *Chemical Abstracts* in the CA File

Name	Section Code	Section Numbers
Biochemistry	BIO/SC	1-20
Organic Chemistry	ORG/SC	21-34
Macromolecular Chemistry	MAC/SC	35-46
Applied Chemistry & Chemical Engineering	APP/SC	47-64
Physical, Inorganic, & Analytical Chemistry	PIA/SC	65-80

CA Keyword Indexes requires the use of alternative words, acronyms, and phrases, as does a subject search in *SCI*'s Permuterm Subject Index. Many libraries do not retain the Keyword Indexes when the printed issues of *CA* are bound, preferring to rely on the controlled-vocabulary subject indexes for the six-month volume and five-year collective indexes. *CA* Keyword Index terms or phrases are selected from the titles or the texts of the abstracts. The following should be kept in mind when using the Keyword Indexes or searching key words in the CA File's Basic Index.

- The singular form is used except for certain commonly used Greek and Latin plurals, for example, spectra, bacteria, data, and algae.
- Abbreviations used in abstracts and in index entries are also used in keywords.
- Words formed by adding prefixes to abbreviations are also abbreviated, for example, photodegrdn.
- Acronyms appearing on the list of abbreviations or defined in the Index Guide may also be used.
- Element names, not symbols, are used.
- Radical names, not abbreviations, are used.
- Words that are normally hyphenated occur as two (or more) words.
- Noun forms of a word are preferred to the adjectival forms, for example, Fungicide dithiocarbamate rather than Fungicidal dithiocarbamate.
- Names of Greek letters are used rather than their symbols.
- Roman numerals may be used. Valence is indicated by Arabic numerals.
- Keyword phrases are not sentences; there are no syntactic words (articles, prepositions, conjunctions), and punctuation is not used.

The **General Subject Index** to *Chemical Abstracts* [230] contains subject references to virtually every topic in *CA* which is not indexed in the Chemical Substance Index. That is, if the work being abstracted does not deal in whole or in part with a compound or compounds as individual entities, the indexing for it is likely to be found in the General Subject Indexes. This includes works dealing with general classes of chemical substances like esters or alcohols, applications, techniques, apparatus, chemical processes, and various other subjects. In Figure 3-2, General Subject Index entries leading to the abstract of the articles would be found for the terms "photolysis," "energy level transition," "bond cleavage," and "kinetics, reaction." Note that the index terms for a given document from both the General Subject Index and the Chemical Substance Index are part of the complete record in the *CA* database.

Shortly after issue number 26 of a *CA* volume has been published, the volume indexes to *Chemical Abstracts* appear. These now include the General Subject Index and the Chemical Substance Index. (Prior to 1972, there was only a **Subject Index** to *CA* which covered both chemical compound names and the terms for all other subjects.) Ten *CA* volumes (two per year) are now published during every five-year collective index period for *Chemical Abstracts*. (From its beginning in 1907 until 1957, a *CA* collective index period was ten years.) Upon completion of a collective index period, all of the volume indexes (including author, molecular formula, patent and other indexes) are now cumulated into the five-year collective index volumes. The collective indexes are enormous, and since they occupy so much space on the shelves, most libraries simply discard the volume indexes once the collective volumes are received.

An online searcher does not have to wait for the end of a volume or collective index period in order to access the controlled-vocabulary subject terms which are assigned in the Chemical Substance Index and the General Subject Index. All of the indexing of a particular document is done before it enters the database. Hence, as soon as the record becomes part of the CA File, the Keyword Index terms, the controlled-vocabulary index terms, and the title words are all immediately accessible to the searcher for searches of the Basic Index (unlabeled) and of the labeled subject fields (/CV and /TI). Registry numbers, the unique numbers assigned by CAS to all chemical substances, are part of the Basic Index of the CA File, but *no* controlled-vocabulary Index Names for chemical substances which are found in the printed Chemical Substance Index are included in the CA File. Why? Because the Index Names of compounds are part of the records in the online Registry File and can only be searched there. (The treatment of chemical names in *CA* is discussed in section 7.2.) In the CV field, the CA File record only has registry numbers (and from 1987 some common names of compounds). Of course, it is possible to search common names of chemical substances as subjects in the Basic Index of the CA File. In most cases, however, it is preferable to use the registry number, which is also included in the CA File Basic Index.

The CA File contains bibliographic records of primary and secondary sources indexed from January 1, 1967 to the present, and the CAOLD File contains pre-1967 records consisting of compound indexing (registry numbers) tied to abstract numbers. The registry number thus provides a link between these two files and the Registry File. In the CAS ONLINE files on STN International, it is very easy to form an answer set for a substance or a group of related substances by searching the Registry File with the techniques discussed in chapters 7-8 and to use the L# assigned to that answer set as input to a search in the CA File and the CAOLD File. Up to 30,000 registry numbers in a single answer set can be transferred from the Registry File in this manner. (Alternatively, a known registry number could be used.) In the CA File, the set of registry numbers may be searched in combination with other terms or parameters which the searcher inputs. These might be subject words or abbreviations for concepts like toxicity, preparation, uses, or section codes, etc. In Figure 3-2, the registry numbers are extensively used in the controlled-vocabulary indexing of the records.

In November 1987, a structure search for isatin in the Registry File yielded from 6 to 345 compounds depending on whether an exact match, family search, or substructure search of the full Registry File was run. (Do not worry about the distinctions among the three different types of structure searches at this time.) When duplicate compounds were eliminated and the answer sets were crossed into the CA and CAOLD Files, the results in Table 5.4 were obtained.

TABLE 5.4
Results of File Crossover from the Registry File to the CA and CAOLD Files for Sets Containing Isatin and Related Compounds

Type of Registry File Search	COMPOUNDS: Unique Full Registry File Results	DOCUMENTS: CA File Results	CAOLD File Results	Total CA References
1. exact	6	667	17	684
2. family	25	29	6	35
3. substructure	314	406	79	485
Total	345	1102	102	1204

In Table 5.4, the six compounds in set 1 are not included in set 2; neither are the 31 compounds in sets 1 and 2 included in the results of set 3. Thus, a total of 1204 documents were identified through crossing the answer sets into the bibliographic databases. The history of the search session is obtained in Figure 5-6 with the command **DISPLAY HISTORY**.

=> **DISPLAY HISTORY**

```
                              FILE REG
         L1                   STRUCTURE
         L2           0       S L1 EXACT SAMPLE
         L3           6       SEARCH L1 EXACT FULL
         L4          31       SEARCH L1 FAMILY FULL
         L5         345       SEARCH L1 SSS FULL
         L6          25       S L4 NOT L3
         L7         314       S L5 NOT L4

                              FILE CA
         L8         667       S L3
         L9          29       S L6
         L10        406       S L7
                              FILE CAOLD
         L11         17       S L3
         L12          6       S L6
         L13         79       S L7

                              FILE CA
         L14          2       S L8 AND (59/SC OR 60/SC)
```

FIGURE 5-6
History of a Search Session in the CAS ONLINE Files on STN International.

Note in Figure 5-6 that the results of L8 (the documents obtained in the search for the six isatin variants in the CA File) were combined in the final step with the section codes for section 59, "Air Pollution and Industrial Hygiene," and section 60, "Waste Treatment and Disposal." (Section cross-references 59/SX, 60/SX could also have been used.) Two references were retrieved, and one of those is reproduced in Figure 5-7.

AN   CA90(26):209518c
TI   Odor and volatile compounds in liquid swine manure. III. Volatile and odorous components in anaerobically or aerobically digested liquid swine manure
AU   Yasuhara, Akiio; Fuwa, Keiichiro
CS   Chem. Phys. Div., Natl. Inst. Environ. Stud.
LO   Ibaraki, Japan
SO   Bull. Chem. Soc. Jpn., 52(1), 114-17
SC   60-2 (Sewage and Wastes)
DT   J
CO   BCSJAB
IS   009-2673
PY   1979
LA   Eng
AB   Volatile and odorous components were isolated from anaerobically digested liq. swine manure by direct solvent extn. and flash distn. under reduced pressure. The concns. of indole [120-72-9], oxindole [59-48-3], DMSO [67-68-5], phenol [108-95-2] and a few carboxylic acids increased, and o-aminoacetophenone [551-93-9] decreased during digestion. Alk. components were not so important for the odor. Offensive odor was formed by the mixing of phenols and carboxylic acids. Complete aeration resulted in a remarkable decrease of lower carboxylic acids.
KW   odor volatile swine manure
IT   Wastes
        (liq. swine manure, odors and volatile components in digested)
IT   Odor and odorous substances
        (of swine manure)
IT   Manure
        (liq., swine, odorous and volatile components in digested)
IT   59-48-3 62-53-3, uses and miscellaneous 64-19-7, uses and miscellaneous 65-85-0, uses and miscellaneous 67-68-5, uses and miscellaneous 79-31-2 83-34-1 90-05-1 ***91-56-5*** 97-61-0 103-82-2, uses and miscellaneous 106-445, uses and miscellaneous 107-92-6, uses and miscellaneous 108-952, uses and miscellaneous 109-524, uses and miscellaneous 111-14-8 116-530 117-817 120-729, uses and miscellaneous 123-079 124-072, uses and miscellaneous 142-621, uses and miscellaneous 50152-0 503-74-2 551-93-9 646-07-1 1561-11-1 64828-47-3 64828-48-4 70328-85-7
        (in odorous and volatile components of digested swine manure)

FIGURE 5-7
Isatin Reference in the CA File Found by Searching with CA Section Codes (Displayed in the ALL Format).

5.4 FORMATS: DOCUMENT TYPES

In a computer search of the *CA* database, a user may even search on the **format** of publication (document type). In STN International's CA File, the formats are coded as shown in Table 5.5. In the printed *CA*, a B or P immediately before the abstract accession number in the indexes shows that a book or patent has been found. Examples are 96:B67903x and 106:P156491d. The other four format codes are not used in the

TABLE 5.5
Codes Used in the Document-Type Field in the CA File on STN International and in the Printed *CA*

CA Code	CA File Code	Document Type
B	B	books (including textbooks, handbooks and encyclopedias) and audiovisual materials
	C	conference proceedings
	D	dissertations
	J	journal articles
P	P	patents
	T	technical reports

printed *CA* indexes. Contrast these codes with those used for source items in the *Science Citation Index* as listed in Table 5.2.

For online searching, the document type codes can be very effective in narrowing a large answer set, perhaps by using a strategy like => **SEARCH L# NOT P/DT** or => **SEARCH L# AND J/DT**. The first strategy would eliminate all patents, whereas the second would limit the answer set to journal articles.

5.5 SEARCHES OF DATABASES AND REFERENCE WORKS DEVOTED TO A DOCUMENT TYPE

In science and technology many documents which contain primary literature are published in formats other than journal articles. A large subject-oriented database like *Chemical Abstracts* will abstract and index all types of primary literature, whereas the multi-disciplinary *Science Citation Index* now restricts its coverage to the new literature appearing in the world's most important journals. There are other databases and reference works which are confined to a particular type of document which is not a journal, but include many different subjects in documents published in that format. Thus, there are specialized reference works or databases restricted to conference proceedings or to dissertations or to patents or which cover only technical reports. Since all of them can be searched by subject, selected examples will be discussed in this chapter. Patents deserve a fuller treatment and are the subject of chapter 6. The fact that these secondary sources concentrate on a single format makes it likely that the scope of their coverage of the chemical literature appearing in those formats will be broader than that given to the same category of document by Chemical Abstracts Service.

5.5.1 Dissertations

For dissertations, the Comprehensive Dissertation Index (CDI) database [571], a product of UMI, can be searched online or as a CD-ROM database. The database includes the contents of printed products such as *American Doctoral Dissertations* [573] and *Dissertation Abstracts International* [572]. While the database does not contain all of the abstracts, it can be searched by subject and covers the entire print file back to 1861. Be aware that universities began their participation in the UMI program at different times, so it cannot be assumed that all U.S. chemical dissertations can be found by searching CDI.

5.5.2 Conference Proceedings

In the area of conference proceedings, one of the major sources is *Index to Scientific and Technical Proceedings* (*ISTP*) [575]. Like *Science Citation Index*, *ISTP* is published by the Institute for Scientific Information. It is a multi-disciplinary index to conference proceedings which have been published in any format, provided the majority of the material is published in the proceedings for the first time. Conference proceedings are another form of primary literature which in recent years has tended to become more and more widely used. In some cases the publication of the item in the conference proceedings represents the only place in the primary literature where that research is presented.

Since 1978, the *Index to Scientific and Technical Proceedings* has covered complete published papers presented at meetings of many major scientific societies. It is estimated that about half of the total published proceedings are indexed each year in *ISTP* and that this represents 75-90 percent of the significant conference literature. The entries for the proceedings become part of the printed *ISTP* within eight weeks of their receipt at ISI. ISTP can be searched online on ORBIT from 1982 onwards.

There are six major sections to *ISTP*. The Contents of Proceedings section contains the complete information about the published proceedings. This includes the contents of the work, the name of the conference, its location, date, up to ten sponsors, the complete bibliographic citation of the book or journal title which contains the proceedings, all authors, the addresses of the authors who are listed first in the papers, and the starting page number of each item. If the proceedings were published as a book (the usual case), there may also be a series title and volume number, up to nine editors, the publisher, city, and year of publication and other miscellaneous information like the price, Library of Congress number, and International Standard Book Number (ISBN).

There is also a Permuterm Subject Index which is used in essentially the same manner as the PSI for *Science Citation Index*. Many of the conventions used in the production of *Science Citation Index* also are applied to the *Index to Scientific and Technical Proceedings*. Foreign language titles are translated into English. Variant spellings are replaced with standard American forms, and the same stop list and semi-stop list is in effect as for *SCI*.

Broad subject access to the conferences is provided in the Category Index, about 200 general topics linked to the conference title and accession number in the Contents of Proceedings section. Terms such as cancer, computer applications, and electricity are found in the Category Index.

Other indexes to proceedings include the *Directory of Published Proceedings (Series SEMT)* [909] and *Conference Papers Index* [908].

5.5.3 Reports: NTIS

The largest single database devoted to technical reports is NTIS [570], the database of the U.S. National Technical Information Service. Available online through many vendors and as a CD-ROM product, NTIS includes U.S. research reports published or funded by over 200 federal agencies. Since the majority of publicly-funded chemical research in the United States comes from agencies such as the National Science Foundation and the National Institutes of Health, much valuable chemical information is found in the database and its printed counterpart *Government Reports Announcements and Index* [570]. Also included are federally sponsored translations.

In addition to its own database, NTIS provides search results from many different databases through its *Published Search Catalog* [578]. Over 3000 searches performed by experts on over two dozen databases (not restricted to research reports) are included. Each search has a fixed price. An example is "Nuclear Magnetic Resonance Coils," containing citations from the INSPEC database (a physics, computer science, and electrotechnology database).

5.6 BIBLIOGRAPHIES

The results of a search in a bibliographic database or printed abstracting or indexing journal is a bibliography. The entries in the NTIS *Published Search Catalog* are all bibliographies on many different topics. A **bibliography** is a collection of references which have in common a certain theme. Very often the unifying theme is a subject, but sometimes a bibliography is compiled for a certain author or company or for other reasons. The list of references (citations) at the end of a journal article or book is also referred to as a bibliography.

Larger printed bibliographies commonly appear first as a comprehensive book covering a range of years. Since bibliographies take a long time to compile, they are usually outdated by at least a year or two at the time they are published. Consequently, supplements may be issued at regular or irregular periods of time. In contrast to abstracting or indexing databases or printed works, bibliographies usually do not cover a large span of time. A typical bibliography might include primary literature which appeared in a five or ten-year period. Furthermore, it is very rare for a bibliography to become a database. An exception is Ohno and Morokuma's *Quantum Chemistry Literature Data Base* [579] which was first published in 1982 for the period 1978-1980. Supplements have appeared each year in the *Journal of Molecular Structure* (*Theochem* section). The database itself, QCLDB [580], can be licensed from the producer, the Japan Association for International Chemical Information.

Bibliographies obviously do not exist on every topic, but when found can save a researcher a great deal of time. One way to discover the existence of a bibliography is to search a library's card or online catalog by looking for subject entries such as: ORGANOMETALLIC COMPOUNDS—BIBLIOGRAPHY. Bibliographies are sometimes produced by industrial chemical companies and distributed free of charge. An example is the *Annotated Bibliography on the Use of Organolithium Compounds in Organic Synthesis* [1101] which was published by the Lithium Corporation of America.

Search **BIBLIOGRAPHY** in the Chemistry Reference Sources Database to see additional examples.

Whatever the source of the bibliography, it presents the user with enough information to locate the primary material in a library or to obtain it from a document delivery service. In that sense, bibliographic databases and bibliographies have in common the fact that they leave the searcher with another task: to seek out the documents which *may* contain answers or needed information.

5.7 CD-ROM DATABASES FOR BIBLIOGRAPHIC SEARCHING

A medium which has moved very quickly into the world of bibliographic searching is the **CD-ROM**, compact disk-read only memory. CD-ROM databases and search software can be accessed on a stand-alone microcomputer equipped with a CD-ROM player. The advantage of such a device from the librarian's point of view is that the cost of searching is fixed. A subscription rate can be paid for an entire year, much as for a printed abstracting or indexing service. A disadvantage is that the normal CD-ROM searching station is designed so that only one person at a time can access the database on one machine. Furthermore, the time periods accessible through the CD-ROM products typically are not as comprehensive as for the online database counterparts. Nevertheless, the number of databases which are being produced in CD-ROM format is likely to increase.

Unfortunately, the number of CD-ROM bibliographic databases which were available for chemistry searching was rather limited at the end of 1989. Some of the files in which chemists might be interested are the SCI CD-ROM edition [1753], MEDLINE [225] for the National Library of Medicine database, Dissertation Abstracts Ondisc [571], NTIS [570] for government reports, and CD-CHROM [1888]. The Compact Cambridge Life Sciences Collection [634] has backfiles available to 1982 covering areas such as biochemistry, biotechnology, chemoreception, microbiology, and toxicology. In addition, the Wilson databases Applied Science and Technology Index [521] and the General Science Index [610] can be obtained on compact disks. The CD-ROM databases are not updated as frequently as their online counterparts. Typically, an updated cumulation is provided quarterly or every six months for a CD-ROM, whereas online database updates are usually bi-weekly or monthly, sometimes even weekly. CD-ROM versions of encyclopedias are also available, including the *Kirk-Othmer Encyclopedia of Chemical Technology* [916], the *Encyclopedia of Polymer Science and Engineering* [917], and the *Academic American Encyclopedia* [454] (the Grolier Electronic Encyclopedia).

5.8 TEXT DATABASES

Text databases have the potential to become the largest databases of all. Indexing and abstracting databases merely contain abbreviated statements of the information content of primary documents, but a text database at the least has all of the words found in the narrative text of an article, patent, or other document. As noted earlier, however, text databases at this time do not include the graphic representations of such things as mathematical symbols, chemical structural formulas, and the like, and most do not include tables of data.

It is best to think of text databases of primary journals as complements to bibliographic databases. The text databases are good for finding certain types of information which is usually not abstracted or indexed in the secondary databases. This includes such things as data for physical and chemical properties of substances, supporting information like the descriptions of experimental procedures, synthetic methods, and other types of information. Such information might be neglected by an abstracter or indexer if it is judged to add nothing new to the knowledge in the field. Nevertheless, it is often the type of information which is most needed by a researcher. Hence, the text databases serve an important function for the end-user. They are also useful where new chemical names, jargon, new terminology, or acronyms are the subject of the search.

Browsing the context in which the search term is found can be extremely helpful, so the vendors have generally provided mechanisms to allow the display of the specific parts of a record where the search key is found. A table which shows the frequency with which the term occurs and lists the sentences or paragraphs where it is found is one device employed by the text database vendors. Often the term which caused the hit is highlighted on the video monitor when the sentence or paragraph is displayed. Another technique which allows the searcher to quickly judge the relevance of the reference is to display a certain number of words before or after the hit term, thus providing a kind of KWIC display.

The normal Boolean operators for combining logical sets do not work very well in searching a text database. There are simply too many terms and too many opportunities in a longer document for false drops to occur when an AND search strategy is used. Normally, it is the proximity operators (ADJ, WITH) and more specialized search commands which are employed. In the latter category are commands to instruct the search software to consider a hit to be a good one only if the search terms all occur in the same sentence or at least in the same paragraph. Otherwise, we might find the first search term coming from the abstract, the second from the experimental details section, a third from the conclusion, and so on. In order to provide such search capabilities, it is necessary to construct index files which indicate the exact position of each word or term relative to others in the document. These indexes elevate the cost of text databases since they require several times the storage space of the original file.

Compared to bibliographic databases, text databases of primary documents do not go back very far in time. Typically, the files start with coverage in the 1980s, with the oldest only extending back to the early years of that decade. There is still a great lack of standardization in the coding used to create text databases of primary journals, although there is some movement toward a **standard generalized mark-up language** (SGML). The adoption of such a standard would permit data from the hundreds of suppliers of chemical text databases to be loaded more easily by a database vendor. Graphics are a major problem, since printers often do not include graphic material in the database used to create the printed product. Even if the majority of them did so, there is relatively little software to search and display graphics.

5.8.1 Chemical Journals Online: The CJO File on STN

Chemical Abstracts Service, which prints the journals of the American Chemical Society, planned from the time it developed a journal text composition system to use the data as a text database as well as to generate the printed journals. Hence, the coding of records which they developed was easily adapted to text searching by STN International.

The files of all American Chemical Society primary, peer-reviewed journals are now available on STN as file CJACS [548]. STN intends to have 21 separate files of Chemical Journals Online (CJO) by 1993. At that time over 160 of the world's most important chemistry journals should be searchable as text databases. The journals listed in Table 5.6 were on the system by the end of 1989.

TABLE 5.6
Chemical Journals Online (CJO) Text Database Files on STN International

Abbreviation	Description	Earliest Coverage
CJACS	eighteen journals of the American Chemical Society	1982
CJRSC	ten journals of the Royal Society of Chemistry	1987
CJWILEY	five polymer journals published by Wiley	1984
CJAOAC	Journal of the Association of Official Analytical Chemists	1987
CJVCH	international edition of *Angewandte Chemie*	1988
CJELSEVIER	four journals of Elsevier Science Pub., B.V.	1990

Let's take a closer look at the CJACS file and the STN Messenger system search software. Messenger has a number of features which make it quite suitable for text database searching. These include:

- proximity operators
- hit term highlighting
- hit display formats
- left-hand and right-hand truncation
- KWIC display option (20 words before and after the hit term)
- interactive browse command (DISPLAY BROWSE).

The proximity operator W implies that the word to the left of the W must precede the word to its right. No such order is implied with the A (adjacency) operator. The L (link) operator says that the terms must be in the same information unit (sentence, paragraph, or field). The S and P operators are more specific, requiring the terms to be in the same sentence or the same paragraph respectively.

The Basic Index on the CJACS file contains terms from the textual units shown in Table 5.7. The main text fields in the CJACS file are the TX, AB (abstract), BA (author biography), and RE (references and footnotes) fields. All of the text fields are proximity searchable.

Another important search capability in text databases of primary journal articles is citation searching. Since the references in the bibliography are searchable, data from a known citation could be used as a search key. An example would be => **S BOYD/ RE AND 1984/PY AND 24/VL**. Yet another search option is to use registry numbers

TABLE 5.7
Fields Included in the Basic Index of the CJACS File on STN International

Field Code	Meaning
TI	titles of the articles
TX	paragraphs of the articles
CP	captions of figures
TT	titles of the tables
RN	Chemical Abstracts Service registry Numbers
CN	names of chemical substances
SM	supplementary material

found in the CAS ONLINE Registry File as input to the CJACS file. This type of file crossover is very easy on STN International simply by using the L# for the answer set of the registry numbers. Finally, the **DISPLAY BROWSE** command lets you search backward or forward a specified number of fields for the occurrence of a term.

The display options for the records in the CJO file include those shown in Table 5.8. The last option, **DISPLAY ALL**, is only used when the searcher wants to print the entire record. This is not usually done except in those rare cases where the document is needed immediately and no other document delivery mechanism is fast enough. Text files of journals are simply too expensive at the present time to regularly compete with the kinds of alternative document delivery options described in Chapter 13.

TABLE 5.8
Display Options in the STN Chemical Journals Online (CJO) File

Command	Meaning
DISPLAY HIT	displays the fields where the hit terms occur
DISPLAY KWIC	displays up to 20 terms before and after the hit term
DISPLAY OCC	displays the number of occurrences of the hit terms and the displayable fields where they occur
DISPLAY BIB	displays the Accession Number (AN), Journal Source Title (SO), Article Title (TI), Author (AU) and Corporate Source (CS)
DISPLAY ALL	displays all of the displayable fields

Figure 5.8 shows a record from the CJACS file obtained by using the command **DISPLAY BIB**.

AN	86:98 CJACS
SO	Industrial & Engineering Chemistry Process Design and Development, (1986), 25(1), 151-155. CODEN: IEPDAW ISSN: 0196-4305
TI	M2 Forming-A Process for Aromatization of Light Hydrocarbons.
AU	(1) Chen, Nai Y. (*): (2) Yan, Tsoung Y.
CS	(1,2) Mobil Research and Development Corporation, Princeton, New Jersey 08540

FIGURE 5-8
Text Record from the CJACS Database Displayed in the BIB format.

DISPLAY OCC gives a table which leads the searcher to the appropriate place in the text, for example:

L1 ANSWER 2 of 71

field	count
TI	2
TX(1)	14
TX(2)	7
TX(7)	5
CP(1)	2
CP(2)	4
TT(1)	2

This indicates that the search term occurs in four fields, the title (TI), the text(TX), the figure captions (CP), and the table titles (TT) fields.

5.8.2 Other Text Searching in Bibliographic and Primary Journal Text Databases

Abstracts in bibliographic databases offer the searcher a greater range of retrieval possibilities than do index terms alone. In the STN implementation of the CA File, the abstract text can be searched using the field qualifier /AB. The texts of all abstracts published since January 1970 (and some earlier) were accessible for searching and display in the CA File by the end of 1989. Since the *CA* abstracts often contain data, it is possible to extract the information from the abstracts. For example, a strategy using the proximity operator (W) such as the following could be used:

=> SEARCH FREE/AB(W)ENERG?/AB(1W)ACTIVATION/AB

A total of 1039 references containing phrases such as "free energy of activation" were found with this strategy in November 1987. When combined with the set number for the isatin compounds, abstract number 93:25540j from Figure 3-2 was retrieved, yielding actual values for the free energy of activation for homolysis of two compounds. A number of other databases now include the abstracts as searchable fields. Examples are BIOSIS (*Biological Abstracts*), PHYS (*Physics Briefs*), and *Ceramics Abstracts*.

    BRS has mounted the Medical Science Research (formerly IRCS Medical Science) database. It contains the texts of all articles published in the journal since 1981. These are biomedical research papers of clinical significance. Each paper is about 1000 words in length and gives introductory, methodological, and research information. The texts of all printed counterparts appeared originally as one title, *IRCS Medical Science*, in both print and microfiche formats. Separate sections were also published individually at various times. In 1987, the publisher changed from the International Research Communication System to Elsevier Applied Science, with a corresponding title change to *Medical Science Research*. Sections of potential interest to chemists include biochemistry, environmental biology and medicine, and pharmacology, among others. As with the journals on the CJO file, it is possible to do a type of citation searching on the references which appear in the bibliographies of the papers. Most subject searching would utilize the proximity operators on BRS.

It is expected that more and more texts of primary journals of interest to chemists will become available as databases in the coming years. Eventually, even the graphics and tabular data are sure to be included.

5.9 SUMMARY

A critical part of a search plan for a subject search is the selection of terms to use as input to either a printed or computer-based reference tool. In many cases, the controlled-vocabulary terms which must be used are governed by a subject-term authority list, and some of those lists and thesauruses are noted in this chapter. Relevant terms can also be assembled in an online search through the techniques of truncation (masking) and neighboring (expanding). Another type of searching for subjects is done in key-word indexes which use natural language. In order to effectively search key-word indexes, synonyms, abbreviations, and related terms must all be used. Natural-language subject indexes can be found in the *Science Citation Index*, *Current Contents*, and the *Index to Scientific and Technical Proceedings*. *Chemical Abstracts* has both key-word and controlled-vocabulary subject access.

The classification scheme of an abstracting or indexing journal (or for that matter, of a library) offers another option for subject searches. Searches of comprehensive databases produce bibliographies consisting of many different primary document types. Other reference works and databases provide subject access to a single primary document type, such as patents, technical reports, conference proceedings, or dissertations.

As newer computer storage media and databases are developed, the options for subject searching increase. One of those options is to search a CD-ROM version of a database. Another is to search a text database or the abstracts of bibliographic databases. It has been difficult to extract from bibliographic databases information such as numeric data or analytical procedures used in the experiments. Likewise, the use of brand names has yielded mixed results in bibliographic databases. Such difficulties can be overcome to a certain extent by searching text databases such as CJO, the STN text file of chemistry journals.

5.10 SELECTED READINGS

Hawkins, Donald T. "A Review of Online Physical Sciences and Mathematics Databases. Part 2: Chemistry." *Database* **1985**, *8*(2), 31-41.

Love, Richard A. "CHEMICAL JOURNALS ONLINE (CJO)—The New Full-Text Database Through STN International." *ONLINE '86 Conference Proceedings*; Online, Inc.: Weston, CT, 1986; pp 149-151.

CHAPTER 6

PATENTS

6.1 INTRODUCTION

The patent literature is very important in chemistry. This can be judged by the fact that about one-sixth of the nearly half a million documents referenced in *Chemical Abstracts* each year are patents. Approximately 1,000,000 patents are published annually for all types of inventions, with a substantial portion of those for chemical inventions. Although many patents are essentially duplicates of the patents issued for the same invention in other countries, a relatively small percentage of the information contained in patents (estimated at 10-30 percent) is ever published in any other document format. Thus, there is much unique information in patents.

Patents are issued by many countries, and the patent laws of those countries vary a great deal. Nevertheless, there are certain general concepts which are relatively uniform. Among these is the use of patent classification schemes. Another is the use of sequential numbers as accession numbers, in most cases, continuing the sequence from year to year.

The voluminous literature of patents is very specialized. Although some subject-oriented abstracting and indexing services include patents among the primary documents covered, this is by no means true of all of them. In fact, there are specialized services devoted exclusively to the information contained in patents. These services must be used if comprehensive searches of the patent literature are to be performed.

101

6.2 DEFINITION OF A PATENT AND THE CRITERIA FOR JUDGING PATENTABILITY

Technically speaking, a **patent** or more properly, letters patent, is a legal document granted by a government to an individual or organization. In the United States, a patent is granted only to an individual, the **patentee**, who may sell or assign the rights to another person or company, the **assignee**. The actual document which is examined in reaching the patentability decision is properly called a **patent application**, the full text of which is known as a **patent specification**. This is the required information which the inventor must submit in order for the appropriate agency to decide to issue the letters patent. However, it is common for chemists and librarians to refer to patent specifications simply as patents, and that is how the term should be understood most of the time in this book.

The inventor submits the required patent application on a certain date, which is then known as the **priority date**. In so doing, the inventor reveals the secret of the invention. If the invention survives the long examination period, and the letters patent is granted, the inventor secures certain rights. Among those is the right to exclude others from using the invention for a profit for a certain period of time. In most countries, the period is 15-20 years. In the United States for a normal utility patent, the period of protection is 17 years.

The inventor also has the right to sell the invention for a profit. In some countries, the inventor must take steps to license others to use an invention after a relatively short period of time (usually 3 years or less) if the invention has not been made available to the public by the inventor or the assignees. That is not the case in the United States, however. In effect, the patent specification, the public disclosure of the invention, is what the state gets in exchange for the protection granted by the patent. In return, the state encourages useful applications of the invention.

Of course, the mere filing of a patent application does not guarantee success. There are certain criteria which must be met. In general, the invention must be some type of new and useful process, machine, manufactured item, composition of matter, or some new and useful improvement on an invention previously known. A patent will not be granted for a useless invention. The inventor must state in the patent specification what is the purpose of the invention. The specification must conclude with one or more **claims**, which describe the subject matter regarded by the inventor as the invention. The claims form the legal description and boundaries of the invention. In terms of the legal requirement, the invention must satisfy three criteria:

- novelty
- utility
- invention (unobviousness).

"Novelty" is relatively straightforward. There can be no prior use of the invention. It must not have been previously known to people outside the inventor's employing organization. This includes demonstrations and market trials in most countries. Additionally, novelty may preclude any kind of publication about the invention. However, in the United States, the inventor may publish a document about the invention, provided that publication occurred no longer than one year before the priority filing date of the application.

"Utility" is the criterion which says that the invention must have a useful purpose, usually some kind of application in industry. There cannot be an anti-social purpose behind the invention. It cannot be intended to harm someone.

The last criterion is perhaps the most difficult to satisfy. "Invention" or "unobviousness" judges the degree of inventiveness. The invention cannot be something that could have easily been seen as an obvious extension of an existing invention by someone who is "skilled in the art". It has to have genuinely overcome some technical difficulty, solved some long-standing technical problem, or resulted in an unexpected advance in the field.

All three of the criteria must be met in order to be successful in obtaining a patent. In chemistry, a patent can be granted for a new composition of matter (a drug, catalyst, polymer, petroleum formulation, etc.), a process (synthesis, industrial process, analytical method, etc.), or machine or equipment used in laboratories or chemical plants. In the U.S., you can even patent a new variety of plant or animal (except human beings). The field of molecular biology has opened up a whole new realm of patent possibilities through the techniques of genetic engineering. Patent laws vary tremendously from country to country and are constantly evolving. Hence, skilled legal help is needed to assess the patentability of an invention.

6.3 THE PATENT PROCESS

The period between the filing of a patent application and the date of publication varies quite a bit depending on the country. Since patent protection must be obtained in each country in which the inventor wants to protect the invention, there are likely to be a number of specifications appearing at different times, perhaps in different languages, but all having essentially the same information content.

An important milestone in patent law occurred over a century ago with the creation of the "Paris Convention for the Protection of Intellectual Property of 1883". Under the convention, the filing date in the first country where an application is made is considered to be the filing date in any other country which has signed the convention, provided the application is made in the other countries within 12 months of the first filing date. In other words, it is as if all of the applications were made on the same day. This is the priority application filing date. The resulting group of patents is called a **patent family**, and the patents other than the first applied for are known as **convention** or **equivalent patents**.

In the United States, the patent specification is not published until the actual patent is granted. Thus, it is not uncommon to see products in this country with the designation "patent pending" or "patent applied for". This means that a patent application had been filed, but no patent had been issued by the time the goods were manufactured.

The examination period actually results in the demise of a large number of applications, perhaps as many as one-third of them. At the application stage, the information in the patent specification is confidential. When published, that confidentiality is broken so that others may see if the invention should truly be patented.

There are a number of countries which employ the practice of **deferred examination**. The unexamined application is published 18 months after the priority application filing date. The applicant may then have to ask that the patent office proceed with the official examination of the document.

The United States does not employ the system of deferred examination. However, a U.S. inventor can use the "Disclosure Document Program" of the U.S. Patent and Trademark Office. A paper disclosing an invention and signed by the inventor is sent to the Patent and Trademark Office. It is held there for two years. This does not substitute for the formal application process, but may help in the resolution of a disputed patent application by providing evidence of the date of conception of an invention. Disclosure documents are not accessible for public examination during the two-year period.

In the U.S., the patent specification is published only when the patent is granted. Hence, it is possible to find out about the existence of a U.S. invention first by keeping track of applications in countries which have early publication (that is, 18 months from the first filing date), such as Great Britain and the Federal Republic of Germany. It is common in the U.S. for an applicant to file here first, thus obtaining a U.S. priority date, then file in European countries within the 12 months allowed by the Paris Convention.

Multi-national filing has become easier in recent years since a number of treaties have fostered closer cooperation. The Patent Cooperation Treaty has about 40 countries as signatories. Applicants file in their national office (or designated receiving office) for their country or nationality or residence, but the actual search results and preliminary examinations are done in offices of the International Searching Authority. This consists of the U.S.A., the U.S.S.R., Japan, Sweden, Austria and the Institute International Des Brevets in The Hague. Each of these offices does searches for the member countries, and the international application and search reports are published 18 months after filing. Applicants still must ultimately convert the international application to a national application in the designated countries' national offices, but the search results of the ISA are accepted to greater or lesser degrees by all signatories.

The World Intellectual Property Organization (WIPO) is a United Nations agency with over 100 members. WIPO administers the Patent Cooperation Treaty. They publish *Industrial Property Statistics* which has important statistical summaries about patents filed and granted. WIPO also has representation at the International Patent Documentation Centre (INPADOC), a private searching organization.

The European Patent Convention requires only one application to be filed, generally through the inventor's national office or through the European Patent Offices in the Hague or Munich, with a designation of other member countries where protection is sought. The participating members of the European Patent Convention include most Western European countries, Sweden and Greece. Portugal will become a participant in 1992. Denmark, Norway, Ireland, Finland, Turkey, and Yugoslavia are not yet participating members.

There is also an African Patent Office (OAPI) with about a dozen members, mainly comprised of French-speaking African countries. A similar organization, ARIPO, serves about a half dozen English-speaking African countries. Patents granted by these organizations are honored in the respective member countries.

In the United States, patent applications are made with the U.S. Patent and Trademark Office. In addition to the normal (utility) patent application which results in the grant of protection for 17 years, there are several other types of U.S. patents. A **statutory invention registration**, formerly called a **defensive publication**, actually gives no positive protection to the applicant, but stands as an official publication which would prevent others from patenting the invention. There are also plant patents which may cover a new species of plant for 17 years. Finally, a design patent can be obtained for 14 years.

6.4 THE VALUE OF PATENT INFORMATION: REASONS FOR CONDUCTING A PATENT SEARCH

There is a tendency in academic chemistry departments to pay less attention to the patent literature than in industry. To a certain extent this might be expected. Academic chemistry libraries rarely attempt to collect chemical patents, preferring instead to order them on an individual basis for the chemist who has uncovered a patent in a literature search. Perhaps it is the fact that chemical patents may not seem as precise and clear as the journal literature which dampens the enthusiasm of academic chemists for patents. It is true that the inventor may state a range of temperatures for a reaction condition or list several catalysts without specifying which led to the optimum conditions for a reaction. In that sense, the patent literature of chemistry is vague compared to other primary literature. Nevertheless, there is a great deal of value to the chemist in patents, as the following paragraphs show.

Patents can suggest the research projects being undertaken by competitors and the types of technology being employed. Patents should be searched when a state-of-the-art search is conducted. The inventors usually choose to discuss the "prior art" in the patent application. Hence, a thorough literature search must be conducted before submitting the application. The most important of the relevant literature will be covered. In addition, the examiners will append a list of references cited.

Another reason for keeping up with the patent literature is to identify possible infringement of an earlier patent. These searches can be limited to the last 20 years. A search for novelty of an invention must be exhaustive, however, going back as far in time as possible. Since patent families often include equivalent patents in several languages, a search for the entire patent family may turn up a version in a language known to the chemist. Thus, patents serve as a cheap translation source. Forecasters of economic trends and directions in research also find the patent literature to be a valuable source of information. It was mentioned in chapter 4 that databases and printed reference works which cover patents have author entries for both the patentees and assignees, both personal and company entries.

Finally, there are subject classification schemes with highly specific subject categories for patents. These make it relatively easy to identify all relevant patents in a certain field. Examples of classification codes are the British Patent Classification, the United States Patent Classification, and the International Patent Classification. The IPC appeared in its 4th edition in 1985. It has 8 major sections divided into 118 classes (designated by numbers) and 620 subclasses (designated by letters). The U.S. Patent

Classification is the oldest system in continuous use. There are about 400 broad subject classes numbered 2 to 935 (with gaps for expansion) and about 115,000 very detailed subclasses in the U.S. Patent Classification. Of value in converting from the U.S. to the IPC code is *Concordance: United States Patent Classification to International Patent Classification* [569] (1985).

6.5 THE MAKE-UP OF A PATENT SPECIFICATION

The classification numbers form an important part of modern patent specifications. All Patent Cooperation Treaty countries publish the International Patent Classification code on the first page of the document. Other bibliographic details are also found on the cover page of the patent. Included are:

1. A patent or document number: a number assigned by the issuing country. In the United States, this is a sequential number which now is in the 4,000,000s. Other published specifications often carry two numbers, an application number and a publication number. For example, British patent application number 8013383 is the same as British publication number 2049570A. The "A" indicates that it is the unexamined version. When the patent is finally granted, the A will be changed to a B and the document reissued with the same number.
2. The date of issue of the patent and the priority filing date.
3. The patent title: This may be somewhat vague and frequently does not give a clear idea of the intent of the invention. Consequently, abstracting and indexing services may re-write the title.
4. The abstract
5. The name of the applicant
6. The list of documents referred to (especially earlier patents)
7. For many patents, one or more drawings
8. The text of the document: This will have a full disclosure of the invention, including its legal description or boundaries in the claims section. It may also include a discussion of the "prior art" if appropriate. For a chemical patent, examples of the new substances will be given.

The look of patent documents has become more uniform in recent years as more countries adopt the standard of the World Intellectual Property Office. This is available in WIPO's *Patent Information and Documentation Handbook.*

6.6 PATENT REFERENCE TOOLS AND PATENT AGENCIES

In this section are discussed some of the major printed reference works or databases devoted to patents. If the database has a printed counterpart, but bears a different name, it is discussed with the relevant printed work. The vendors which mount the database on their systems are listed with the respective record in the Chemistry Reference Sources Database.

Some of the general subject abstracting and indexing services cover the patent literature. Among them are the Institute of Paper Chemistry's *Abstract Bulletin*, *Derwent Biotechnology Abstracts*, *World Aluminum Abstracts*, *Chemical Abstracts*, and the R. H. Chandler publications *Catalysts in Chemistry* and *Organometallic Compounds*. Some broad discipline-oriented services which might be expected to include patents in their databases in fact do not. *Physics Abstracts*, for example, covers no patents. Not all A&I services use the standard bibliographic description of patents recommended by the World Intellectual Property Organization.

The most successful patent searches are carried out in databases or printed products specifically designed to cover patents. Companies or organizations like Derwent, INPADOC, and IFI/Plenum are well known for their high-quality coverage of patents.

6.6.1 The United States Patent and Trademark Office

By the year 2000 the United States Patent and Trademark Office's search files will contain an estimated 50 million patent documents. The prospect of having to cope with that much paper has precipitated an intensive automation effort in the Patent Office. The Automated Patent System (APS) is being designed to equip the Patent Office with the computer-based tools needed to handle such a large mass of information. Text searches of all U.S. patents issued since 1975 can be conducted using APS. Ultimately the system should be capable of searching the 115,000 categories of the traditional U.S. Patent Classification Code as well as abstracts of foreign patents. Furthermore, the retrieval of graphics along with text is a high priority. Most of the software for the project is being developed by Chemical Abstracts Service.

A related system, the Classification and Search Support Information System (CASSIS) [1890], is now accessible through over 60 patent depository libraries in the United States, about half of them public or state libraries and the other half university libraries.

Among the search capabilities which CASSIS provides at the present time are:
- search of key words in classification titles
- search of key words in patent abstracts and titles, from 1969 to date
- search of patents assigned to a particular assignee, from 1969 to date.

The user may choose to display:
- original and cross-reference classifications of a known patent
- all patents assigned to a given classification
- classification titles
- patent title and/or company name.

Classification information for all U.S. patents from 1790 to the present is covered in CASSIS. There is no cost for searching CASSIS at a patent depository library. The availability of CASSIS is sure to expand now that the Patent and Trademark Office has produced a CD-ROM version of the system.

Traditional printed products issued by the U.S. Patent and Trademark Office are also available at the patent depository libraries, as well as being held by many non-depository libraries. These include the *Index of Patents Issued From the United States Patent and Trademark Office* [1891] and the *Official Gazette of the United States Patent and Trademark Office* [1161]. The *Official Gazette* is a weekly publication containing summaries of normal patent specifications, usually called **utility patents**, those patents which do not cover designs, plants, or animals. Specifications are listed in the three broad categories, General and Mechanical, Chemical, and Electrical. Each entry, which gives the principal claim and a diagram if appropriate, is classified within the broad groups in numerical order by class and subclass. The *Index of Patents* has an alphabetical list of patentees and an index to the subjects of the invention. It is published annually in two volumes.

6.6.2 Derwent Patent Publications, Databases, and Products

The commercial company which offers the broadest range of products and services in the patent area is Derwent Publications Ltd. Not only does Derwent market a number of specialized patent services for the chemical industry, it also has a very good document delivery service for patents of all major industrialized countries. Derwent includes in its coverage the patents of 31 countries plus published applications from the European Patent Office and Patent Cooperation Treaty signatories. In addition to patents, important items from the periodicals *Research Disclosures* and *International Technology Disclosures* are found in its files. Like other abstracting or indexing services which handle patents, Derwent covers in depth only the first patent of a family which is received. This is called the **basic patent**. It may be the patent from the country in which the priority filing was made, but that is often not the case. Other patents in the family (the equivalents) are cross-referenced to the basic patent whenever Derwent receives them.

For comprehensive coverage of chemistry patents, the WPI database [554] may be searched. WPI is extensively indexed from the complete patent specification. The coverage includes pharmaceuticals from 1963, agricultural chemicals from 1965, plas-

TABLE 6.1
Derwent Classification Sections for Chemical Patents in the WPI Database and the *Chemical Patents Index*

A	(Plasdoc)	Polymers, plastics
B	(Farmdoc)	Pharmaceuticals
C	(Agdoc)	Agricultural chemicals
D		Food, disinfectants, detergents
E	(Chemdoc)	General chemicals
F		Textiles, paper
G		Printing, coating, photographic
H		Petroleum
J		Chemical engineering
K		Nucleonics, explosives, protection
L		Refractories, glass, ceramics
M		Metallurgy

tics from 1966, and all chemistry from 1970 to the present. Thus, the chemical records in the database constitute a considerable fraction of the nearly 4 million total records. Subject searching of WPI is possible for both title and abstract words. Furthermore, the International Patent Classification codes provide a type of subject access.

Derwent has devised its own classification scheme for patents, with 21 sections labeled by letters of the alphabet. Chemical patents comprise sections A to M, and the subject matter of the non-chemical patent areas falls into sections P to X. Chemical patents account for nearly half of those included in Derwent's coverage in WPI and are divided into twelve sections, as shown in Table 6.1. If the patent falls logically into more than one section, it is included in all relevant sections. Available free from the publisher is "The Derwent Classification for All Technology".

In terms of printed products, Derwent's coverage of chemistry is found in the *Chemical Patents Index* [555] (formerly, the *Central Patents Index*). Subscribers choose the sections of the *CPI* in which they are interested. Subscribers to the printed product receive preferential rates when searching the WPI database (and WPIL, the newer Derwent patent information from 1981, including abstracts). Furthermore, they have access to certain indexing features which are not available to non-subscribers.

The *Chemical Patents Index* is issued in a series of publications. The Alerting Bulletins come in twelve different weekly editions arranged either in country order or in classified order. The Basic Abstracts Journals of the *CPI* contain detailed summaries of the basic patent specifications in the corresponding Alerting Bulletins which were issued 1-2 weeks earlier.

Derwent has developed several methods to retrieve information on chemical compounds based on fragmentation codes (see section 7.3.4.). Such codes exist for polymers and chemicals which fall into *CPI* sections B, C, and E (pharmaceuticals, agricultural, and general chemicals). Mastery of the Derwent codes requires a great deal of experience. Hence, Derwent has created a microcomputer program, TOPFRAG [359], which facilitates the drawing of chemical structures on a microcomputer and their conversion to Derwent fragmentation codes. These can then be uploaded to a Derwent database. Only subscribers to the *Chemical Patents Index* with full subscriptions to sections B, C, or E or who satisfy certain other requirements are eligible to buy and use TOPFRAG. Another microcomputer product from Derwent is PATSTAT Plus [1892], designed to analyze the results of a patent search and present them in the form of tables, graphs, or charts to identify trends.

Access to non-chemical patents in a printed Derwent product is provided through the *General & Mechanical Patents Index* and the *Electrical Patents Index*.

6.6.3 INPADOC, the International Patent Documentation Center's Database

INPADOC [557] is the most nearly comprehensive worldwide patent database, covering over 50 patent offices. (Also included is information on the legal status of data from ten countries.) Produced by the International Patent Documentation Center, INPADOC has over 14 million bibliographic records for patents extending back even before 1920. In common with Derwent's WPI database, INPADOC is one of the few databases to provide online access to patent family data. (This information is supplied to Chemical Abstracts Service by INPADOC for the printed *CA* patent indexes, but the online database of *CA* does not include patent family data). The INPADOC database is searchable on a number of vendors' systems, including STN International, ORBIT, and DIALOG. Only title words can be searched as subject terms.

The database corresponds to the microfiche publication *INPADOC Patent Gazette* and the INPADOC Legal Status Service. INPADOC is strictly a bibliographic database with no abstracts. The *Gazette* portion of the file covers the time period 1968 to the present, but legal status information begins with 1978.

On STN is a current awareness file INPAMONITOR [1893], with about 120,000 records from the last six weeks. INPAMONITOR is updated weekly. Searchable fields include inventor names, assignees, International Patent Classification codes, as well as titles and complete publication and application data in most cases.

6.6.4 IFI/Plenum's Patent Services

The IFI/Plenum Data Company has been indexing U.S. chemical patents for over four decades. Thus, chemical patent coverage from 1950 is available in their CLAIMS database [559]. All other U.S. patents are in the database from 1963. Over 1.7 million U.S. patents are now included.

IFI has developed a system of indexing chemical patents which employs "uniterms". This provides a depth of indexing for U.S. chemical patents which is unrivaled in such a large database. CLAIMS has compound fragment and role indexing for all chemical patents. Access to the CLAIMS Comprehensive database, the file which includes the indexing of chemical compounds, is restricted to subscribers. The file includes two-digit role qualifiers for searching concepts like polymers. In the STN versions of the IFI/Plenum databases (files IFIUDB and IFICDB), CAS registry numbers have been added to about 85 percent of the 14,500 compounds in the Uniterm Compound Registry System. Direct L# crossover of registry numbers from the CAS ONLINE Registry File to the IFIUDB and IFICDB files is now possible.

Several vendors, among them ORBIT and STN International, also mount the IFI Reassignment and Reexamination Database. The records for the U.S. patents in this file are for patents which have been reassigned because ownership has been transferred since 1980. Also included since 1981 are U.S. patents which have been reexamined due to questions concerning their patentability.

6.6.5 Chemical Abstracts Service's Coverage of Patents

Patents constitute one of the largest groups of primary documents found in *Chemical Abstracts*, accounting for about one-sixth of the total documents abstracted in *CA* each year. Patents from 27 national patent offices as well as EPC and PCT patents are included in *CA*'s coverage and now number in excess of 1.3 million items in the CA File. The depth of indexing of the patents found in *CA* does not approach that given to chemical patents in the CLAIMS database. CAS indexes in *CA* only those compounds which were made in the examples cited in a patent, not all compounds covered by generic claims. Nevertheless, patents have always been a part of the *CA* database, so it is possible to search them all the way back to 1907 in *Chemical Abstracts*. However, CAS did not initially provide the extensive coverage for patents which was given to other forms of literature. Beginning around 1960, that situation changed, and *Chemical Abstracts* is now one of the best sources for information on chemical patents.

There was originally no numerical index to the patents found in *Chemical Abstracts* prior to 1937. The Special Libraries Association eventually published *Patent Index to Chemical Abstracts, 1907-36* [935], which brought the coverage up to the year when *CA* patent indexes began.

The *CA* patent indexes were originally constructed in two parts—a Numerical Patent Index and a Patent Concordance. Like other abstracting and indexing services, CAS abstracts from the first patent of a family which it receives. These basic patents were indexed in the Numerical Patent Index by country and cross-references were made in the Patent Concordance from later patent equivalents to the abstracts for the basic patents. There was no easy way to tie together all of the equivalent patents in the earlier patent indexes.

Beginning in 1981, a new patent index for *CA* was introduced. Appearing in both the weekly and volume/collective indexes, the expanded Patent Index contains in a single printed file the new patents abstracted by *CA* and the equivalents that have been received for basic patents which were abstracted earlier. Each time a new equivalent or related patent is received, it is listed in the Patent Index with all of the other members of the patent family. Furthermore, the new listing includes the priority relationships between the members of the family and indicates the stages of patent examination. In the case of regional and international documents, the countries in which the patent documents are applicable are also listed. An example of a *CA* Patent Index entry is shown in Figure 6-1.

```
4623632 A, 106:35904g
     CN 86/106891 A
        (Nonpriority)
     CN 86/107521 A
        (Nonpriority)
     EP 262238 A1 (Designated
        States: AT, BE, DE, FR,
        GB, IT, NL, SE;
        Nonpriority)
     EP 264468 A1 (Designated
        States: AT, BE, DE, ES,
        FR, GB, GR, IT, NL, SE;
        Nonpriority)
     JP 63/111943 A2
        (Nonpriority)
     US 4619906 A
        (Continuation-in-part),
        106:69139b
     US 4652689 A (Division;
        Related)
     US 4746764 A
        (Continuation-in-part;
        Related)
     ZA 86/07044 A (Nonpriority)
     ZA 86/07755 A (Nonpriority)
```

Figure 6-1
Patent Index Entry for a U.S. Patent Covered by *Chemical Abstracts*

Patents are also indicated in the other indexes of the printed *Chemical Abstracts* by a "P" placed before the abstract number for each entry which refers to a patent. Thus, an author or subject search might uncover an entry such as 105:P20537v or 100:P56896x. In an online search of the *CA* bibliographic database, the document-type field can be used to include or exclude patents. For example, => **SEARCH L9 AND P/DT** would limit a search to patents in the STN International CA File.

6.6.6 Other Patent Reference Sources

The American Petroleum Institute's APIPAT [587] database covers the petroleum refining and processing patents from nine countries since 1964. There are also a number of patent databases devoted to particular countries, for example CHINAPATS [1895], JAPIO (Japanese Patent Abstracts in English) [560], or PATDPA (German Patent Database) [553]. The legal aspects of patents give rise to some very specialized services, such as LitAlert [1896], a database with information about U.S. patents and trademarks which are the subject of infringement legislation. Lexpat [1897] has the full text of U.S. patents from 1975 on. Since patents are frequently cited in bibliographies of journal articles, it is possible to use a patent as the cited reference in both the *Science Citation Index* and SciSearch. In the database, the reference is searched on Dialog as **?S CP=ENTRY**, where ENTRY contains at least a country code followed by a patent number, for example, **?S CP=GB 2088880**

6.7 TRADEMARKS

A trademark is a design which is associated with a specific product. Unlike a patent, which expires after a certain period of time, a trademark enjoys the possibility of indefinite renewal in most countries. Legal protection is usually granted for the trademark itself and for associated tradenames for the product.

Since trademarks are designs, it is very desirable for reference works or databases associated with them to be able to reproduce the actual design. The TRADEMARK-SCAN database [218] on DIALOG does just that. Trademarks which are either designs only or those which have an image record in the database can be displayed. However, in order to view the design online, the DIALOGLINK software [1898] is required. Intended as a fast screening tool for checking the availability of new product or service names, the TRADEMARKSCAN database covers all active registered and pending trademarks on file in the U.S. Patent and Trademark Office. In most of the U.S., however, trademark rights are conferred by use, so it is not necessary to register a trademark. The producer, Thomson & Thomson, has linked together variant spellings (for example, TediBar would be included in a set defined by TEDDY AND BEAR), and it is planned to include a comprehensive design element search (for example, to be able to retrieve all trademarks which include a bear in the design).

6.8 SUMMARY

Patents form a very substantial and important part of the chemical literature. Just as there are separate databases and reference works devoted to such primary literature types as reports, conference proceedings, and dissertations, so there exist a number of such tools for patents. The products of organizations or companies which specialize in patent literature (Derwent, INPADOC, IFI/Plenum, and others) as well as products which include patents among other types of literature (for example, *Chemical Abstracts*) are discussed in this chapter.

6.9 SELECTED READINGS

Rimmer, Brenda M. "Abstract Journals: A Survey of Patent Coverage." *J. Doc.* **1988**, *44*(2), 159-165.

Walker, Richard D. *Using Patent Literature: A Comprehensive Guide*; Oryx Press: Phoenix, AZ, 1988. [1899]

Simmons, Edlyn. "Patent Family Databases." *Database* **1985**, *8*(1), 49-55.

Maynard, John T. *Understanding Chemical Patents: A Guide for the Inventor*; American Chemical Society: Washington, DC, 1978. [567]

Eisenschitz, Tamara S. *Patents, Trade Marks, and Designs in Information Work*. Croom Helm: London, New York, Sydney, 1987. [1900]

Kaback, S. M. "Access All the Information in Patents." *Chemtech* **1985**, *15*(3), 146-151.

CHAPTER 7

ONLINE CHEMICAL DICTIONARIES AND OTHER SOURCES FOR CHEMICAL COMPOUND SEARCHES

7.1 INTRODUCTION

The vast majority of subject searches in chemistry involve chemical compounds. There are numerous ways to code chemical compounds for effective retrieval, based on such information as molecular formulas, compound names, numbers of rings, etc. The use of such codes in both printed and online sources is the subject of this chapter. Another technique, structure searching, is the topic of the next chapter.

A search for information about a chemical compound may be for many different types of information: its properties, how to make it, its occurrence, uses, etc. Chemists have developed elaborate systems for coding, writing about, and talking about chemical compounds. Some of these systems provide a unique and unambiguous avenue to the desired compound. Others allow us to isolate compounds which are related in some manner to the one we are seeking.

7.2 COMPOUND NAME SEARCHES

It is very common to use chemical names for searching. Over the years, a number of indexes and other reference tools based on chemical nomenclature have been developed. (See section 14.5 for standard works on chemical nomenclature.) Chiefly printed or other formats of reference sources which do not require a computer to use are dealt with in sections 7.2 and 7.3. Online chemical dictionaries are presented in section 7.4.

7.2.1 *Chemical Abstracts* Index Guide

The Index Guide to *Chemical Abstracts* [176] is one of the first sources which should be consulted when beginning either a manual or computer-based subject search for a chemical compound. It contains cross-references and indexing policy notes to guide the searcher to the controlled vocabulary for both chemical substance name searches and other types of subject searches. The Index Guide began to be issued in the 8th collective index period (1967-71). Each five-year collective index period has its own Index Guide volume, which serves to control the vocabulary used in the subject indexes during that period. Thus, in a retrospective search which crosses the boundaries of the collective index periods (1967-71, 1972-76, 1977-81, etc.), the terms which were used at the time the printed product was produced are the terms which must be used. This means that two entirely different controlled-vocabulary names may be required in the printed *CA* to search for information on one chemical compound, depending on the time an article or other primary work was indexed. Indexing by registry numbers instead of CAS Index Names solves this problem in the online CA File database.

A compound name found in the Index Guide often has a SEE reference to lead the searcher to the CAS Index Name. For example, in the Index Guide for the 10th collective index period (1977-81), the example in Figure 7-1 can be found. Note the number in brackets in Figure 7-1. It is the CAS **registry number** which is unique to that compound. In the printed *CA*, registry numbers cannot be used for searching. However, they are very valuable in online searching. The closest online counterparts to the portions of the *CA* Index Guide which deal with chemical compounds are the online chemical dictionaries discussed in section 7.4. Registry numbers are further defined and described in section 7.3.6.

7.2.2 *Chemical Abstracts* Chemical Substance Indexes

No abstract accession numbers are found in the Index Guide. The user must go to the appropriate subject index to find references of interest in *Chemical Abstracts*. Prior to 1972, information on all topics, including chemical compounds, was found through the *CA* Subject Indexes. Since January 1972, information on classes of compounds, such as flavonoids, and on other subjects has been placed in the *CA* General Subject Index [230]. Subject searches for a single compound are now covered in the Chemical Substance Index of *CA* [232].

Flavan
See 2H-1-Benzopyran, 3,4-dihydro-2-phenyl- [494-12-2]

FIGURE 7-1
See Reference for a compound Name in the *Chemical Abstracts* Index Guide

Note the inversion of the preferred compound name (the *CA* Index Name) for Flavan in Figure 7-1. The Index Name is 2H-1-Benzopyran, 3,4-dihydro-2-phenyl-. All compounds entered in the *CA* indexes which cover chemical substances are filed alphabetically by the name of the **Heading Parent**, the basic structural skeleton of a compound, including any principal functional group. This means that subject index entries for Flavan file in the *B* section of the relevant index as Benzopyran, not in the *D* section under Dihydro, nor in the *F* section under Flavan. For purposes of alphabetizing, the basic compound comes first, followed by derivatives. Almost all *CA* Index Names are systematically constructed from terms which represent fragments of a chemical structure. Thus, the inverted entries in the *CA* chemical substance indexes serve to group together structurally-related substances.

Numbers for locants, symbols designating stereochemistry, and other italicized characters come into play in alphabetizing the names only when the alphabetic characters do not serve to distinguish two distinct compounds. If a substance is named only with a number, it comes before the alphabetic sequence of names. Alphanumeric substance names, such as 9AAP or SQ 1089, are interfiled according to the alphabetic characters they contain.

Certain compounds have a lot of primary literature published about them in a given time period. Such substances, which number about 600 are called **qualified substances** by CAS. A list of them can be found in *Qualified Substances in the CA File* [203] (1985). In order to facilitate subject searching for such a qualified substance, CAS classifies all of the abstract references for that compound into seven basic categories in the Chemical Substance Indexes as shown in Table 7.1. It is important to include the abbreviations in column 2 of Table 7.1 when formulating an online search strategy on one of those topics. The abbreviations are used in preference to the complete words in both indexing and in the texts of the abstracts in *CA* for all substances, not just for the qualified substances.

TABLE 7.1
Standard Subject Divisions for Qualified Chemical Substances in *Chemical Abstracts*

Subject	Abbreviated Online As:
analysis	anal
biological studies	biol study
occurrence	
preparation	prepn
properties	prop
reactions	
uses and miscellaneous	uses and misc

7.2.3 Other Sources of Compound Names

There is a CAS microform compilation, *Registry Handbook—Common Names* [191] (annual), which is very useful. A name is considered "common" in this set if the name is relatively uncomplicated. This does not necessarily imply that the substance is a common one. Many chemical substances have literally dozens of names. The *Handbook* does not claim to have all of them for every substance. Nevertheless, an average of about 1.5 names for each of the more than 530,000 unique substances covered in the set is found in the Name Section of the *Handbook*. Each compound name is tied to the CAS registry number for the substance. The Number Section of the *Handbook* lists variant names for the substance plus the molecular formula, if known. Furthermore, names which were used in one or more collective index periods of *Chemical Abstracts* from the 6th collective index period onward (1957-) are designated in the Number Section. This gives an easy way to select alternative terms for substances which have undergone name changes. The great number of chemical substances included in this work makes it by far the largest printed source of non-systematic chemical names available today.

Giese's *Beilstein's Index: Trivial Names in Systematic Nomenclature of Organic Chemistry* [160] (1986) contains all of the trivial names used as roots of names in the chemical substance indexes of major reference works such as *Chemical Abstracts* and *Beilstein's Handbook of Organic Chemistry* plus trivial names allowed by IUPAC rules. The work also includes root-modifying prefixes used mostly with steroid and terpene names, for example, Cyclo-, Homo-, Nor-, etc.

Another useful list of chemical compounds is found in the National Library of Medicine's *Medical Subject Headings—Supplementary Chemical Records* [200] (annual). The printed version contains approximately 23,000 chemical compounds mentioned in journals covered by *Index Medicus* and the computer database MEDLINE. The online counterpart, a component of the NLM controlled-vocabulary thesaurus *MeSH* (Medical Subject Headings) [224], has supplementary chemical records for at least 38,000 chemicals, enzymes, biologicals, drug combinations, and plant products. The file grows by about 300 new chemical records each month. These records form a subset of MeSH which became searchable in MEDLINE in June 1980. Among other fields, each record includes the generic name, CA Index Name and registry number,

TABLE 7.2
Selected Printed Sources of Chemical Compound Names

American Drug Index [216] (annual)
Chemical Synonyms and Trade Names [198] (1987)
Encyclopedia of Chemical Trademarks and Synonyms [214] (1981)
The Merck Index [177] (1989)
Organic Chemical Drugs and Their Synonyms [201] (1987)
Pharmacological and Chemical Synonyms [197] (1985)
Registry of Toxic Effects of Chemical Substances (RTECS) [211] (1988 plus supplements)
SOCMA (Synthetic Organic Chemical Manufacturers Association) Handbook [210] (1965)
Thesaurus of Chemical Products [186] (1986)
USAN and the USP Dictionary of Drug Names [206] (1986)

related registry numbers, and synonyms. The supplementary chemical records do not rank as full MeSH descriptors for searching the MEDLINE file. so each of the records is mapped to one of the broader related names which can be searched in MEDLINE, for example, norbornanes.

A frequent problem which occurs in chemical compound reference questions is that a name for a compound is known, but it is a trivial name, a brand name, a drug name, or even an abbreviation or acronym. The searcher often needs to find a structural drawing, molecular formula, or more formal name for the substance. The printed works discussed above (or their online counterparts) solve many such problems. In addition, the sources listed in Table 7.2 may also be of use in such searches.

Containing almost 90,000 compounds, *RTECS*, the *Registry of Toxic Effects of Chemical Substances* [211] has many common and brand names. The printed product is supplemented by quarterly microfiche, and there are online and CD-ROM versions of RTECS. The *Merck Index* [177] can also be searched online. Another database which may prove helpful at times is Trademarkscan [218].

7.3 SEARCHING FOR COMPOUNDS WITH CODES

From the time that Berzelius standardized the chemical symbols for the elements in 1813, there have been many systems developed to represent chemical compounds with codes. Common to most of these codes is the use of the Roman alphabet, Arabic numerals, and standard punctuation marks. In some cases, other special characters are utilized.

7.3.1 Molecular Formulas

Chemists have long used letters as symbols for the chemical elements and combined them with numbers to designate molecular formulas of chemical compounds. Over the years, several systems to standardize the order of the parts of the molecular formulas in indexes have been invented. Once the rules of the systems are understood, the indexes can be used for retrieval of information on a compound of interest. A molecular formula does not necessarily depict a unique molecule. There may be a number of isomers which satisfy a given formula. Hence, formula indexes usually combine names of the compounds with the formulas. It is expected that the users of the index will be familiar enough with chemical nomenclature to be able to distinguish the isomers.

Three systems of arranging elements in molecular formula indexes have been widely used. All of them are named for the people who codified the rules. In the Hoffman system, priority numbers are assigned to the elements, and the highest priority numbers are first in the formula. The Hoffman system covers only inorganic compounds. A similar system was adopted for arrangement of compounds in *Gmelin's Handbook of Inorganic Chemistry*. Another system is the Richter System, which is only for organic compounds. According to the Richter system, the formulas are first divided by the number of carbon atoms, then are arranged by the frequency of occurrence of the elements following this scheme:

C H O N Cl Br I F S P, then all others in alphabetical order. The number of atoms of each element in the formula is first considered, and formulas with the same number of atoms of identical elements are thus grouped together. The Richter System was used in some of the older handbooks and abstracting journals which included formula indexes. Among them were *Beilsteins Handbuch der organischen Chemie* and *Chemisches Zentralblatt*.

The system which is most widely used today is the Hill System, which covers both organic and inorganic compounds. Under the Hill System, the elements are arranged alphabetically within a given molecular formula. However, if carbon is present, C always comes first, followed by H if hydrogen is also present. The arrangement of an index of formulas ordered by the Hill System is then alphabetical, with the numbers of atoms of the elements coming into play if necessary. In alphabetizing, all numbers of a single-letter element are considered before a two-letter symbol starting with the same first letter. Thus, all hydrocarbons come before calcium-containing compounds which have Ca as the first symbol in Hill System order. The lowest number is listed first as the elements are encountered in a list of similar molecular formulas. The Hill System arrangement can give rise to some odd-looking formulas when compared to the normal order used by chemists. This is evident in the formulas in Figure 7-2 which are arranged in Hill formula order.

7.3.2 *Chemical Abstracts* Formula Indexes

The weekly issues of *Chemical Abstracts* do not have formula indexes. *CA* volume and collective indexes include formula indexes starting with volume 14 (1920), arranged according to the Hill System. There is a twenty-seven-year collective formula index covering v. 14-40 (1920-46), a ten-year collective formula index for v. 41-50 (1947-56), and five-year collective formula indexes thereafter.

The *Chemical Abstracts* Formula Index [231] does not in any sense replace the subject indexes for chemical substances. However, incompletely described compounds can sometimes be found only in the Formula Indexes. On the other hand, the voluminous *CA* entries for the 600 or so qualified substances can only be found in the relevant subject indexes. The sheer bulk of the entries for these especially common substances forced Chemical Abstracts Service to direct the searcher from the Formula Index to the appropriate subject index entries rather than duplicate the entries in both places. (See Figure 7-3.) In fact, cross-references to the *CA* Chemical Substance Index are not limited to the qualified substances. Whenever more than twenty abstract entries would be required for any substance in a volume (six-month) Formula Index or more than fifty entries per substance in a collective (five-year) Formula Index, a cross-reference to the Chemical Substance Index is inserted instead. This explains why searches of the CAOLD file for common substances are usually not very productive. The CAOLD file is created from the *CA* formula indexes.

Only the subject index entries for the names of substances (both qualified substances and all other compounds) have descriptive phrases (**text modifications**) attached to them for such things as properties, effects, use of a substance, or other

$Al_6Ca_5O_{14}$
B_2O_3
B_2Zr_3
BrH
CCl_4
$CHCl_3$
CHNO
C_2Ca
C_2H_4
C_2H_4BrCl
$C_2H_5AlBr_2$
$C_5H_8O_2$
$C_8H_5NO_2$
$C_{15}H_{24}N_2$
$C_{22}H_{24}FN_3O_2$
CaO_3Ti
ClH
H_2O_4S
H_4Sn
O_3PbRb_2
$O_8P_{14}Zn_7$
$SnZr_4$

FIGURE 7-2
Selected Molecular Formulas in Hill System Order and Alphabetized as in a Formula Index.

specific aspects relating to the substance. Hence, the subject indexes provide much more in-depth indexing than do the formula indexes.

For a given formula, isomers are arranged alphabetically by preferred name (Index Name), the inverted names printed in the *CA* subject indexes. Thus, the *CA* Formula Index helps to find the Index Name of a specific chemical compound without having to generate the name from the nomenclature rules. Of course, the user does have to be able to recognize the right compound when isomers exist. Once the Index Name has been found, additional references can be located in the subject indexes. The Subject Indexes (more recently, the Chemical Substance Indexes) can also be used to find references to compounds which are chemically related to the desired compound (for example, esters of acids). Such entries will be listed there after the Index Names for the parent compounds. In the Formula Indexes, entries for related compounds are likely to be widely separated.

A typical *CA* Formula Index entry from the 10th Collective Index period (1977-81) is shown in Figure 7-3.

$C_8H_5NO_2$
1H-Indole-2,3-dione [91-56-5]. See Chemical
   Substance Index
   sodium salt [3486-31-5], 90:6180p;
      91:157670v; 94:209034z

FIGURE 7-3
Chemical Abstracts Formula Index Entry.

For salts, addition compounds, and mixtures, the molecular formulas for the components are arranged separately, with the major component first. Ratios for salts and addition compounds are specified when known. If unknown, a lower case x before the second or any subsequent formula is used. Examples are:

$$C_{15}H_{24}N_2 \cdot 2ClH \qquad C_{22}H_{24}FN_3O_2 \cdot xH_2O_4S$$

No ratios are used for mixtures. These are examples of **dot-disconnected formulas**. Copolymer formulas show the formulas of the monomers individually, and, again, no ratios are indicated. The formula combination for copolymers or the single formula for a homopolymer is listed in parentheses with a subscript x.
Examples are:

$$(C_5H_8O_2)_x \qquad (C_4H_8 \cdot C_2H_4)_x$$

To find the formulas in the *CA* Formula Index, look up the main component only. The name under the main component formula such as "dihydrochloride," "sulfate," "polymer with . . ." indicates that this is a salt, an addition compound, a copolymer, etc. In the case of polymeric repeating units, the formula is written in parentheses followed by a subscript n, for example,

$$(C_8H_8)_n \qquad (C_2H_4O)_n H_2O.$$

The last case shows that end groups are included in the formula as a summation of their individual formula units. However, only the polymeric repeating units have their formulas listed in the printed Formula Index. The end groups are not shown.

Now that computer-readable versions of the *Chemical Abstracts* and other formula indexes are available, it is possible to use the molecular formula as an online search term. On the STN International system, the *CA* Registry File can be searched using the molecular formula field (See 7.4.4). Online searches are entered in the Hill formula order, with the numbers typed on the same line as the letters. Complete dot-disconnected and polymer formulas are displayed in the online Registry File. BRS ties the molecular formula to the CAS registry numbers in its CHEM file, the bibliographic file of *Chemical Abstracts*.

7.3.3 Ring Indexes

Many chemical substances include in their structures rings which can be used to index the compounds. In the 1920s, rules for numbering ring systems were devised which were applicable to all systems of simple or fused rings. Chemical Abstracts Service adopted the rules for the Third Decennial Index (1927-36). In 1940 the American Chemical Society published the first edition of *The Ring Index*, followed by a second edition in 1960. When the first Index Guide to *Chemical Abstracts* appeared in 1969, a section on ring systems (covering also stereoparents and cage systems such as polyboranes and metallocenes) was included. In 1977 CAS published the *Parent Compound Handbook* which replaced the Index Guide section. By then over 40,000 ring, stereo, and cage parents had been identified.

The *Ring Systems Handbook* [212] (1988) is the latest effort to collect and codify rings and related compounds. It replaces all previous ring indexes. Nearly 60,000 ring and cage systems were included in the basic edition which is supplemented each six months with new ring systems and updates to previous entries. It is planned to reprint the entire *Ring Systems Handbook* every three to five years. Included for each entry in the *Ring Systems Handbook* are the CAS registry number, the structural diagram, the *CA* Index Name, the molecular formula, and the Wiswesser Line Notation (see section 7.3.5 below).

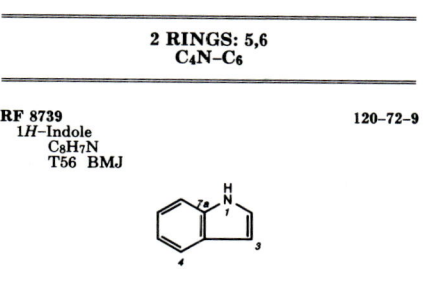

Figure 7-4
Typical *Ring Systems Handbook* Entry

Access to the entries is either by molecular formula (via the Ring Formula Index), by *CA* Index Name, or by ring analysis. The Ring Formula Index is a molecular formula index in Hill System order which ignores hydrogen counts. The *Ring Systems Handbook* provides an easy route to the Index Name for the *CA* Chemical Substance Indexes. Entries in the main body of the work, the Ring Systems File, are in ring analysis order, such as:

$$\text{2 RINGS: 5,6}$$
$$C_4N\text{-}C_6$$

The entry is found by considering the smallest set of smallest rings in the compound and figuring the molecular formulas of each independently (ignoring hydrogens). They are then arranged in ascending order from the smallest to the largest ring. Cage systems are placed at the end of the Ring Systems File.

Chemical Abstracts itself has for many years included an Index of Ring Systems. The Index of Ring Systems is published with each Formula Index for a volume (six-month) or collective (five-year) index period, beginning with the 7th collective index period (1962-66). Prior to then, such information was found in the Introduction to the Subject Indexes. Substances containing rings are ordered by the number of constituent rings, as in the *Ring Systems Handbook*. Thus, the number of component rings, the sizes of those rings, and the elements comprising them are all of the information needed to find a ring compound. As with the *CA* Formula Index, once the *CA* Index Name is known, the searcher can consult the Chemical Substance Indexes to find more detailed information about the ring systems and related compounds. Online search capabilities involving compounds which contain rings are discussed in sections 7.4.1 and 7.4.2.

124 CHEMICAL INFORMATION SOURCES

Substructure searches, covered in Chapter 8, offer another possibility for retrieving ring compounds.

7.3.4 Fragment Codes

Molecular formulas do not completely represent the topology of a molecule and are thus ambiguous. Another ambiguous coding system for molecules is fragment codes. **Fragments** are parts of molecules which can be named or otherwise described and, when taken collectively, designate all parts of a molecule. Fragments correspond to functional groups, skeletal groups, etc., but give no information about the relative positions of the groups. Fragment codes were developed earlier in this century for manual indexing of compounds. When computer searching of large structure files became possible, it was necessary to partition the files into smaller units for efficient searching. Fragment codes were used for this purpose and became known as **screens**. The screens could be single elements or large partial structures. The important thing to know about such systems is that the set of fragment codes or screens must be determined *before* indexing of the substances occurs. A fragment code for epichlorohydrin is shown in Figure 7-5. In that code, 5 could represent a halide-containing compound, 12, an ether, 21 for chloro-, 62 for an epoxide, and 71 could designate a 3-atom ring. If all of these separate fragment numbers were combined in an AND statement, there is a good chance that the search would retrieve epichlorohydrin. However, there is no guarantee that epichlorohydrin would be the only compound retrieved.

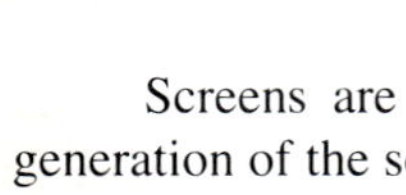

CH_2—CH—CH_2—Cl (with epoxide O bridging)

Fragment Code: 5-12-21-62-71

FIGURE 7-5
Fragment Code for Epichlorohydrin.

Screens are still used in computerized substructure searching, but automatic generation of the screens is usually built into the systems. Occasionally, a substructure search will require the manual input of one or more screens to the search strategy. For assistance on such occasions, the *CAS ONLINE Screen Dictionary* [189] (1981) is available from CAS.

7.3.5 Wiswesser Line Notation and Other Linear Notations

There have been three editions of rules for the **Wiswesser Line Notation (WLN)**. The first edition in 1954 led a number of chemical companies to adopt WLN as a practical way of uniquely and unambiguously coding chemical structures. A second edition followed in 1968, and a third, *The Wiswesser Line-Formula Chemical Notation*

(WLN) [205] in 1975. That no fourth edition has been published reflects a major shift away from the use of the WLN in recent years. Nevertheless, it is still a versatile and useful system for coding chemical substances.

The Wiswesser Line Notation code is amazingly compact, built on only 41 characters, including a blank space, &-/*0123456789, and the upper-case alphabet, A-Z. Wiswesser changed a few of the traditional symbols for the elements (for example, E for Br and G for Cl), invented a few other symbols (like Q for the -OH group), and constructed the system of rules for manipulating the symbols. The resulting system visually depicts the structure of a molecule in a linear character string as meaningful to those who understand the code as any regular chemical structural drawing. Wiswesser thus provided a system which was extremely flexible, standardized, and allowed substructure searching with normal sorting routines on a computer.

Dozens of chemical companies adopted the system to organize their internal files, and a number of printed reference works appeared with WLN indexes. Among the latter were the *CRC Atlas of Spectral Data and Physical Constants for Organic Compounds* (1975), the *Parent Compound Handbook*, and *Index Chemicus*. The WLN-based Chemical Substructure Index [192] to *Index Chemicus* continues to be produced as a microfiche serial set.

The compactness of the WLN code can be judged from the WLN for epichlorohydrin: T3OTJ B1G, where the T3O indicates a heterocyclic ring, defined with 3 elements (2 carbons and an oxygen) within the ring. The second T indicates that the ring is saturated with hydrogens wherever possible, and the B1G designates a chain with one carbon found at the B position relative to the oxygen. Attached to the carbon chain is a chlorine atom, and hydrogens are understood to occupy the other available points on that carbon.

One of the most successful implementations of WLN is the CROSSBOW (Computerized Retrieval of Organic Structures Based on Wiswesser) [227] system developed by ICI (Imperial Chemicals Industries). The search system utilizes a three-step process involving screens, WLN, and connection tables. (Connection tables form the basis of the system used by Chemical Abstracts Service and are covered in the next section.) As more and more companies sought to convert their WLN files to CAS connection files, programs were developed to make the transformation. A program called DARING [226] is available from the Fraser Williams Company to accomplish the conversion.

Other people have devised linear notations, among them, Hayward, Skolnik, and Dyson. However, none of those systems gained as many supporters as did WLN, despite the fact that a variant of the Dyson system received official sanction from IUPAC, the International Union of Pure and Applied Chemistry.

7.3.6 Connection Codes and CAS Registry Numbers

The coding system which is most widely used today is based on atom connection tables. The major difference between connection tables and fragment codes or linear notations is that no pre-classification of the component parts of molecules is required. For a full connection table a square matrix is drawn containing the same number of

```
    5
    O
   / \
C—C—C—Cl
1  2  3  4
```

Atom	Atom No.	Connections	Bond	Type
C	1	-	1-2	1
C	2	1	2-3	1
C	3	2	3-4	1
Cl	4	3	—	—
O	5	1,2	1-5	1
O	5	1,2	2-5	1

FIGURE 7-6
Connection Code for Epichlorohydrin.

rows as atoms in the chemical structure to be coded. All standard atomic symbols are used, and the bonds are assigned a code, perhaps numbers. The connection table then contains information about the way each atom is attached to those surrounding it. For any given compound, several connection tables could be constructed since there are no rules for numbering atoms. In the compact variant of connection tables, each connection is specified only once. Any atom is chosen as number 1, hydrogens are ignored, then the remaining atoms are numbered. Only the connections to lower-numbered atoms are coded. A compact connection table for epichlorohydrin might then be listed as shown in Figure 7-6.

Connection tables can be formed and input by relatively untrained clerical labor. Once the data have been entered, all possible structures can be scanned using atom-by-atom searches. Fragment codes can automatically be generated from connection tables, and it is possible to convert back and forth between connection tables and other structure codes if necessary.

In a registry system, each compound has only one identifier, and a new identifier is assigned to a new compound only when there is no other compound exactly like that one in the database. An example of a classified registry system is that of the Enzyme Commission. In that system, enzymes are assigned numbers such as this one for Ribokinase: E.C. 2.7.1.15. Another system which is based on connection tables, but does not classify the compounds is the Chemical Abstracts Service Registry System.

The development of the Chemical Abstracts Service Registry System is a major accomplishment in the history of chemical information science. Chemical Abstracts Service uses a number known as the **registry number** to identify each unique substance in the database. Registry numbers look somewhat like Social Security Numbers, with five to nine digits. Each registry number is divided into three groups of the format 987654-32-1. The first set of digits may vary in length from two to six. A different registry number is assigned for each stereochemical or positional isomer. Acids and bases have registry numbers, but their salts have different registry numbers which probably look quite different, that is, are in a totally different range of numbers. Hence, related compounds cannot be gathered into a set in an online search by inputting a range of registry numbers (with very rare exceptions). In other words, the CAS Registry Number System does not classify similar compounds together. However, the registry numbers do provide unique access points for the particular compounds they represent. The CAS registry numbers are widely used and can be found in both primary and secondary sources. In the latter, there is sometimes an index to the entries based on the registry numbers, as in the *Dictionary of Organic Compounds*.

In an online search it is very important to obtain the registry number when seeking information about a compound. However, there are some compounds with registry numbers which are not linked to any abstracts in the online *Chemical Abstracts* file. This situation arose in part because CAS assigned registry numbers to compounds which came from sources other than the primary literature abstracted in *CA*. These included the *Colour Index* and the *Merck Index*. Likewise, registry numbers are assigned to mixtures, with each component getting its own registry number. Thus, it could occur that there is nothing new in the literature of the last few decades for a substance which has long been known. If that compound has recently been incorporated into a new mixture, it would be assigned a registry number, but the link to an abstract would be through the registry number for the mixture. Occasionally registry numbers are replaced. This may happen when a structure is incorrectly or partially determined, then a correct full structure is later published.

The file of compounds in the CAS Registry System has surpassed 10,000,000 in number, with about 350,000 new entries each year. The main value of CAS registry numbers lies in their use as search terms in computer databases. It is much, much easier to input a registry number than a *CA* Index Name for a compound. In addition, a number of databases besides *Chemical Abstracts* can be accessed with the registry number. For example, many journals which have the texts of their articles online, such as the CJACS database [548] of American Chemical Society journals, include registry numbers. In addition, the National Library of Medicine makes extensive use of registry numbers in its online databases. Over 50,000 CAS registry numbers have been added to the BIOSIS Previews/RN database [234] exclusively on STN International for records which were entered into the database from July 1980. About 1.5 million of the records in the database have at least one registry number. Thus, cross-file searching among the Registry and CA Files and BIOSIS Previews/RN is straightforward. See Table 8.1 for additional databases which can be searched with the CAS registry number.

Registry number searching of the online *Chemical Abstracts* database is available through a number of vendors. Depending on the vendor used, there are variations in what can be done with the registry numbers. DIALOG, ORBIT, STN International, and Questel allow searching by registry number from 1967. STN International has begun to provide registry number links to abstracts before 1967 in the CAOLD File [233]. The output consists of a list of abstract numbers for records which contain information about the compound. The abstracts must then be found in the printed *CA*.

BRS allows registry number searching only from 1977, but BRS does not link the registry numbers to the modifying phrases (text modifications) in the indexing, for example, "reaction of, with isatin". The **text modifications** are not controlled-vocabulary terms. Nevertheless, they pinpoint the specific aspect of a compound which is discussed in an article or other abstracted work. Thus, they lend considerable flexibility to searches for chemical compounds. On STN International, text modification terms are part of the Basic Index in the CA File.

BRS has chosen to link the molecular formula directly to the *CA* bibliographic records. On most systems the record in the bibliographic file of *Chemical Abstracts* contains only the registry numbers, with no links to molecular formulas, names, etc. This makes it difficult to identify the compound of interest when looking at the full record. However, beginning with v. 106 (January 1987), CAS is adding common names for many substances directly to the registry number in the index term field of the records. In the record for abstract 93:25540j in Figure 3-2, the first compound listed in the CV (Controlled Vocabulary) field has registry number 57765-60-3. It is discussed in the article in the context of its reaction with isatin. Note that from the record alone it is not possible to know that isatin is the compound below it with registry number 91-56-5. However, the newer 1987 record in Figure 3-2 clearly identifies the most common compounds with their names as well as the registry numbers.

As a further aid in identification, STN International now provides the option of SELECTing registry numbers from a record to form an answer set in the CA File which can be input to the Registry File to identify the substances. In this manner, the registry numbers do not have to be typed in individually. Registry numbers which go the other way, from the Registry File to the CA File and result in a hit are now highlighted on STN International when the Index Term field is printed. This makes it easy to see from the text modifications what aspect of the compound is being discussed. The registry numbers are highlighted with three asterisks on either side, as in ***91-56-5*** in Figure 3-2. By linking the registry numbers with text-modification terms using the field-specific operators, it is possible to perform very precise searches for compounds on some systems.

On certain systems registry numbers for compounds whose preparation is discussed in an abstracted work are coded with a symbol designating the preparation of the substance. For instance, in order to retrieve articles on the preparation of Flavan on DIALOG, the searcher would enter:

? **Search rn = 494-12-2p**

and on STN International, simply:

=> **Search 494-12-2p**

Questel and ORBIT require the use of the abbreviation PREPN AND-ed to the registry number in order to perform the same search.

Thus, we see that that the treatment of registry numbers in *Chemical Abstracts* online files varies from vendor to vendor. It is essential to become familiar with how the vendor of choice provides access to information on chemical compounds via registry numbers. In the next section, we will learn how to use the *CA* textual data about compounds in online chemical dictionaries. In chapter 8, the use of the *CA* connection tables and screens for true substructure searching by inputting actual chemical structures is presented.

7.4 ONLINE CHEMICAL DICTIONARIES

As noted before, online chemical dictionaries are not online versions of books which are dictionaries of chemical terms. Those works, such as Hawley's *The Condensed Chemical Dictionary* [199], define a broad range of chemical concepts including, but not limited to compounds. In contrast, an **online chemical dictionary** is a file which ties together nomenclature of various types, registry numbers, and perhaps other information such as molecular formulas, ring information, etc. For the most part, the online chemical dictionaries exist as companion files to bibliographic or numeric databases and make searching for information on a compound considerably easier. This is done through links between the dictionary files and the databases. The CAS registry number is invariably the link between the files. Vendors have developed relatively easy methods to switch the registry numbers from one file to another, including the MAPRN command on DIALOG and the PRINT SELECT RN command on ORBIT. On the STN International system, it is possible to transfer the registry numbers from an answer set formed in the Registry File to the CA File of bibliographic records. As many as 30,000 registry numbers may be present in the L# formed in the Reg File. Once the CA File is entered, the search for all of those substances is performed simply by the command =>S L#

7.4.1 Online Chemical Dictionaries on DIALOG and ORBIT

DIALOG uses the number of times a compound is associated with different records in their CA Search bibliographic file as the basis for developing their online chemical dictionaries. There are two DIALOG chemical dictionary files, Chemsearch [595] and Chemname [594]. Chemsearch has all substances in the CAS Registry System from 1965 onwards. Chemname, a subset of Chemsearch, is limited to those substances cited in the bibliographic file two or more times since January 1, 1967. In addition, DIALOG files 308 through 312 provide the option of limiting a search by collective index period. In the past, there was a separate file on DIALOG for compounds which have *no* CA bibliographic file entries linked to those registry numbers from 1967 onward. This file, called **CHEMZERO**, had over a million entries! All of those compounds are now part of Chemsearch.

DIALOG's online chemical dictionary files are indexed by fragments of the *CA* Index Name, element count, periodic index or group name, periodic transition row, stereochemical descriptors, molecular element, and molecular formula. They also include various access points by ring data. The "element count" search option can be very useful because it allows the searcher to specify no carbon atoms, thereby limiting the search to inorganic compounds. Carbon is the only element for which an element count of zero is allowed, however. Other elements which are searched with the EC qualifier on DIALOG must have a number of 1 or greater in the search strategy. All metallic elements can be gathered with the strategy ?**S EC = Mnnnn** and all halides with the strategy ?**S EC = Xnnnn**, where M = metals, X = halides, and nnnn is a four-digit, zero-filled number. The strategy ?**S EC = Annnn** limits the answer set to substances containing a specific number of non-hydrogen elements.

Substantial capabilities for ring searches have been included in the DIALOG files. For example, it is possible to specify the ring analysis, the number of or arrangement of the elements in the ring, the total number of elements in a ring system, as well as the number of rings and sizes of the rings.

ORBIT's online chemical dictionary files are CHEMDEX2/3/4 [520] (1976- , covering registry numbers 56700-46-0 to the present) and CHEMDEX (1965-75, for registry numbers 36-88-4 through 56700-44-8). These files can also be searched by fragments of *CA* Index Names, element counts, stereochemical information, molecular formulas, and ring descriptors. There is some difference in the way ORBIT and DIALOG handle this information for indexing purposes. It is not possible to do an element count search on ORBIT. However, ORBIT does allow a range of numbers of a specific element to be searched, for example, less than three sulfur atoms.

A search using chemical name fragments is subject to the same chances of error which any fragment search entails. The fragments may be combined in quite unexpected or unwanted ways. Proximity operators and string searching can reduce the number of false drops in such searches. (See section 3.6.6.)

7.4.2 The National Library of Medicine's Online Chemical Dictionary

The National Library of Medicine's CHEMLINE file [593] includes information on over 500,000 substances indexed in NLM's databases. CHEMLINE does not allow searches by element count, ranges of numbers of elements, or classes of compounds. However, it is possible to perform name searches on both common and systematic names, name fragment searches on systematic chemical names, and molecular formula searches (including variants which allow retrieval of salts, mixtures, and fragments of molecular formulas). Information about rings can also be searched, such as the number of rings, size, elemental analysis, and component line formula, or fragments of these last three. The main purpose of CHEMLINE is to assist in searching other NLM databases, such as TOXLINE [219] or RTECS [211], for information on chemical compounds. Hence, the CHEMLINE record includes a Locator field which identifies the NLM databases containing information on the substance.

A much smaller group of chemical substances can be searched in either the MeSH Vocabulary File [224] or directly in the MEDLINE bibliographic database [225] (See NLM's *Medical Subject Headings—Supplementary Chemical Records* [200]). Either file may be searched by the name of the substance, by CAS registry number, related registry number, or Enzyme Commission Number, by a synonym or name fragment of the substance name, and by MeSH headings for compounds.

Searches in NLM files are considerably cheaper than those in files based on CAS products. Therefore, if a search of a biologically active compound is needed, the CHEMLINE database may be a good first choice.

7.4.3 The Chemical Information System's Online Chemical Dictionary

The Chemical Information System (CIS) was originally made available through a joint effort of the National Institutes of Health and the Environmental Protection Agency.

By 1985, it had been "privatized," and various components are now available from different vendors. One of them is Chemical Information Systems, Inc., a subsidiary of Fein-Marquardt Associates.

The heart of the CIS is SANSS, the Structure and Nomenclature Search System [222]. In this section are discussed only those parts of SANSS which are not related to its substructure search capability. A discussion of the structure searching features of SANSS is included in section 8.3.1.

Like NLM's CHEMLINE, SANSS serves as the hub of the CIS databases, directing the searcher to databases which have information on the substance. A number of these files are numeric databases which contain actual data such as infrared spectra, mass spectra, C^{13} NMR spectra, x-ray crystallographic data, etc. SANSS can be searched by name, CAS registry number, molecular formula, molecular weight, or fragments of names, structures, or formulas.

7.4.4 STN International's Online Chemical Dictionary

STN International offers a number of options for searching chemical substances in the CAS ONLINE Registry and CA Files. Most of the capability to search for chemical substances is found in the Registry File [223]. Discussion of the substructure searching facets of this file will be deferred to Chapter 8.

In 1985 STN International added many of the chemical dictionary searching features offered by DIALOG, ORBIT, and other vendors. Substances can now be searched in the Registry File by common and trade names, chemical name fragments, complete *CA* Index Names, and molecular formulas. Ranges of occurrences of particular elements, and specific molecular weights or a range of molecular weights offer further options. CAS registry numbers and component registry numbers are also searchable.

The CAS ONLINE Registry File contains the complete store of over 10,000,000 chemical substances found in the CAS Registry System. The compounds are not segmented into files by date on the STN International system. The close relationship between the Registry File and the CA File (which contains the bibliographic data in one file covering 1967 on) was taken fully into consideration in the design of the search system. An answer set formed in the Registry File consists of a set of compounds identified by registry numbers. All of the registry numbers in that set number can be searched in the CA File, the CAOLD File, and certain other databases on STN International simply by listing the set number (L#) as the search key, provided the set does not contain more than 30,000 registry numbers. Thus, STN International has a very easy method of moving from an online chemical dictionary search into a bibliographic search. STN definitely provides the greatest range of search dates with the possibility of locating pre-1967 references in the CAOLD File, which is exclusively available on STN International.

An excellent set of manuals for CAS ONLINE searching was issued in 1985. The multi-volume set entitled *Using CAS ONLINE* [235] includes five physical volumes. Volume III of the set is devoted to "Dictionary Searching". CAS has also published the *Search Aid for Name Searching: Frequently Posted Name Segments in the Registry File* [204] (1985). This booklet helps determine at which point a compound name should be broken when selecting fragments for searching. For example, inputting "methoxy" as a fragment in the Registry File yields zero postings. Examination of the lists in the *Search Aid* reveals that the proper input should be:

=> **Search METH(W)OXY**

The Basic Index in the Registry File contains fragments from the CAS Index Names, synonyms, and molecular formulas. Thus, character strings (with no intervening spaces) which represent names or fragments of names of compounds or formulas can be searched there. Note that a name search for a single compound is best entered in the Registry File using the Chemical Name field label (/CN), as discussed below.

There are special inverted files for records in the Registry File, including the Chemical Name index. These are listed in Table 7.3, and, along with the Basic Index, comprise the searchable fields in the Registry File.

TABLE 7.3
Inverted Files for the CAS ONLINE Registry File

Field Name	Field Code
Basic Index	none
Compound Class Identifiers	/CI
Chemical Name (Index Name or Synonym)	/CN
Component Registry Number	/CRN
Formula Weight	/FW
Heading Parent	/HP
CAS Registry Number Locater	/LC
Complete Molecular Formula	/MF
Total Number of Components in the Molecular Formula	/NC
Periodic Group Codes	/PG
Registry Number	/RN
Element Count	none

Substances in the CAS ONLINE Registry File records are classified into the eleven compound categories in Table 7.4. These compound class identifiers can be used in a search of the Registry File with the field label /CI, as in **=>S L# AND PMS/CI** to limit a search to polymers. Most of these identifiers are self explanatory, but some need better definition. Compounds classed as registered concepts (CTS) or generic registrations (GRS) would not normally be considered registerable substances by CAS, perhaps because they are not unique chemical substances or part of the structures are unknown. However, in compliance with a request from another organization, such as a government agency, they were registered. For example, the Toxic Substance Control Act (TSCA) governs certain of these GRS and CTS substances which are carefully controlled and regulated by the government from the point of manufacture to disposal. Substances which have more than 253 non-hydrogen atoms have no structure diagrams in the Registry File; they are manually registered substances, since there are no connection tables for them in the Registry File.

TABLE 7.4
Compound Class Identifiers in the CAS ONLINE Registry File

Class Name	Code
Alloy	AYS
Coordination Compound	CCS
Registered Concept	CTS
Generic Registration	GRS
Incompletely Defined Substance	IDS
Manually Registered Substance	MAN
Mineral	MNS
Mixture	MXS
Polymer	PMS
Radical Ion	RIS
Ring Parent	RPS

The Chemical Name field allows single or multi-word common or trade names and CAS Index Names to be entered as search terms. Examples are:

=> **Search Isatin/CN**
=> **Search Acetylsalicylic acid/CN**
=> **Search 2H-1-Benzopyran/CN**

Note that punctuation is included if appropriate. As with molecular formulas, subscripts are typed on the line, as are superscripts. Greek characters are spelled out completely (for example, .DELTA.) with periods before and after. Brackets become parentheses, and any symbol or word which has meaning to the search software must be **masked**, that is, enclosed in single or double quotation marks. The symbols thus affected are:

/) (> < = ? ! $ ' and "

The words are AND, OR, and NOT.

It should be noted that Enzyme Commission numbers are considered to be part of the Chemical Name field. Thus a search for Ribokinase by E.C. number would be entered:

=> **Search E.C. 2.7.1.15/CN**

A mixture, copolymer, etc. is generally given a separate registry number from that of its components. However, each component also has a registry number which is searchable in the component registry number field (CRN).

A search for a formula weight may either be for an exact numerical value, a range of values, or values greater than or less than a certain number. If the latter options are chosen, the field identifier *precedes* the value, as in:

=> **Search FW > 10000**

Otherwise, it is in the normal end position, as with:

=> **Search 100/FW**

For mixtures and other multi-component substances (excluding polymers, which cannot be searched by formula weight), each component's formula weight is calculated separately. Hence, no attention is paid to the ratios of such components.

The Heading Parent field contains the complete **Heading Parent**, the part of the name which comes first in the inverted listing of Index Names in the *Chemical*

Substance Index. Since substances with more than 253 non-hydrogen atoms cannot be searched by substructure, the HP field is useful for searching such things as large analogs of naturally occurring peptides, as with:
=>S bradykinin/HP
Names used as Heading Parents are not separately displayed in the records, but are part of the chemical names section of the records.

The Registry Number Locater field lists the databases on STN International which contain records indexed by the registry number.

Molecular formulas are searched in Hill System order. Recall that molecular formulas are also included in the Basic Index of the Registry File. In the /MF inverted file, it is assumed that the *complete* molecular formula is being searched for either single-component or multi-component substances. Of course, the rules for multi-component substances which were given in section 7.3.2 apply when the molecular formula for such substances is searched in the Registry File.

The total number of components can be used in the search, qualified with the field code /NC. The number of dots in the formula plus one determines the number to enter. Thus, for $C_{15}H_{24}N_2.2ClH$, the number of components is two. It is also possible to search for a range of components, for example, => **Search 3-5/NC** In addition, compounds having fewer than or more than a certain number of components can be searched, provided the field code is entered before the < or > sign, as in: => **Search NC < 5**

The Periodic Group codes must be obtained from a chart in the "Dictionary Searching" manual (volume III of *Using CAS ONLINE: The Registry File* [235]). These codes cannot be displayed as part of the full record. The Periodic Group codes can be very helpful when searching for inorganic substances, especially when combined with Element Counts.

The registry number itself, if already known, can be used to search the Registry File. Unlike the CA File, the Registry File does not include registry numbers in its Basic Index. Thus, a search for a known registry number must include the field label /RN, for example, => **Search 91-56-5/RN**

Element Count searches employ no field codes. Instead, the number of elements desired is separated from the symbol for the element by a "/" to form the search statement. For example, => **Search 2/BA** locates all compounds with 2 Barium atoms in the molecular formula.

7.5 DISPLAY OPTIONS IN THE REGISTRY FILE

The default option for displaying a record in the Registry File is IDE (Substance Identification Information), which displays such things as the registry number, Index Name, other names, the structure diagram, the number of references in the CA File, and an indication of the existence of records in the CAOLD File. (See Figure 7-7.) Up to 50 names for a substance are included in the default IDE format. The final field in the record displayed in Figure 7-7 is the File Segment (FS) field. It designates a substance which could be displayed in three dimensions. About 40 percent of the

```
=>    s isatin/cn
L1              1 ISATIN/CN

=>    d ide
```

L1   ANSWER 1 OF 1
COPYRIGHT (C) 1989 AMERICAN CHEMICAL SOCIETY

RN   91-56-5
CN   1H-Indole-2,3-dione (9CI)   (CA INDEX NAME)
CN   o-Aminobenzoyl formic anhydride
CN   2,3-Diketoindoline
CN   2,3-Dioxoindoline
CN   2,3-Indolinedione
CN   Isatic acid lactam
CN   Isatin
CN   Isatinic acid anhydride
CN   Indole-2,3-dione (8CI)
CN   Pseudoisatin
CN   2,3-Dioxo-2,3-dihydroindole
CN   Isatine
CN   Tribulin
DR   84788-92-1, 5815-00-9
MF   C8 H5 N O2
CI   COM
LC   BEILSTEIN, BIOSIS, CASREACT, CHEMLIST, CIN, CSCHEM, EINECS, HODOC, TSCA
FS   3D CONCORD

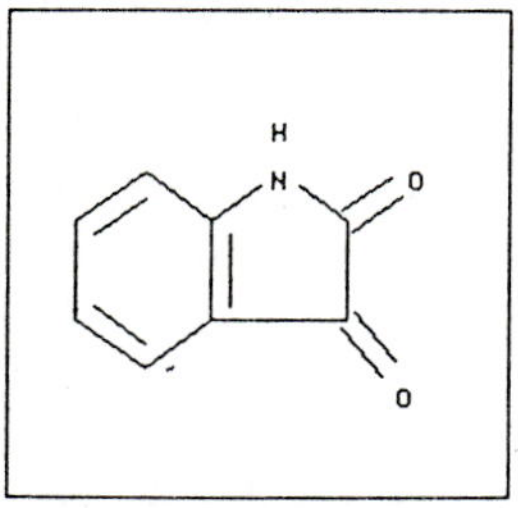

REFERENCES IN FILE CAOLD (PRIOR TO 1967)
770 REFERENCES IN FILE CA (1967 TO DATE)

=> d cost

COST IN U.S. DOLLARS	SINCE FILE ENTRY	TOTAL SESSION
CONNECT CHARGES	5.60	5.90
NETWORK CHARGES	0.80	0.88
DISPLAY CHARGES	3.69	3.69
FULL ESTIMATED COST	10.09	10.47
	DISCOUNT	
CA SUBSCRIBER PRICE	-0.67	-0.67

IN FILE 'REGISTRY AT 18:15:33 ON 17 DEC 89

Figure 7-7
CAS ONLINE Registry File Record Displayed in the IDE Format on a Type 2 (Graphics) Terminal

substances in the Registry File were capable of such display by the end of 1989. In order to view the 3-D structure display, two pieces of software are required: STN Express [534] and the Alchemy molecular modeling software [1901]. The FS field is also searchable by entering either "3D" or "CONCORD" as labeled field search terms, for example, =>S 3D/FŞ

A number of other options for viewing the results of a Registry File search are available. In order to see all chemical names associated with a substance, the FIDE command is used, as in =>**DISPLAY L1 2 FIDE**
If only the CA Index Name is desired the command is =>**DISPLAY IN**
Chemical Abstracts Service attaches to a record in the Registry File as many as ten of the most recent bibliographic references from CA File entries which include that registry number. Thus, one of the DISPLAY options in the Registry File is to see the substance information with part or all of the bibliographic records, for example,
=>**DISPLAY IDE BIB**
Bibliographic records are *not* searchable in the Registry File. They can only be displayed, and a command which asks to see the attached bibliographic records will display all of them in the format requested. Options for such a display include abstract numbers only (AN) or abstract numbers with: bibliographic information (BIB), abstracts (ABS), indexing information (IND), or all of that information (ALL). The ALL display format also displays all substance names associated with the record.

7.6 SEARCHING WITH CHEMICAL NAMES IN THE CA FILE

It is usually preferable to search for a single compound first in the CAS ONLINE Registry File, then use the answer set in a further search of the CA File. Giving the **SELECT CHEM** command in the Registry File extracts the CAS registry numbers and chemical names from the display fields for entry into all STN files which use them as search terms. Of course, there are times when it is desirable to search directly in the CA File with a compound name. Since the Basic Index in the CA File contains single words from the titles, keywords, controlled-vocabulary terms, and text modifications, many chemical names are found in the Basic Index. Before entry in the Basic Index, all punctuation is removed from chemical names. Then each component of the name which is separated by a space is entered into the appropriate alphabetical sequence of the Basic Index. Thus, Indole-2,3-dione would be indexed at four points in the Basic Index:

Indole 2 3 dione

Likewise, common names or names referring to families of compounds, such as isatin or indoles, will have indexing entries in the CA File's Basic Index. Furthermore, the chemical class names may be controlled-vocabulary terms which could be searched precisely with the field label, as in: => **S PEPTIDES/CV** Consult the Index Guide when in doubt about the use of a class name as a controlled-vocabulary index term. (CA Index Names are not searchable in the CV field of the CA File's Basic Index; they can only be searched as chemical names (with the /CN field label) in the Registry File.) The mere mention of a chemical substance in a document does not mean that it will necessarily be indexed in *Chemical Abstracts*. There must be something new reported about it in the document, or it must be a new chemical substance. Class names are used in *CA* indexes whenever three or more individual members of a recognizable class are described in the document. (See 11.4.1 for more information.)

7.7 ELEMENT TERM (/ET) SEARCHES ON STN INTERNATIONAL

STN International has made it relatively easy to search for chemical substances in the engineering (COMPENDEX), physics (PHYS, INSPEC), energy (ENERGIE, ENERGY), and materials (METADEX) databases on the STN system by creating the element term (/ET) field. Substance information is extracted from the titles and abstracts of records on these and other databases. Possibilities for searching in the ET field include:
- molecular formulas
- materials descriptions
- alloys
- intermetallic compounds and metal systems
- nuclear reactions and dopings
- positive and negative ions
- isotopes
- elementary particles.

A search strategy for retrieving references on aluminun-zinc-magnesium alloys and intermetallic compounds which could be used in cross-file searching of the relevant databases is: => **SEARCH AL*MG*ZN/ET**

Element symbols in such searches are arranged in alphabetical order and separated by asterisks.

7.8 SUMMARY

Chemists have several options for searching which are not available in other disciplines. These include a variety of compound names and codes for chemical compounds, such as fragment codes, linear notations, and molecular formulas. Online chemical dictionaries, while expensive to search on most systems, provide a wide range of access points for chemical compound searches. Another alternative is to search the structural drawing directly in an online system. That option is discussed in the next chapter.

7.9 SELECTED READINGS

Warr, W.A. "Available Systems of Structure Representation." In *Chemical Nomenclature Usage*; Lees, R.; Smith, A.F., Eds.; Halsted Press: New York, 1983; pp 124-131.

Buntrock, Robert E. "Pitfalls in Chemical Searching." *Database* **1989**, *12*(1), 112-114.

Krumpoic, Miroslav; Trimakas, Diana; Miller, Connie. "Searching Chemical Abstracts Online in Undergraduate Chemistry." *J. Chem. Educ.* **1989**, *66*(1), 26-29.

CHAPTER 8

STRUCTURE SEARCHING

8.1 INTRODUCTION

In the previous chapter we learned how to use various facts about a chemical substance (its name, molecular formula, number of rings, etc.) to retrieve information on that substance. A few of those search techniques even allow us to gather into one set all compounds which share certain characteristics, for example, those with 2 barium atoms or those with formula weights greater than 10,000.

Another way to search for a compound or a class of related compounds is by inputting a structural diagram or a partial structural diagram. **Structure** or **substructure searching** leads to the identification of all compounds in a given file having certain structural features. Structure searching systems allow the user to assess the novelty of a compound and to correlate the structure with certain chemical or biological activities. In addition, structure searching systems can be linked to chemical reaction databases to seek model compounds or look for specific reaction conditions. Substructure searches might also help identify competitive products or lead to a market for a product.

Some of the online chemical dictionaries discussed in chapter 7 include structure searching capabilities. Among them are STN's CAS ONLINE Registry File and Chemical Information Systems, Inc.'s SANSS file. Another online vendor, Questel, Inc., was the first company to produce a commercial product capable of true structure searches of millions of compounds using data from the full CAS Registry System. The Questel DARC system, which is based on a different theoretical concept from that employed in searches of the STN Registry File, has found widespread use throughout the world.

More and more programs have been developed for structure searches on stand-alone computers within a given company, educational institution, or other organization. Among them are Molecular Design Ltd.'s MACCS (Molecular Access System) and ORAC Ltd./Maxwell Communication Corporation's OSAC (Organic Structures Accessed by Computer), which run on large mainframe or supermini computers. Even a microcomputer can now be used for structure searching.

8.2 STRUCTURE SEARCHING ON STAND-ALONE COMPUTERS

Chemical companies have long used computers to organize information about the compounds in which they are interested. The searching of chemical structures using WLN or other linear codes was well established by the time the CAS Registry System began in 1965. The last few decades have seen the development of an increasing number of structure searching systems which are based on connection tables, especially those which use data from the CAS Registry System.

8.2.1 Microcomputer-Based Systems

The most significant limitation to structure searching software for microcomputers has been the amount of memory which could be devoted to the connection codes for the chemical compounds. The usual search process, which involves an atom-by-atom, bond-by-bond approach, slows considerably as the size of the file of compounds stored on a microcomputer increases. However, as microcomputer data storage capacities and processing speeds increase and as new programming techniques are developed, the number of compounds which can be searched on a microcomputer increases dramatically. For example, the HTSS [249] software package (formerly, TREE) can search a database of 15,000 structures on a hard disk in about 20 seconds. Other software for structure searching on a PC includes ChemSmart [79], ChemBase [73], and S4 [1902]. Developed by the Beilstein Institute, S4 can search a file of hundreds of thousands of compound structures on a CD-ROM in 30 seconds or less.

8.2.2 Structure Searching Systems on Larger Stand-Alone Computers

If a larger computer, such as a DEC VAX or other equipment of similar capacity is available, the number of compounds which can be searched is considerably greater. It is common nowadays to find such systems which can search databases containing several hundred thousand structures.

8.2.2.1 MACCS

One of the most widely-used structure searching systems is MACCS [250], the Molecular Access System. MACCS-II allows the product to be customized to search files in excess of 500,000 compounds. MACCS-3D introduces the capability to store, search, and retrieve 3-D chemical models and related data. By installing Molecular

Design Limited's ChemBase on PCs within a given company, MACCS files associated with a particular project can be downloaded and manipulated at individual workstations. This saves time on the organization's mainframe computer. Searches of various databases which cover reaction chemistry are also possible through the related software REACCS [251].

Subsets of the Index Chemicus database (the full file of which contains over 4 million compounds reported since 1962) can be obtained for in-house use with the MACCS software. Molecular Design Limited also markets a version of the Fine Chemicals Directory [252] and the MACCS-II Drug Data Report (containing 10,000 models) which can be searched with MACCS. Both of these come in two- or three-dimensional versions. Information from a structure search on the Fine Chemicals Directory includes suppliers of fine chemicals, their catalog numbers, as well as grades of purity, and variants which are salts or isomers.

MACCS-II offers several options for structure searching in addition to the traditional substructure search. One is similarity searching, which finds compounds in the database that have several features in common with the query molecule. Another option with MACCS is R-group searching, a technique which is useful for requiring certain functional groups to be present or absent at certain positions in a molecule.

MACCS can retrieve stereochemical information about a compound which contains an asymmetric carbon atom. The four possibilities are: no stereochemistry (that is, none specified), a racemic mixture, the R enantiomer, and the S enantiomer. If a compound exhibits stereochemistry about a double bond, three variants can be handled by the MACCS software: *cis*, *trans*, and "either". In the last case, either a mixture of *cis* and *trans* forms are present or the configuration is unknown. The designation of stereochemistry in a search query is possible only in the graphic search mode in MACCS. The other search option, which consists of a linear character string designating the molecule, does not permit the input of stereochemical features as search parameters.

8.2.2.2   OSAC

OSAC (Organic Structures Accessed by Computer) [253] was developed at the University of Leeds by the same group which created ORAC (Organic Reactions Accessed by Computer) [254]. The system is menu-driven, with online help built in. It runs on any DEC VAX or MicroVAX under the VMS operating system.

Like MACCS, the OSAC system provides an easy method to draw structures on a computer's video monitor. The menu displays onscreen the most common ring substructures and substituents. These can be assembled to produce the desired compound. Furthermore, commonly-used structure fragments can be stored and recalled as needed. Up to 32 databases can be created with OSAC, each consisting of chemical structures and related information. Search options are presented on a Query Menu page.

The chemist may draw a needed structure and perform a search for an exact match or for compounds whose substructure matches the data input. Data in tabular form can also be accommodated in OSAC. Options for both the display of data (the forms menu) and the display of query (the query menu) can be customized by the user. The system also has a utility package for defining a thesaurus and a data-definition utility, allowing tailor-made databases to be created.

8.3 STRUCTURE SEARCHING ON REMOTE SYSTEMS

The greatest flexibility of online searching in general is obtained through searches on remote systems which combine the power of structure searching with access to other databases. Once a particular structure or group of related structures has been isolated, there are links provided (usually, via the compound's CAS registry number) to numerous bibliographic, text, numeric, and directory databases. Some of the databases which can be accessed on various systems are listed in Table 8.1.

**TABLE 8.1
A Sample of Databases Which Can Be Searched With the CAS Registry Number**

Name	Type
Analytical Abstracts	bibliographic
EMBASE (Excerpta Medica)	bibliographic
Agrochemicals Handbook	numeric
Heilbron (Dictionary of Organic Compounds)	numeric
CJACS (American Chemical Society Journals)	text
Kirk-Othmer Encyclopedia of Chemical Technology	text
CAS ONLINE Registry File	dictionary
SANSS (Structure and Nomenclature Search System)	dictionary
Janssen Catalog of Chemical Products	directory

8.3.1 SANSS

Oldest of the commercially-available structure search systems which are based on the CAS Registry System is SANSS, the Structure and Nomenclature Search System [222]. SANSS is designed for structure searching of the compounds indexed in the databases provided by Chemical Information Systems, Inc. The decision was made in late 1984 to turn over to the private sector what was then known as the National Institutes of Health-Environmental Protection Agency Chemical Information System (CIS). Fein-Marquart Associates, through its subsidiary Chemical Information Systems, Inc., then began to provide public access to the databases. The system has been upgraded and improved in the succeeding years to include several new components.

Various search capabilities are included in SANSS, the hub of the CIS system. SANSS provides a gateway to all of the CIS databases. The structure of the desired compound can be drawn on any terminal or microcomputer using the SANSS commands. This can then be used to search the database. If a registry number is known, the command **SSHOW** n (n = registry number) will display the structure and nomenclature information. For most of the chemical substances, the CIS database records include the CAS registry numbers, structures, chemical names, CAS Index Names, molecular formulas, and a list of sources of further information. Thus, approximately 350,000 substances are to be found in the database. The collection is particularly valuable for structure searches of compounds which are fairly common and for which data on their toxic properties, spectra, etc. exist. Although alphanumeric depictions of the chemical compounds (as seen in Figure 8-1) are clear and sufficient for most purposes, graphic depictions of molecules are also possible on the CIS. The SuperStructure software [1904] is a graphics program for creating structures on a PC which can be uploaded and searched online in SANSS.

SANSS does not allow substitution to be blocked at certain sites on the molecule by specifying hydrogen atoms at those sites. This is a feature of some structure searching systems. Likewise, SANSS lacks the capability to specify variable groups at certain sites, a feature of the DARC and CAS ONLINE systems. It can, however, specify variable atoms attached to a particular node.

The cost of searching the Chemical Information System is relatively low com-

Option? **sshow 91-56-5**

Entry 1 CAS RN 91-56-5

   CIS Sources of Information

      2 - CIS, EI Mass Spectrometry
      6 - CIS, Cambridge X-Ray Crystallography: ISATIN
      7 - CIS, Merck Index: 4951
     22 - NBS, Crystal Data File
     32 - NIOSH/CIS, RTECS: NL7873000
     51 - CIS, TSCAPP (TSCA Plant and Production)
     60 - CIS, WMSSS (Wiley Mass Spectral Database)
    124 - CIS, CI Mass Spectrometry
    130 - CIS, FRSS (Federal Register Search System)
    214 - CIS, PHYTOTOX (Plant Toxicity)
11 Non-CIS References Available

$C_8H_5NO_2$

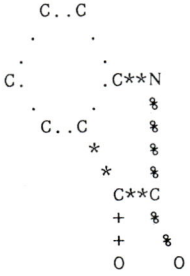

1-H-Indole-2,3-dione (9CI)
Indole-2,3-dione (8CI)
o-Aminobenzoylformic anhydride
Isatic acid lactam
Isatin
10 more names available

FIGURE 8-1
Chemical Information Systems, Inc. SANSS File Record.

pared to searches of STN International's Registry File or Questel's DARC. Thus, it is a good first choice if the compound of interest is a common compound or one that has significant commercial applications.

8.3.2 DARC and Markush Searches

The DARC structure searching system [255] was introduced in the United States in 1981. It caused quite a stir in the chemical information community, and soon Chemical Abstracts Service was moving full steam ahead to catch up with the French database vendor Telesystemes-Questel (Questel for short).

For compounds registered by Chemical Abstracts Service, DARC offers four types of searches. These are an exact structure match, a substructure search for related compounds, a generic search which allows the definition of variable groups at certain points on the molecule, and a Markush structure search which is especially important for determining the patentability of a substance. The DARC software accommodates both graphic and textual input and display of the answers. Compounds registered by CAS since the beginning of 1967 are available for searching. However, polymers are searched in a separate file, POLYCAS [245].

Three main commands are used to draw the molecules. GR, the graph command, draws the basic outline of a molecule. AT, the atoms command, specifies what noncarbon elements are located at various nodes on the molecule. BO, the bonds command, defines the types of bonds between molecules. Other specifications for the molecule include the designation of a ring system as either isolated or fused to other rings, plus the capability to specify the maximum acceptable number of atoms in the retrieved compound, or to limit the number of rings or the number of components.

The search on DARC takes place in two stages. First, the "RE" fragment search identifies from the millions of compounds in the database certain compounds for closer examination. Up to 50,000 structural candidates might be retrieved at this stage. Second, a detailed examination of the candidate structures is made in the "AA" stage, the atom-by-atom, bond-by-bond comparison to the stated requirements for the structure or structures of interest. It is useful to think of the RE stage as a bunch of Boolean "AND" statements for fragments of a molecule. At this stage the computer really does not know how the fragments are attached to each other. It is the AA stage which determines that. The AA step requires a tremendous amount of computing power. Hence, there are limitations placed on the total number of candidate structures which can be processed in stage 2. What happens when the RE search retrieves too many compounds? With DARC, it is possible to split up the set of compounds retrieved in stage 1 and look at the compounds which have been assembled in the RE search.

One of the most powerful features of the DARC structure searching system is the Markush DARC software and related database. A **Markush structure** is one which can represent thousands or even hundreds of thousands of specific chemical structures. The concept is particularly important in the process of patenting chemical substances. Essentially, a Markush structure consists of a basic, constant structure with variable groups attached. Each variable group can have a number of different values which can vary with their points of attachment or the number of times they occur. The groups may be defined using terms such as -alkyl, -aryl, -alkoxy, etc. or groups such as $-NH_2$, $-CH_3$, -ethyl, etc. By using generic structures, an almost infinite number of chemical substances can be claimed in a patent.

A generic structure query searches for a basic structure and attached variable groups. Databases are now appearing which claim to be generic. That is, the database itself consists of generic structures. What has been available up to now are searches with specific compounds as answers to a search query involving some variability in the groups at certain sites on a structure. Derwent is currently involved in the creation of a system which allows generic queries to be matched against a generic database. In cooperation with Telesystemes-Questel and INPI (the French Patent and Trademark Office), Derwent has created the databases MPHARM [1905] and WPIM (World Patents Index Markush) [1906] which both contain Markush structures capable of being searched with the Markush DARC software. The DARC CHEMLINK software [1894] allows the formulation of structural queries offline which can be uploaded for structural searches. CHEMLINK now includes Markush DARC searches.

8.3.3 CAS ONLINE Registry File

Structure searches of the CAS ONLINE Registry File [223] are performed on the STN International system. STN places all of the compounds registered by Chemical Abstracts Service in one huge file, the Registry File. The CAS ONLINE Registry File thus encompasses all of the compounds registered from the inception of the Registry System in 1965. This includes compounds which CAS is retrospectively registering, compounds which were used in indexing pre-1967 records, but have not been used in indexing since that time. Beginning with 1967, the registered compounds were linked to all relevant full bibliographic records produced for the printed *Chemical Abstracts*.

STN has the unique advantage of providing as optional output the abstract text for most of the abstracts published after January 1, 1967. With the retrospective project, links are being established to abstract numbers prior to 1967 through the CAOLD file [233]. Thus, STN alone offers the capability to utilize structure searches to find pre-1967 *CA* records through the CAS registry numbers. For those reasons, we will concentrate on the STN version of structure searching. This does not imply that the Registry File structure searches are any better than those performed with DARC.

8.3.3.1 General Features of Searching in the Registry File

As with the Questel DARC and other systems, the user has the option of true graphics input and output, assuming that appropriate equipment and software are being used. The graphics option has the advantage of showing on the screen and in print a depiction of the compound which is very similar to that which chemists are used to working with. Contrast the structure image produced by the graphics option in Figure 7-7 with the image of the same compound as printed on a standard text printer in Figure 8-2. The difference is striking, and one might conclude that, given the choice, most chemists and information specialists would always opt for the graphics output. However, with the technology currently employed for online searching, the graphics display requires the user to "copy and clear page" at certain points in the search. This means that a signal must be manually input to continue receiving information. In order to print the compound as a graphics image, the graphics portion of the data may be downloaded. After the online session is complete and the user has logged off, the image can then be printed. Alternatively, a slow "screen dump" of the graphics image to a printer or plotter can be performed while still connected to the STN system. There are no such delays with the standard text output. Once a display of the answer has been requested, it can run from start to finish as text (ASCII) output. The answer can be continuously printed on paper and displayed on the video monitor while the search session is in progress. (Another alternative is to eliminate the structure diagram from the display of the records.)

The searcher chooses the type of output desired upon logging on to STN. This is done immediately after the password is entered, when the user must respond to:

TERMINAL (ENTER 1, 2, 3, OR ?):

The response for text output is **3** and for the most widely-used option for graphics output, **2**, which implies Tektronix 4010 graphics emulation.

Structure searching in the CAS ONLINE Registry File is similar to that in the Questel DARC system in certain respects. STN offers four types of searches for compounds registered by Chemical Abstracts Service. These are an exact structure match, a family search, a closed substructure search, and a substructure search for all related compounds. One of the most important points to grasp is that the output from a structure search in the Registry File of a particular structure is progressively larger as we move from an exact search to a family search to a closed substructure search and finally to a full substructure search.

The exact search retrieves the substances which match exactly the data input for the Registry File search. It is not possible to specify stereochemistry in the structural drawing on STN. Therefore, more than one substance may be found in the answer set for an exact search. That is because this option retrieves not only a match for the unsubstituted structure which was input, but also any stereoisomers, isotopically labeled substances, ionic substances, radical substances, or even homopolymers which match the structure. It is important to recognize that the CAS Registry System assigns a unique registry number to stereoisomers, labeled compounds, etc., and that they are *not* tied to the registry number for the basic compound. Thus, a structure search for an exact match of a substance is a sure way to pull together into one set all of the variants which are in the database.

A family search will also find those substances which match the structure input with no substitution. In other words, the output for an exact search is a subset of the output for a family search. The family search also retrieves any multi-component compounds which match that structure. Thus, the answer set will also contain salts, mixtures, copolymers, or addition compounds if they are present in the database.

A closed substructure search is similar to a family search. However, in a CSS, variable nodes, system defined nodes, and generic groups are allowed. It is also possible to allow further substitution at a particular position by using the **CONNECT** command.

The substructure search results in the largest potential set of compounds, for it retrieves all of the substances in the previous three options and others if present. Thus, any substance in the database which contains the *substructure* will be included in the answer set when this option is chosen.

8.3.3.2 How to Create a Structure and Search in the Registry File

There are several ways to create a structure in the Registry File, including starting with a known substance, using a pre-drawn structure, or creating the structure from scratch. When the command **STRUCTURE** is given, the system responds with the following prompt:

ENTER NAME OF STRUCTURE TO BE RECALLED (NONE):

As is true throughout the STN system, the option in parentheses is the default option, which can always be selected by simply entering a period. To do so in this case would tell the system that we want to create a structure from scratch. We might choose to enter something besides a period, for example, the L# (identifying number) for a structure already created in this search session. We could also start with a registry number for a known compound which is similar to the one we want to build. Finally, the searcher may select a model from a file of 38 ring systems which CAS has provided. These include rings containing 5-12 atoms and a number of complex ring systems

TABLE 8.2
Some Pre-Drawn Ring Systems for Use in Creating Structures in the CAS ONLINE Registry File

Code	Unsubstituted Structure
n (n = 5-12)	ring of size n
ACENAP	acenaphthene
ADAMAN	adamantane
ANTHRA	anthracene
BILINE	biline
CAROTE	carotene
CEPHAL	cephalosporins
FLUORN	fluorene
IBOGAM	ibogamines
INDENE	indene
MORPHN	morphine
NAPHTH	naphthalene
NORBRN	norbornane
PENICL	penicillins
PHENAN	phenanthrene
PHORBN	phorbine
PORPHN	porphines
PORPHY	porphyrazine
PROSTA	prostaglandins
PURINE	purine
SEQTER	sesquiterpenes
STEROD	steroids

which would be difficult to draw from scratch, such as fused ring systems, coordination compound fragments, and boron cages. Examples include the steroid ring, the adamantane ring, and others listed in Table 8.2. To use one of these codes as a model, enter **STRUCTURE ABBREV** or **STRUCTURE n,** where **ABBREV** is one of the six-character codes in Table 8.2, and **n** is a number 5 through 12. Structure Codes for coordination centers and boron cages are also available. See *Using CAS ONLINE: The Registry File* Volume IIA Appendix 2 [235] for those codes.

Once the starting option has been selected, the system prompt for building structures appears:

ENTER (DIS), GRA, NOD, BON OR ?:

This prompt includes the three basic commands for creating structures in the Registry File. **GRA**, the graph command, draws the outline of the substance, initially with all bonds unspecified and all nodes designated as C. For example, **GRA R6** creates a six-membered ring (rings of 3-8 atoms can be built in this manner); **GRA R65** creates a six-membered ring fused with a five-membered ring. If followed by **GRA 8 C1**, a chain of one atom in length is added to the ring at atom number 8. The GRA command is followed by the command **DIS** (for display), causing the structure to appear on the screen with C at all numbered nodes (points where atoms are to be defined) and all bonds depicted as undefined or unspecified.

NOD, the node command, allows an atom to be changed to any element or further defined as a group or variable. Possible substitutions are listed in Table 8.3.

All four of the generic group symbols, Ak, Cy, Cb, or Hy, can be further characterized as shown in Table 8.4. The symbols in Table 8.4 are used in conjunction with the command **GGC**, the Generic Group Category, to define nodes specified as Ak, Cy, Cb, or Hy.

Table 8.5 shows the shortcut symbols which can be used in addition to the symbols shown in Table 8.3 when the NOD command is given. The symbols in Table

TABLE 8.3
Valid Node Symbols in the Registry File

Symbol	Substitution Allowed at the Site
element symbol	only that element
X	any halogen (F, Cl, Br, I, At)
M	any metal[1]
-	excludes the element or group which follows the minus sign
Q	any element except C or H
A	any element except H
Gk (k = 1-20)	user-defined variable groups[2]
Ak	any carbon chain
Cy	any cyclic group
Cb	any carbocyclic group
Hy	any heterocyclic group
shortcut symbols	See Table 8.5.

1. CAS defines a metal as any element except Ar As At B Br C Cl F H He I Kr N Ne O P Rn S Se Si Te Xe.
2. A special command, either VAR (variable) or REP (repeating) is used to define Gk groups.

8.5 are used in the same manner as symbols for elements or other symbols. Exact hydrogen counts are assigned where hydrogens are present in the structures represented by the symbols.

TABLE 8.4
Specifications for Generic Group Symbols in Structure Building in the Registry File

Abbreviation	Meaning
LIN	linear
BRA	branched
SAT	saturated (all single bonds)
UNS	unsaturated (at least one bond not single)
LOC	low carbon (6 or fewer carbons)
HIC	high carbon (more than 6 carbons)
LOQ	low hetero (only 1 non-carbon atom)
HIQ	high hetero (more than one non carbon atom)
MCY	monocyclic
PCY	polycyclic

TABLE 8.5
Shortcut Symbols for Use in Building Structures in the CAS ONLINE Registry File

SYMBOL	MEANING	SYMBOL	MEANING
C(O)CH3	acetyl	M-C6H4	m-phenylene
CBR2	dibromomethyl	ME	methyl
CBR3	tribromomethyl	MEO	methoxy
CCL2	dichloromethyl	N-BU	n-butyl
CCL3	trichloromethyl	N-BUO	n-butoxy
CF2	difluoromethyl	N-PR	n-propyl
CF3	trifluoromethyl	N-PRO	n-propoxy
CH	methyne	NH	amine
CH2	methylene	NH2	amine
CH3	methyl	NH3	ammonia
CHO	formyl	NO2	nitro
CI2	diiodomethyl	O-C6H4	o-phenylene
CI3	triiodomethyl	OH	hydroxy
CN	cyano	OPO3H2	phoshpate
CO2H	carboxy	OSO3H	sulfate
COOH	carboxy	P-C6H4	p-phenylene
COSH	thiocarboxy	PH	phenyl
CS2H	dithiocarboxy	PHO	phenoxy
CSSH	dithiocarboxy	PO3H2	phosphono
ET	ethyl	S-BU	sec-butyl
ETO	ethoxy	S-BUO	sec-butoxy
I-BU	isobutyl	SH	mercapto
I-BUO	isobutoxy	SO2	sulfonyl
I-PR	isopropyl	SO3H	sulfo
I-PRO	isopropoxy	T-BU	tert-butyl
		T-BUO	tert-butoxy

The bond command, **BON**, defines the type of bonds to be included. For example, **BON ALL SE** would make all bonds in the compound single exact. See Table 8.6 for other possibilities.

TABLE 8.6
Bond Codes Used in the CAS ONLINE Registry File

Code	Bond Type
u	unspecified
n	normalized
se	single exact
s	single exact or normalized
de	double exact
d	double exact or normalized
t	triple

A **normalized bond** is one which is found in a ring like benzene. CAS also designates as normalized the bonds in a tautomer or combinations of rings and tautomers. Normalized bonds are specified in single rings if the ring has an even number of atoms and contains alternating single and double bonds all the way around the ring. For fused rings, the outside path around the rings must contain an even number of atoms and the bonds must alternate between single and double.

The situation with tautomers is more complex. The following environment must exist in order for the bonds in a tautomer to be designated n.

$$H1—2=3 \qquad 1=2—3H$$

where:

- A. The central atom of the three non-hydrogen atoms is connected to any two of the following: N, O, S, Se, or Te. (1 and 3 could be the same atom.)
- B. The central atom is: C, N, P, As, Sb, S, Se, Te, Cl, Br, or I.
- C. At least one hydrogen, hydrogen isotope, or charge is on 1 or 3.

Included as tautomers are unsubstituted or monosubstituted amides or certain acids. In order to get all acids or all amides which satisfy the rules in A and B in a substructure search, the bond between 2 and 1 should be designated simply S, and between 2 and 3, D, or vice versa.

There are certain assumptions which the system makes about the structures when a substructure search is run. The first is that an atom drawn in a chain is only to be in a chain and may never be part of a ring. (Likewise, an atom drawn in a ring is assumed to be only a component of a ring and never part of a chain). A second assumption is that a ring may have other rings fused onto it. Another assumption is that in a substructure search, substitution may occur at any point in the molecule where it is chemically possible. (This can now be overridden by choosing the closed substructure search option (CSS) which blocks substitution at all possible nodes.)

The ring atoms are always going to be part of a ring, but a chain atom in some compounds may also be part of a ring. If such compounds are of interest, **NSPEC** or

NSP, the node specification command allows that adjustment to be made. For example, **NSP 8 RC** would allow atom 8 to be either in a ring or in a chain.

It is possible to block further rings from being attached to the ring system in a substructure search by giving the command **RSPEC** or **RSP**, the ring specification command. Failure to use this option implies to the system that compounds which have the ring system embedded as part of a larger ring system are acceptable. RSP eliminates that group of compounds. It is also possible to specify that one or more specific rings in the system be isolated, with rings which are fused onto other rings being acceptable. To accomplish this, simply enter **RSP R n** where "R" designates a ring with node number "n".

Another very useful option is to specify the exact or minimum number of hydrogens at particular nodes using the command **HCO**. This prevents or limits substitution. The command might be: **HCO 8 E3** or **HCO 8 M2** or **HCO 8 E0**. In the first case, the atom at node number 8 must have exactly three hydrogen atoms attached to it. In the second case, there must be at least two hydrogens at node number 8, but more are acceptable. In the final case, maximum substitution is called for since no hydrogens are allowed.

There are other attribute commands to do such things as specify a charge, the mass, or the valence of a node, designate abnormal mass for hydrogens attached to a node, or specify a delocalized charge on a set of nodes. For their use, consult volume IIA of *Using CAS ONLINE: The Registry File* [235]. When the structure has been created, the command **DIS SIA** will show the structure image (SI) and any special attributes (NSPEC, RSPEC, etc.). At that point, the command **END** is given, and the system assigns an L# to the structure which was created.

There is one very important step in Registry File structure searching which must never be overlooked. That is to run a sample search *before* running the search on the full file of structures. When the command **SEARCH L#** is given, the following prompt appears:

ENTER TYPE OF SEARCH:(SSS), CSS, FAMILY, OR EXACT:

followed by:

ENTER SCOPE OF SEARCH: (SAMPLE), FULL, RANGE:

Always do a sample search first. A sample search is quickly run on 5 percent of the total compounds in the Registry File.

A structure search occurs in two stages. The first is a rapid overview of the file based on **screens** (fragments and other search keys) without regard to how these are put together in the molecule. The second stage is iteration, an atom-by-atom, bond-by-bond look at the candidate molecules found in step one. The iterations consume large amounts of computer processing time. Consequently, an upper limit of 50,000 candidate compounds from step one has been built into the CAS ONLINE Registry File search system when searching online. The SAMPLE search is used to project whether that limit will be reached in a FULL file search. Once the sample search is complete, the message we hope to see is: FULL FILE PROJECTION: COMPLETE. That signifies that no more than 50,000 iterations should be necessary and that the maximum number of projected answers in the full file search will not exceed 30,000 in an online search. (Up to 50,000 answers can be assembled in a batch search.) If the message is INCOMPLETE or UNCERTAIN, then the structure must be refined and tested with another sample search. Only after the prediction of a COMPLETE search is received should the search be initiated in the full file or a range of registry numbers be indicated for the search. The sequence of events necessary for a successful structure search in the Registry File is outlined in Table 8.7.

TABLE 8.7
Outline of the Search Sequence in the Registry File

1. Enter the **FILE REG** command.
2. Initiate structure creation with the **STRUCTURE** command.
3. Build the outline of the structure using the **GRA** command.
4. Specify the non-carbon atoms with the **NOD** command.
5. Specify the bonds with the **BON** command.
6. Specify additional requirements with **RSP, NSP, HCO**, etc.
7. Do a final display and check of the structure with the **DIS SIA** command.
8. Terminate structure building with the **END** command. (An L# is assigned).
9. Begin the search in the Registry File with the **SEARCH** command. (SEARCH L#)
10. Specify the type of search by indicating **SSS, CSS, FAMILY**, or **EXACT**.
11. Specify the scope of the search by indicating **SAMPLE, FULL**, or **RANGE**. (Always do a sample search first.)
12. Display the answers by entering the **DISPLAY** command and the L# of the answer set. (**DIS L# 1-n**)

The basic display options in the Registry File are: **IDE**, the default option which shows the Structure Image and various facts about the molecule; **BIB**, which displays the ten most recent bibliographic records from the CA File which have that registry number in their indexing; **ABS**, the abstracts associated with those records; and **ALL**, all of the above, plus the indexing for the CA File records.

Once the answer set has been formed, it may be combined with other search keys in the Registry File or searched in the CA File or other databases on STN which use registry numbers as search keys. It is important to realize that it is the registry numbers of the compounds in the answer set which will be searched in the basic index of the CA File or matched to registry numbers in the other files on STN International.

In the Registry File itself, the technique of combining dictionary terms such as the molecular formula, formula weight, and name fragments can significantly refine a structure search which has a large number of answers. The answer set from the structure search is matched with the desired features, for example, => **SEARCH L10 and FW>1000**

Let us now refer to Figure 8-2 to see how a structure is built on a type 3 (text) terminal. Note that the five-membered ring forms a tautomer with nodes 7-8-10.

The search in Figure 8-2 produced six answers in the Registry File, including the normal isatin molecule and five others which were either ionic or isotopically labeled substances. In contrast, a family search of the full file produced 31 substances, and a substructure search gathered 345 compounds into the answer set.

Space does not allow a more thorough presentation of Registry File structure searching in this book. There are other commands and other search techniques which are used by more experienced searchers. Especially when the full file projection is

STRUCTURE SEARCHING   **153**

incomplete or uncertain, the technique of screen searching can be employed. Chemical Abstracts Service has produced a number of useful guides and other compilations to assist in searching the Registry File. Search **INDX = "REGISTRY FILE"** in the Chemistry Reference Sources Database in order to find them.

⇒**file reg**

```
file reg
COST IN U.S. DOLLARS                               SINCE FILE        TOTAL
                                                     ENTRY         SESSION
FULL ESTIMATED COST                                   0.40            0.40

FILE 'REGISTRY' ENTERED AT 18:00:16 ON 02 NOV 89
COPYRIGHT (C) 1989 AMERICAN CHEMICAL SOCIETY

STRUCTURE FILE UPDATES: HIGHEST RN 123482-23-5
DICTIONARY FILE UPDATES: 28 OCT 89 (891028/ED) HIGHEST RN 123463-03-6
⇒structure
ENTER NAME OF STRUCTURE TO BE RECALLED (NONE):.
ENTER (DIS), GRA, NOD, BON OR ?:gra r65,dis

            2           7
            C           C
  1    ?   ?   3   ?    ?    8
      C?      ? C?      ?C
       ?        ?        ?
       ?        ?        ?
       C        C????????C
  6   ?        ?                9
         ? C?    4
            5

ENTER (DIS), GRA, NOD, BON OR ?:gra 8 c1,9 c1,dis

            2           7
            C           C         C    10
  1    ?   ?   3   ?   ?  8   ?
      C?      ? C?     ?C    ?
       ?        ?        ?
       ?        ?        ?
       C        C????????C
  6   ?        ?              ?
         ? C?    4         9   ?
            5                  C    11

ENTER (DIS), GRA, NOD, BON OR ?:nod 7 n,10 11 o,dis

            2           7
            C           N         O    10
  1    ?   ?   3   ?   ?  8   ?
      C?      ? C?     ?C    ?
       ?        ?        ?
       ?        ?        ?
       C        C????????C
  6   ?        ?              ?
         ? C?    4         9   ?
            5                  O    11
```

FIGURE 8-2
Structure Search for Isatin in the CAS ONLINE Registry File on a Type 3 (Text) Terminal.

154 CHEMICAL INFORMATION SOURCES

ENTER (DIS), GRA, NOD, BON OR ?:**bon r 1-2 n,4-9 3-7 8-9 s,9-11 de, 7-8 8-10 n,dis**

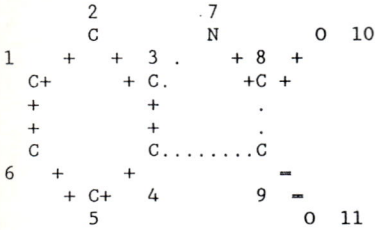

ENTER (DIS), GRA, NOD, BON OR ?:**rsp, hco 7 e1, dis sia**

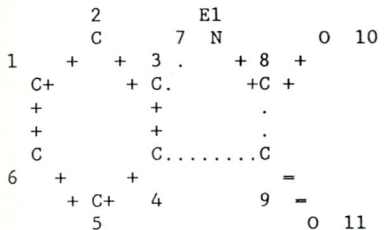

NODE ATTRIBUTES:
HCOUNT IS E1   AT   7

GRAPH ATTRIBUTES:
RSPEC 1
NUMBER OF NODES IS 11
ENTER (DIS), GRA, NOD, BON OR ?:**end**
L1   STRUCTURE CREATED
=>**search L1**
ENTER TYPE OF SEARCH: (SSS), CSS, FAMILY, OR EXACT:**exact**
ENTER SCOPE OF SEARCH: (SAMPLE), FULL, OR RANGE:**.**
SAMPLE SEARCH INITIATED 18:03:46
SCREENING
SAMPLE SCREEN SEARCH COMPLETED —   4 TO ITERATE
100.0% PROCESSED   4 ITERATIONS   0 ANSWERS
SEARCH TIME: 00.00.25

FULL FILE PROJECTIONS: ONLINE   **COMPLETE**
                         BATCH   **COMPLETE**
PROJECTED ITERATIONS: 4 TO   199
PROJECTED ANSWERS: 0 TO   0

L2   0 SEA EXA SAM L1

=>**search L1**
ENTER TYPE OF SEARCH: (SSS), CSS, FAMILY, OR EXACT:**exact**
ENTER SCOPE OF SEARCH: (SAMPLE), FULL, OR RANGE:**full**
FULL SEARCH INITIATED 18:05:02
FULL SCREEN SEARCH COMPLETED - 100 TO ITERATE

FIGURE 8-2
Structure Search for Isatin in the CAS ONLINE Registry File on a Type 3 (Text) Terminal.

```
100.0% PROCESSED          100 ITERATIONS                                    6 ANSWERS
SEARCH TIME: 00.00.21


L3                6 SEA EXA FUL L1

=> display 6

ENTER (L3), L# OR ?:.
ENTER DISPLAY FORMAT (IDE):.

L3      ANSWER 6 OF 6

RN    91-56-5
CN    1H-Indole-2,3-dione (9CI)    (CA INDEX NAME)
CN    o-Aminobenzoylformic anhydride
CN    2,3-Diketoindoline
CN    2,3-Dioxoindoline
CN    2,3-Indolinedione
CN    Isatic acid lactam
CN    Isatin
CN    Isatinic acid anhydride
CN    Indole-2,3-dione (8CI)
CN    Pseudoisatin
CN    2,3-Dioxo-2,3-dihydroindole
CN    Isatine
CN    Tribulin
DR    84788-92-1, 5815-00-9
MF    C8 H5 N O2
CI    COM
LC    BIOSIS, CASREACT, CHEMLIST, CSCHEM, EINECS, TSCA
```

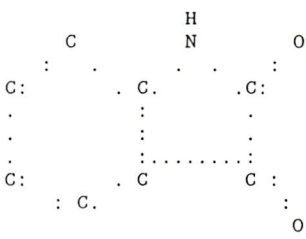

REFERENCES IN FILE CAOLD (PRIOR TO 1967)
763 REFERENCES IN FILE CA (1967 TO DATE)

=> logoff y

COST IN U.S. DOLLARS	SINCE FILE ENTRY	TOTAL SESSION
	37.16	37.56
DISCOUNT AMOUNTS (FOR QUALIFYING ACCOUNTS)		
CA SUBSCRIBER PRICE	-0.30	-0.30

STN INTERNATIONAL LOGOFF AT 18:08:20 ON 02 NOV 89

FIGURE 8-2
Structure Search for Isatin in the CAS ONLINE Registry File on a Type 3 (Text) Terminal.

8.4 STRUCTURE INPUT TO THE REGISTRY FILE AND OTHER DATABASES VIA FRONT-END SOFTWARE

Learning to create structures with the Registry File's commands is not easy. For the infrequent user of such systems, there is another option. Sophisticated graphics front-end software packages now exist to assist with structure searching on remote databases as found on DIALOG, ORBIT, STN International, etc. Among the packages are MOLKICK [1908], HTSS [249], SuperStructure [535], STN Express [534], DARC CHEMLINK [1907], and ChemConnection [1909]. See *Chemical Structure Software for Personal Computers* [1889] (1988) for other software packages.

MOLKICK is a universal graphics query program which works with various data communications software packages. With MOLKICK, structures are drawn and stored offline in a microcomputer, then uploaded as a structure query to the vendor's computer. Furthermore, since MOLKICK is a memory-resident program, structures can be modified or formulated during the search session. Drawing of the structure is accomplished either with a mouse or with MOLKICK's pull-down menus for atoms, bonds, screens, and input choices. The user selects the appropriate format in which to store the structure (CAS, DARC, etc.), then uploads the query to the appropriate host.

HTSS, Hierarchical Tree Structure Search, is available both as a database management PC package and as a VAX or IBM mainframe computer product. The software takes advantage of a windowing capability to display on the same screen a window with the starting structure and an adjacent window with the structure(s) resulting from the search. SuperStructure is software which facilitates structure searches on the Chemical Information System.

STN Express is a package which assists the user in the creation of both textual and structure search questions. As with other such packages, the STN Express software for structure queries incorporates the rules for tautomers and normalized bonds. Structures can be constructed from pull-down menus or drawn with a mouse. There are templates for common structures, as shown in Table 8.8.

TABLE 8.8
Templates Available With STN Express

Structure	Structure
aromatics	biline
cycloalkanes (3-7)	carbohydrates-linear
cycloalkanes (7-16)	carbohydrates-cyclic
cycloalkanes (3-8) fused	carotenes
cycloalkanes (5-6) fused	cephalosporin
bridged rings	coordination centers
nitrogen-containing rings (5)	crown ethers
nitrogen-containing rings (6)	nucleosides
oxygen-containing rings	nucleotides
sulfur-containing rings	penicillin
N-O-S heterocycles	porphyrins
alkaloids	prostaglandins
amino acids	steroids

The DARC CHEMLINK software searches single-compound databases on Questel, such as EURECAS, as well as Markush formula databases, such as MPHARM and WPIM. The menu includes the standard DARC query and the MARKUSH-DARC query. The third option within CHEMLINK's opening menu is to connect to the Questel computer via the Emu-Tek graphics telecommunications software [349] which is built into CHEMLINK. CHEMLINK can also be ordered with INFOLOG G software [1910] instead of Emu-Tek.

The preceding four packages were designed to work on IBM PC and compatible microcomputers. ChemConnection is for the Macintosh. Like MOLKICK, ChemConnection is a graphics software package only. Hence, chemical structures can be formulated offline, but not textual questions. Also like MOLKICK, ChemConnection does not include telecommunications software. However, it does run with all of the common Macintosh communications packages.

8.5 STRUCTURE SEARCHING OF BEILSTEIN

By 1989 the Beilstein database [1911] had begun to be loaded on the search systems of the major vendors. Beilstein covers the literature of organic chemistry from 1830 through 1979. The retrospective registration of chemical compounds by Chemical Abstracts Service had reached back to 1957 by the time the Beilstein database became available. Thus, very comprehensive structure searching of organic chemical compounds is now possible far back into the 19th century by utilizing structure searches of both the CAS ONLINE Registry File and the Beilstein database.

The *Beilstein Handbook of Organic Chemistry* [669, 670], from which the database was created, has a long and illustrious history. The first group of compounds to be loaded into the database consisted of heterocyclic compounds. By the end of 1989, over 3 million heterocyclic substances were in the database. More detailed information on the contents and coverage of *Beilstein* may be found in sections 10.4.2. and 11.3.1. In this section, we will concentrate on access to the information in the database through structure searches.

Major vendors with intentions to provide access to the Beilstein database on their systems include DIALOG, ORBIT, and STN International. On DIALOG structure searches of Beilstein are conducted through the use of MOLKICK software and the DIALOG telecommunications software package DialogLink [1898]. ORBIT, on the other hand, has chosen HTSS (Hierarchical Tree Structure Search) software as its link to the Beilstein file.

On STN International, the searcher has the option of running a structure search in the CAS ONLINE Registry File and crossing over the L# of the answer set to Beilstein. At the end of 1989, CAS registry numbers were linked to about 75 percent of the substances in the Beilstein database. However, CAS registry numbers cannot be used in the Beilstein database as the starting point to build an existing structure by inputting a registry number. A Beilstein registry number must be used in that case, and these are quite different from CAS registry numbers. Aside from that, the process of creating a structure and searching by structure on STN is exactly the same in

Beilstein as in the CAS ONLINE Registry File. Various other search options and the record structure of the Beilstein file are discussed in 10.4.2.3.

8.6 OTHER STRUCTURE SEARCHING SYSTEMS

The database ChemQuest [359], containing over fifty catalogs from chemical suppliers in the U.S.A., Europe, and Japan, has an optional graphics mode of searching on ORBIT. Both exact structures and substructures can be used as input for the search.

Another structure software product is Derwent's TOPFRAG [358]. TOPFRAG is an acronym for Topological to Fragmentation Code Conversion Program. Derwent has used at least three versions of its fragmentation codes to index chemical substances in its patent files. Thus, it has become increasingly complicated for users of those files to select the correct codes for searching. The program converts the graphic depiction of the molecule to appropriate search strategies which can then be uploaded to search Derwent files on ORBIT, DIALOG, or QUESTEL. Only the exact structure is used as input.

8.7 SUMMARY

In addition to using the alphanumeric, textual search parameters from the last chapter (such as the molecular formula, molecular weight, complete compound name or name fragment) to conduct a search, structural depictions of molecules can be used as search input. Both microcomputer and mainframe systems are available for structure searching within a given organization. Telecommunications links to remote databases provide a powerful avenue to the vast stores of information in bibliographic and non-bibliographic databases. Several vendors have developed online structure searching capabilities, among them, Chemical Information Systems, Inc. (SANSS), Questel (DARC), and STN International (Registry Search System). The use of the results of the structure searches as input to the CA File and to other databases is a very powerful feature of online searching in chemistry.

8.8 SELECTED READINGS

Anderson, Susan. "Graphical Representation of Molecules and Substructure-Search Queries in MACCS." *J. Mol. Graphics* **1984**, *2*(3), 83-90.

Ash, Janet E.; Chubb, Pamela A.; Ward, Sandra E.; Welford, Stephen M.; Willett, Peter. "Chemical Structure Search Systems and Services." In *Communication, Storage and Retrieval of Chemical Information*; Ellis Horwood Limited: Chichester; Halsted Press: Chichester, New York, 1985; pp 182-202.

Wagner, A. Ben. "Chemical Substructure Searching: Comparing Three Commercially Available Databases." *Online Rev.* **1986**, *10*(3), 173-183.

Hansen, Peter J.; Jurs, Peter C. "Chemical Applications of Graph Theory. Part 1. Fundamentals and Topological Indices." *J. Chem. Educ.* **1987**, *65*(7), 574-580.

Barnard, John M. "Online Graphical Searching of Markush Structures in Patents." *Database* **1987**, *10*(3), 27-34.

Meurling, Anita. "CAS ONLINE and DARC: A Comparison." *Database* **1990**, *13*(1), 54-63.

CHAPTER 9

SEARCHING FOR INFORMATION INVOLVING CHEMICAL MEASUREMENTS (CONSTITUTIONAL CHEMISTRY)

9.1 INTRODUCTION

Most of the information sources dealt with in this chapter are traditionally associated with analytical chemistry. They are concerned with methods for the detection, identification, or quantitation of a substance. The laboratory techniques to determine the composition and identity of a chemical compound and those used to separate mixtures of chemical substances have generated a large body of chemical literature.

Searches in this area are again subject searches, but they are more difficult than those which involve information about a known chemical compound or those which approach the literature with an author's name. This is because the object of the search is less well defined or broader in scope. Sampling techniques, sample preparation, methods of separation or purification, and methods of identification are examples of the type of information sought in these searches. The information is often found in **methods** books or series which have sifted the most relevant and reliable techniques from the primary literature and have repackaged descriptions of those techniques in various compendia, handbooks, treatises, and so forth. Reference works of this type, often with the word "methods" in the title, are also found in the area of reaction chemistry. For the most part, those are treated in chapter 11. Search the Chemistry

Reference Sources Database for **METHODS** in order to survey the broad range of material which deals with this topic. A guide to relevant literature on this subject is Stuart James's *Using Literature* [289], published in the Analytical Chemistry by Open Learning series.

9.2 COLLECTIONS-I: CONTINUING METHODS SERIES, TREATISES, AND ENCYCLOPEDIAS

Some of the collected secondary works on methods have been published in series which can stretch into many volumes. Occasionally these series have a classified arrangement and resemble a treatise. At other times the editors of a series of methods volumes will simply identify the "hot" topics in the field and structure the series with a view toward the current market. It is not always possible to tell from the title whether a series concerns itself with methods. For example, the *Journal of Chromatography*

TABLE 9.1
Contents of *Techniques of Chemistry*

Vol No.	Part No.	Year	Title
I			Physical Methods of Chemistry
	1A	1971	Components of Scientific Instruments
	1B	1971	Automatic Recording and Control, Computers in Chemical Research
	2A-B	1971	Electrochemical Methods
	3A-D	1972	Optical, Spectroscopic and Radioactivity Methods
	4	1972	Determination of Mass, Transport, and Electrical-Magnetic Properties
	5	1971	Determination of Thermodynamic and Surface Properties
	6	1977	Supplement and Cumulative Index [including] Determination with the Ultracentrifuge, Determination of Viscosity, Mass Spectrometry
II		1970	Organic Solvents
III		1971	Photochromism
IV	1-3	1972-73	Elucidation of Organic Structures by Physical and Chemical Methods
V	1-3	1974-82	Technique of Electroorganic Synthesis
VI			Investigation of Rates and Mechanisms of Reactions
	1	1986	General Considerations and Reactions at Conventional Rates
	2	1986	Investigation of Elementary Reaction Steps in Solution and Very Fast Reactions
VII		1975	Membranes in Separations
VIII	1-2	1975-76	Solutions and Solubilities
IX		1980	Chemical Experimentation Under Extreme Conditions
X	1-2	1976	Applications of Biochemical Systems in Organic Chemistry
XI		1976	Contemporary Liquid Chromatography
XII			Separation and Purification
XIII		1979	Laboratory Engineering and Manipulations
XIV		1978	Thin-Layer Chromatography
XV		1980	Theory and Applications of Electron Spin Resonance
XVI		1981	Separations by Centrifugal Phenomena
XVII		1982	Applications of Lasers to Chemical Problems
XVIII		1984	Microwave Molecular Spectra
XIX			Techniques of Melt Crystallization
XX		1988	Techniques for the Study of Ion-Molecule Reactions

Library [290] contains volumes which deal with topics such as modern techniques for liquid chromatography, the theory and practice of electron capture in chromatography, and liquid chromatography detectors. *Methods in Enzymology* [268], on the other hand, which numbers over 150 volumes, covers a whole range of topics, such as enzyme structure or immunochemical techniques.

Although similar in subject content to the continuing methods series, treatises and encyclopedias in this field have a more limited scope of subject coverage and a fixed number of volumes planned for the material. The arrangement of the material in treatises is quite different from that in encyclopedias, however. There is a classified subject arrangement in treatises and an alphabetical sequence of topics in an encyclopedia.

9.2.1 *Techniques of Chemistry* and *Physical Methods of Chemistry*

The series of volumes published as *Techniques of Chemistry* [675] actually evolved from two earlier series: *Technique of Organic Chemistry* and *Technique of Inorganic Chemistry*. The contents of volumes published through 1989 are listed in Table 9.1. The titles of the volumes indicate the broad range of methods which are used in constitutional chemistry. While some of them are specifically slanted toward organic chemistry, many of the volumes describe techniques which can be used in both organic and inorganic chemistry.

The widespread success of the works published as volume I in *Techniques of Chemistry* led to a new edition of *Physical Methods of Chemistry* [315] (1986-). When complete, the treatise is expected to contain the volumes listed in Table 9.2.

9.2.2 *Treatise on Analytical Chemistry*

The most up-to-date comprehensive treatise on the topic of constitutional chemistry is Kolthoff and Elving's *Treatise on Analytical Chemistry* [500]. The second edition

TABLE 9.2
Contents of *Physical Methods of Chemistry*

Vol. No.	Part No.	Year	Title
I		1986	Components of Scientific Instruments and Applications of Computers to Chemical Research
II		1986	Electrochemical Methods
III	A	1986	Determination of Chemical Composition
	B	1989	and Molecular Structure
IV			Microscopy
V			Determination of Structural Features of Crystalline and Amorphous Solids
VI			Determination of Thermodynamic Properties
VII			Determination of Elastic and Mechanical Properties
VIII			Determination of Electronic and Optical Properties
IX	A-B		Investigations of Surfaces and Interfaces

of this monumental work began publication in 1978. It follows the arrangement of the first edition which appeared in 33 physical volumes between 1959 and 1980. Part I of that edition is devoted to "Theory and Practice". Part II covers "Analytical Chemistry of Inorganic and Organic Compounds," and Part III, "Analytical Chemistry in Industry". The final volume of each part contains a detailed subject index for that part, plus an author index and complete table of contents.

9.2.3 *Encyclopedia of Industrial Chemical Analysis*

The Snell-Ettre *Encyclopedia of Chemical Analysis* [518] was published from 1966 to 1974. The first part (in three physical volumes) is devoted to general analytical tech-

TABLE 9.3
Contents of *Wilson and Wilson's Comprehensive Analytical Chemistry*

Vol. no.	Part no.	Year	Contents
I	A	1959	Analytical Processes
			Gas Analysis
			Inorganic Qualitative Analysis
			Organic Qualitative Analysis
			Inorganic Gravimetric Analysis
I	B	1960	Inorganic Titrimetric Analysis
			Organic Quantitative Analysis
I	C	1962	Analytical Chemistry of the Elements
II	A	1964	Electrochemical Analysis
			Electrodeposition
			Potentiometric Titrations
			Conductometric Titrations
			High-frequency Titrations
II	B	1968	Liquid Chromatography in Columns
			Gas Chromatography
			Ion Exchangers
			Distillation
II	C	1971	Paper and Thin-Layer Chromatography
II	D	1975	Coulometric Analysis
III		1975	Elemental Analysis with Minute Samples
			Standards and Standardization
			Separations by Liquid Amalgams
			Vacuum Fusion Analysis of Gases in Metals
			Electroanalysis in Molten Salts
IV		1975	Instrumentation for Spectroscopy
			Atomic Absorption and Fluorescence Spectroscopy
			Diffuse Reflectance Spectroscopy
V		1975	Emission Spectroscopy
			Analytical Microwave Spectroscopy
			Analytical Applications of Electron Microscopy
VI		1976	Analytical Infrared Spectroscopy
VII		1976	Thermal Methods in Analytical Chemistry
			Substoichiometric Analytical Methods

niques of use in a wide variety of instances and has its own subject index at the end of volume 3. Volume 20 is the index to the second part and also lists the contents of the entire encyclopedia. There are chapters in part 2 on elements and classes or types of chemicals such as tin, dienes, agricultural chemicals, or cosmetics. Thus, raw materials, intermediate substances, and finished products are all found in the encyclopedia. Some chapters in the second part also deal with techniques, for example, bead and flame tests, bioassay, carrier gas and vacuum fusion methods, centrifugal separation, and Fourier transform spectroscopy. The editors of the *Encyclopedia* did not attempt to present a comprehensive overview of all available analytical techniques, leaving that to the sister publication, the *Treatise on Analytical Chemistry*, and the Fresenius *Handbuch der Analytischen Chemie* [497]. Thus, the *Encyclopedia* concentrates on the practice of analytical chemistry and not its theory.

TABLE 9.3
Contents of *Wilson and Wilson's Comprehensive Analytical Chemistry*

Vol. no.	Part no.	Year	Contents
VIII		1977	Enzyme Electrodes in Analytical Chemistry
			Molecular Fluorescence Spectroscopy
IX		1979	Ultraviolet Photoelectron and Photoion Spectroscopy
			Auger Electron Spectroscopy
			Plasma Excitation in Spectrochemical Analysis
X		1980	Organic Spot Test Analysis
			The History of Analytical Chemistry
XI		1981	The Application of Mathematical Statistics in Analytical Chemistry
XII			Thermal Analysis
	A	1981	Simultaneous Thermoanalytical Examinations by Means of the Derivatograph
	B	1982	Biochemical and Clinical Applications of Thermometric and Thermal Analysis
	C	1984	Emanation Thermal Analysis and other Radiometric Emanation Methods
	D	1984	Thermophysical Properties of Solids
XIII			Analysis of Complex Hydrocarbons
	A	1981	Separation Methods
	B	1981	Group Analysis and Detailed Analysis
XIV		1982	Ion Exchangers in Analytical Chemistry
XV		1983	Methods of Organic Analysis
XVI		1982	Chemical Microscopy
			Thermomicroscopy of Organic Compounds
XVII		1982	Gas and Liquid Analyzers
XVIII		1983	Kinetic Methods in Chemical Analysis
			Application of Computers in Analytical Chemistry
XIX		1986	Analytical Visible and Ultraviolet Spectrometry
XX		1985	Photometric Methods in Inorganic Trace Analysis
XXI		1988	New Developments in Conductimetric and Oscillometric Analysis
XXII		1986	Titrimetric Analysis in Organic Solvents
XXIII		1988	Analytical and Biomedical Applications of Ion-Selective Field-Effect Transistors
XXIV		1989	Energy Dispersive X-Ray Fluorescence Analysis

9.2.4 Comprehensive Analytical Chemistry

When it began to be published in 1959, *Wilson and Wilson's Comprehensive Analytical Chemistry* [498] was projected to have five volumes devoted to: I. Classical Analysis; II. Electrical Methods, Physical Separation Methods; III. Optical Methods; IV. Industrial and Other Specialist Applications; and V. Miscellaneous Methods, General Index. After nearly three decades of continuous publication, the set is contained in more than two dozen physical volumes with contents as shown in Table 9.3.

9.2.5 Comprehensive Treatise of Electrochemistry

The *Comprehensive Treatise of Electrochemistry* [925] (1980-) covers the broad field of electrochemistry. The volumes listed in Table 9.4 had appeared by the end of 1989.

TABLE 9.4
Contents of *Comprehensive Treatise of Electrochemistry*

Volume no.	Year	Title
1	1980	The Double Layer
2	1981	Electrochemical Processing
3	1981	Electrochemical Energy Conversion and Storage
4	1981	Electrochemical Materials Science
5	1983	Thermodynamics and Transport Properties of Aqueous and Molten Electrolytes
6	1983	Electrodics: Transport
7	1983	Kinetics and Mechanisms of Electrode Processes
8	1984	Experimental Methods in Electrochemistry
9	1984	Electrodics: Experimental Techniques
10	1985	Bioelectrochemistry

9.2.6 Encyclopedia of Electrochemistry of the Elements

The 15-volume *Encyclopedia of Electrochemistry of the Elements* [808] (1973-84) deals with the descriptive electrochemistry of inorganic and organic compounds. Standard potentials, voltammetric characteristics, kinetic parameters and double-layer properties, electrochemical reactions, and applied electrochemistry are found in each chapter. The applied electrochemistry sections are concerned in part with the isolation or purification of the element. The arrangement of compounds in the encyclopedia generally follows that found in *Gmelin* for inorganic compounds and in *Beilstein* for organic compounds. (See sections 10.4.2. and 10.4.3.)

9.2.7 Methods in Enzymology

Colowick and Kaplan's monumental series *Methods in Enzymology* [268] is a key reference source for libraries serving biochemists. Representative topics include basic theory, sources of equipment and reagents, and methods for DNA sequence analysis.

In v.152 (*Guide to Molecular Cloning Techniques*) is found "Requirements for a Molecular Biology Laboratory".

Subject Access to the volumes of *Methods in Enzymology* is facilitated by a series of indexes which have appeared periodically. Table 9.5 lists the location and coverage of the volumes.

TABLE 9.5
Subject Indexes to *Methods in Enzymology*

Volume No.	Covering Volumes	Dates
33	1-30	1955-74
75	31-32, 34-60	1974-79
95	61-74, 76-80	1979-81
120	81-94, 96-101	1982-83
140	102-119, 121-134	1983-86

9.2.8 Methods of Enzymatic Analysis

Now in its third edition, Bergmeyer's *Methods of Enzymatic Analysis* [274] (1983-87) aims to provide a laboratory manual for the major applications of enzymes to analysis. The treatise presents in a clear and concise manner such topics as fundamental principles (v.1); samples, reagents, assessment of results (v.2); enzymes (v.3-5); metabolites (v.6-9); antigens and antibodies (v.10-11); and drugs and pesticides (v.12). The Cumulative Subject Index is a valuable addition to the set, since individual volumes were not cross-referenced to other volumes as they were published. The work deals both with analytes that can be assayed with the aid of enzymes and the analytical techniques to analyze enzymes themselves in areas such as clinical medicine or biomedical research laboratories. Information on the sensitivity and limitations of the methods can be found in *Methods of Enzymatic Analysis*, as well as experimental "tricks" which actually make the techniques work.

9.3 COLLECTIONS-II: STANDARD METHODS, HANDBOOKS, AND SMALLER WORKS

There are many smaller analytical handbooks and methods volumes. Furthermore, a number of useful compilations of standard methods of analysis have appeared in many editions over the years. Some of the more important ones are discussed in this section.

9.3.1 Standard Methods and Pharmacopoeias

A special class of reference sources, often published in one volume, reports standard methods for various techniques. An example is *Standard Methods for the Examination of Water and Wastewater* [326] (1985), which covers such topics as the determination of metals, inorganic non-metallic constituents, and organic constituents; examination for radioactivity; and test methods for aquatic organisms.

The fifteenth edition of *Official Methods of Analysis of the Association of Analytical Chemists* [305] (1990) incorporates newer methods such as inductively coupled plasma emission spectroscopy as well as revisions of older methods. Since the first edition of this work appeared over sixty years ago, it has been the standard source for techniques with which to analyze commercial agricultural, chemical, food, and pharmaceutical products.

ASTM, the American Society for Testing Materials, each year publishes the multi-volume *Annual Book of ASTM Standards* [1742]. Many test methods are included in the set, which now contains over sixty volumes. Individual titles are frequently revised and may be purchased separately.

A class of reference book known as a **pharmacopoeia** sets the standards of purity for prescription pharmaceuticals and drugs. The designation "USP" indicates that a drug conforms to the standards established by the U.S. Pharmacopoeial Convention as found in *The United States Pharmacopoeia* [956] (now published with *The National Formulary*). The *Pharmacopoeia* section is limited to drug substances and dosage forms, whereas the *Formulary* deals with pharmaceutic ingredients.

The third edition of the *NIOSH Manual of Analytical Methods* [282] (1984) contains methods to establish environmental limits for toxic exposures. Perhaps the largest set of sampling and analytical techniques available for personal monitoring is found here. Methods are given for chemical substances commonly encountered in industry and academic institutions, such as benzene, chloroform, phosphoric acid, etc.

9.3.2 Handbooks and Selected Smaller Works

In its latest edition, *Instrumentation for Environmental Monitoring* [994] (1983-) appears in three volumes, the first covering radiation, the second, water, and the third, air. Each volume discusses the basic problems, methods of sensing and detection, and the available instruments and techniques.

Meites' *Handbook of Analytical Chemistry* [300] (1963), though somewhat dated now, is still a useful resource. It contains sections devoted to both samples and techniques. In addition to presenting fundamental data, the book deals with qualitative analysis, titrimetric analysis, the measurement of pH, and many other methods.

A set which was never completed, but includes several useful volumes is *Methodicum Chimicum* [502]. Volume 1 (in two parts) (1974) covers "Analytical Methods," including purification, wet processes, determination of structure, micromethods, biological methods, quality control, and automation.

There are a number of books on reagents which are useful in analytical work. Among them are the *Handbook of Organic Reagents in Inorganic Analysis* [506] (1976), *Reagent Chemicals* [504] (1986), the *CRC Handbook of Organic Analytical Reagents* [505] (1982), the *Handbook of Analytical Derivatization Reactions* [508] (1979), and the *Handbook of Derivatives for Chromatography* [511] (1978).

Purification of Laboratory Chemicals [311] (1988) assists in the purification of laboratory reagents. Thousands of commercially available compounds can be purified using these methods. In addition to the physical techniques of distillation, recrystallization, drying, chromatography, solvent extraction and molecular sieves which are commonly used, the authors devote many pages to purification of individual organic, inorganic and organometallic chemicals. Furthermore, general methods for purifying classes of compounds are included in case the specific chemical of interest is not covered in the previous chapters.

The original *CRC Handbook of Chromatography* [328] (1973) has been supplemented by special editions devoted to particular classes of compounds, such as lipids [702] (1984) and steroids [886] (1985). The CRC Press has also published many other handbooks of interest to analytical chemists, including *CRC Handbook of Electrophoresis* [516] (1980), *CRC Handbook of Solubility Parameters and Other Cohesion Parameters* [802] (1983), *CRC Handbook Series in Inorganic Electrochemistry* [812] (1980-), *CRC Handbook Series in Organic Electrochemistry* [689] (1977-83), and *CRC Handbook of Basic Tables for Chemical Analysis* [1919] (1989).

Albert and Serjeant's classic book on *The Determination of Ionization Constants* [308] (1984) first appeared in 1962. It is the standard work for those who need to determine an ionization constant. The novice is guided in Chapter 1 to an appropriate method. In Chapter 9 are found ionization constants of 370 commonly prescribed drugs and other biologically active substances. A wealth of other relevant data is found throughout the book.

Shriver and Drezdzon's *The Manipulation of Air Sensitive Compounds* [321] (1986) includes data and information to assist in the design of experiments involving substances which undergo chemical changes when exposed to air. Most of the techniques are intended for use with inorganic and organometallic compounds, but they are generally applicable to gases and air-sensitive solids or liquids.

Techniques in Molecular Biology [866] (1983) contains a selection of the more advanced analytical and preparative techniques frequently used by researchers in molecular biology. Both the theory and relevant practical details of each technique are presented.

An up-to-date manual for inorganic trace analysis is Marczenko's *Separation and Spectrophotometric Determination of Elements* [1737] (1986). Proven methods in the chemical literature up to the beginning of 1985 were incorporated in the book, which concentrates on the separation methods.

9.3.3 Dictionaries and Nomenclature Aids

Among the dictionaries for the field of constitutional chemistry are: *Dictionary of Chromatography (English-German-French-Russian)* [509] (1984), *A Dictionary of Chromatography* [510] (1982), *A Dictionary of Spectroscopy* [1169] (1982), *Multilingual Dictionary of Important Terms in Molecular Spectroscopy* [1176] (1966), the *Dictionary of Electrochemistry* [809] (1984), and *Acronyms and Abbreviations in Molecular Spectroscopy: An Encyclopedic Dictionary* [1920] (1990).

The definitive rules for the nomenclature of analytical chemistry are found in *Compendium of Analytical Nomenclature* [157] (1987). The *Guide to Index Terms in Analytical Abstracts* [355] (1986), while it is a thesaurus for a particular abstract journal, can also be of use in selecting terms to search in other reference sources, such as *Science Citation Index*.

9.4 COLLECTIONS-III: SPECTRAL, STRUCTURAL, AND SEQUENCE ANALYSIS COLLECTIONS

There is a wealth of information available to assist in the determination of the structure of a compound. Spectral data, crystallographic data, and sequence analysis data compilations are very numerous. Some of the most important works are discussed in this section.

9.4.1 Spectral Data Collections

Organic Electronic Spectral Data [1167] (1960-) presents in molecular formula order the organic compounds for which ultraviolet-visible spectra have been measured. The first two volumes covered the literature from 1946 through 1955. Supplements have regularly appeared, and since 1966 are published each year. More than half a million spectra are thus listed in formula order with name, solvent or phase, and wavelength values in nanometers for all maxima, shoulders, and inflections. Data are referenced to the original journal articles.

Lang's *Absorption Spectra in the Ultraviolet and Visible Region* [1166] (1961-) presents actual spectrographs. All of the spectra were measured at various research institutes in Eastern Europe. Both a name and formula index are included. Cumulative indexes have been produced for groups of five volumes as they are completed.

The third edition of the *Eight Peak Index of Mass Spectra* [1190] (1983) is a comprehensive index covering the collections of a number of different organizations, as shown in Table 9.6. In addition, 10,431 spectra are taken from the general mass spectrometry literature. The total of 66,720 mass spectra might include as many as three spectra for a given compound if different experimental conditions prevailed.

The 7-volume *Wiley/NBS Registry of Mass Spectral Data* [1691] (1988) has also appeared in a Wiley CD-ROM edition [1597]. This makes it possible to search a database of over 123,000 spectra on an IBM personal computer. The database is searched with the PBM/STIRS software developed at Cornell University. The same database is found in the Wiley Mass Spectral Search System [1194], a component of the former NIH/EPA Chemical Information System, now available through Chemical

TABLE 9.6
Collections of Mass Spectra Covered by the *Eight Peak Index of Mass Spectra*

Collection	Number of Spectra
Imperial Chemical Industries	4,707
John Wiley [1194]	1,622
NIH-EPA [1218]	36,832
Thermodynamics Research Center [1332]	312
American Petroleum Institute	2,443
ASTM	191
Dow Chemical Company	1,142
Mass Spectrometry Data Centre	9,040
Total	56,289

Information Systems, Inc. (CIS). The PBM/STIRS software is also used with the CIS database to search the mass spectra of pure compounds or mixtures. Peaks are entered, candidate substances are identified, and a direct comparison of the retrieved spectrum and the unknown spectrum is made. (CIS, Inc. also has databases for searching carbon-13 NMR and infrared spectra.)

There is another MS database on the CIS system, the NIST/EPA/MSDC Mass Spectral Database [1603]. That database is the successor to the NIH-EPA mass spectral database and contains about 48,000 electron impact and chemical ionization mass spectra. The collection is also included in *The Wiley/NBS Registry of Mass Spectral Data* [1691]. The original NIH-EPA collection of mass spectra was published as a set in 1978 as the *EPA/NIH Mass Spectral Data Base* [1218], containing spectra for 25,556 substances, with two supplements adding many new substances. A Cumulative Index covers the basic volume and the first supplement through substance name, molecular formula, molecular weight, and CAS registry number indexes. Now a microcomputer version of the current database, which is supplemented with additional spectra from the Mass Spectrometry Data Centre, can be obtained from the National Institute of Standards and Technology. The PC version includes a utility for adding new spectra to the database. Structures are included for 85 percent of the spectra.

Fein-Marquart Associates, Inc., the developer of the Chemical Information System, also have a PC software package for the analysis of mass spectral data. Known as MASCOT [1596], the software is available with search libraries of environmental compounds or forensic drugs.

In addition to collections of mass spectra, the TRC (Thermodynamics Research Center) publishes collections of proton and carbon-13 NMR spectra, as well as ultraviolet, infrared, and Raman spectra. Another company which produces many standard spectral collections is Sadtler Research Laboratories. Sadtler has large collections of infrared (both prism and grating), proton NMR, carbon-13 NMR, and ultraviolet spectra.

Sadtler also markets smaller collections of IR (and other) spectra for inorganic or organometallic compounds, minerals, pharmaceuticals, toxic chemicals, steroids, drugs, etc. The company has microcomputer software for FT-IR, IR, and carbon-13 NMR spectra libraries. Utilizing the capabilities of Microsoft Windows, the Sadtler PC SEARCH software [1599] allows an IBM PC to conduct a full spectrum search or peak search on a spectral library. The packages of spectral libraries cover a wide variety of chemical classes, each with representative spectra. Sadtler also supplies the Coblentz Society IR spectral collection.

Sprouse Scientific Systems' MicroSearch software [1598] offers another option for searching libraries of IR spectra. Sprouse is one of several companies which provide diskettes containing IR libraries of their own spectra and that of others. These include FT-IR libraries of polymers, solvents, and pure chemical compounds. The multi-volume *Sprouse Collection of Infrared Spectra* [1595] (1988-) includes coverage of polymers, solvents, and other compounds of interest in applied spectroscopy.

STN International has introduced the C13-NMR/IR database [1912]. Although accessed through STN, the file is searched with different software than the normal STN system search software (Messenger). Carbon-13 NMR spectra are taken from a core of 20 primary journals, as well as spectra catalogs and unpublished spectra from BASF and other organizations. BASF is also the source of the unpublished IR spectra which are included in the database. The database can be searched by chemical shifts, molecular formulas, chemical name fragments, structures, or substructures. In addition, C13-NMR/IR has the capability to locate carbon-13 NMR spectra through a method of spectrum estimation.

The *Carbon-13 NMR Spectral Data* collection [1254] (1987) edited by Wolfgang Bremser is available in both computer-output-microfiche (COM) and in machine-readable formats [1610]. The COM collection has 58,108 spectra for 48,357 compounds. Other useful smaller collections of spectra include the *Atlas of Carbon-13 NMR Data* [1211] for 3017 compounds, the *CRC Atlas of Spectral Data and Physical Constants for Organic Compounds* [1261] (1975) which has IR, Raman, Ultraviolet, proton and carbon-13 NMR, and mass spectra, *The Aldrich Library of NMR Spectra* [1256] (1983), *The Aldrich Library of FT-IR Spectra* [1251] (1985), the *Merck FT-IR Atlas: A Collection of FT-IR Spectra* [1364] (1988), *The Aldrich Library of Infrared Spectra* [1252] (1981), and *The Coblentz Society Desk Book of Infrared Spectra* [1199] (1982).

9.4.2 Crystallography

In the field of crystallography, a standard work is *International Tables for Crystallography* [842], the second, revised edition of which began to appear in 1987. Successive versions of the tables have moved progressively away from the special topic of X-ray structure determination toward data and text which can be used in all aspects of crystallography. *Volume A. Space-Group Symmetry* will be followed by *B. Reciprocal Space* and *C*, which will contain physical and chemical data of interest to crystallographers. Earlier editions of volumes II-III and supplementary tables are still available in the interim [849].

Six volumes of the third edition of *Crystal Data: Determinative Tables* [850] appeared between 1972 and 1988. Volumes 1 and 2 cover respectively organic and inorganic crystalline materials through 1966. Volume 3 covers organic literature through 1974 and volume 4, inorganic substances 1967-69. Volumes 5 and 6 are devoted to organic and organometallic compounds, plus metallic complexes containing organic ligands from 1975-81.

Wyckoff's *Crystal Structures* [852] (1963-71) is limited to those studies which define the positions of the atoms in a crystal. Both a bibliography and a statement of atomic positions, the work is admirably illustrated. Crystal structures of both inorganic and organic compounds are found in the work.

Structure Reports [1163] (1913/28-) covers the numerous studies by X-rays which produce unit cells and possible space groups. In 1965, the annual volumes split between A. Metals and Inorganic Compounds and B. Organic Compounds.

The Crystallographic Data Centre in Cambridge, England published a comprehensive bibliography on *Molecular Structures and Dimensions* [360], which covers organic and organometallic crystal structures from 1935 onward. Both the bibliographic file and the structural file are accessible in the CRYST database [1607] through Chemical Information Systems, Inc. The database contains over 50,000 bibliographic records and includes both X-ray and neutron diffraction studies. The structural component provides atomic coordinates and cell parameters for organic and organometallic compounds reported since 1960. The tapes and search software can also be leased in the United States through the Medical Foundation of Buffalo, Inc. (Buffalo, New York) at relatively low cost to academic and non-profit users. Another crystal structure database on the CIS is XTAL [1602]. The Inorganic Crystal Structure Data file [1618] contains over 28,000 crystal structures for inorganic substances (compounds with no C-C and/or C-H bonds and with at least one of the nonmetallic elements H, He, B, C, N, O, F, Ne, Si, P, S, Cl, Ar, As, Se, Br, Kr, Te, I, Xe, At, Rn). The CRYSTMET database [1921] concentrates on metallic phases. The Powder Diffraction File [1914] is now available on CD-ROM. The file is a collection of single-phase X-ray powder diffraction patterns in the form of tables of the interplanar spacings and relative intensities which characterize the compounds. Thus, it is a non-structural crystallographic database.

Methods of crystallographic database compilation and searching are found in *Crystallographic Databases* [1913] (1987). This is a useful introduction to the available databases. Included is an overview of crystal structure analysis for the non-specialist.

9.4.3 Biochemical Compounds and Sequence Databases

A number of works exist to assist in visualizing the structure and configuration of complex biochemical compounds. These include the *Atlas of Steroid Structure* [727] (1975-84), *Atlas of Protein Sequence and Structure* [728] (1968-78), *Nucleic Acid Sequences Handbook* [729] (1981), and the *Atlas of Stereochemistry* [759] (1978-86). In addition, there are over a dozen databases which register molecular sequences. These sequences are often cited in the journal literature by a reference to the database and the accession number rather than reproducing a graphic representation of the structure itself. A useful guide is *Directory of Protein and Nucleic Acid Sequence Data Sources* [1915] (1987). Works which introduce the field of computing with protein and nucleic acid sequences are *Computational Molecular Biology Sources and Methods for Sequence Analysis* [1916] (1989) and *Sequence Analysis; An Introduction to Computer Methods* [1922] (1989).

The *Atlas of Molecular Structures in Biology* [865] first appeared in 1973 with the volume "Ribonuclease-S," containing stereodiagrams of the S-form of the enzyme. This was eventually followed in 1981 with a second volume containing stereodiagrams of hemoglobin and myoglobin. Croft's *Handbook of Protein Sequence Analysis* [884] (1980) presents the complete amino acid sequences for proteins and peptides. The peptides included in the work are related to proteins through biosynthesis or are directly derived from proteins. Excluded are cyclic peptides and those containing D-amino acids. *AMSOM; Atlas of Macromolecular Structure on Microfiche* [932] (1976) provides a comprehensive compilation of structural information for biological macromolecules whose three-dimensional structures were well defined at that time. The journal *Nucleic Acids Research* has for several years included a supplement which covers RNA sequences [894] [895] [919].

Computer techniques for the determination of sequences of structures of nucleic acids and proteins are now quite numerous. Results of these investigations are recorded in databases containing millions of sequences. Such databases facilitate comparisons with sequences which have been newly determined and allow the generation and testing of hypotheses about the organization and evolution of molecular sequences. In 1988, the National Library of Medicine introduced a new data element, the Secondary Source ID, into its bibliographic database, MEDLINE. This field links articles in MEDLINE to the molecular sequence databases where the structures of the cited compounds are deposited. Two very large compilations of data in this area are those of the European Molecular Biology Laboratory's Nucleotide Sequence Data Library (EMBL) [1341] and the GenBank Genetic Sequence Data Bank [1342] of the U.S. National Institutes of Health and other U.S. health-related agencies. More than 8,000,000 entries are included in the collections, which have appeared in printed form as *Nucleotide Sequences* [1339] (1987). Data from the *Atlas of Protein Sequence and Structure* [728] are available in the Protein Identification Resource (PIR) database [1340]. Nearly 3000 entries for over half a million amino acid residues are included for completely sequenced proteins and sequences translated from nucleic acids. Search the Chemistry Reference Sources Database for **INDX = "MOLECULAR SEQUENCE DATABANK"** for other databases of this type.

One of the unique features of the sequence databases is their free (or very low) costs. The tapes can be obtained for in-house use. Alternatively, they can be searched online. Until 1990, the EMBL, GenBank, and PIR databases were accessible online through BIONET [1337], a computer network linking the computer files at IntelliCorp's IntelliGenetics Division to academic and nonprofit organizations for a $400 subscription fee. A number of programs for data manipulation were also available to the users of that system, including sequence editor, management, and analysis programs, as well as the major searching programs, IFIND and QUEST. Most of BIONET's services were scheduled to be taken over by GenBank after 1989.

The Protein Identification Resource at Georgetown University provides access to the GenBank and EMBL databases over the Tymnet telecommunications network. Additionally, GenBank can be accessed directly through the distributor's computer system at BBN Laboratories (formerly Bolt, Beranek and Newman, Inc.) in Cambridge, Massachusetts. Their system even has a software clearinghouse component which includes information on some of the available sequence analysis packages. Also of help in this regard is the *Software Directory for Molecular Biologists* [1336] (1986), a guide to selecting computer software for the analysis of molecular sequences. Three-dimensional structures of proteins from residue sequence information can be generated with the molecular modeling software Chem-X [1923].

Chemical Abstracts Service includes on their staff trained specialists who conduct searches for a fee in the GenBank, EMBL, and PIR databases. In addition, CAS has developed a sequence database for protein and nucleotide sequence data reported in patents since 1981. The database, CASSEQ [1917], contains over 7000 sequences. The coverage of sequences in the patent literature is a unique aspect of the database.

With larger storage capacities on microcomputers, some of the sequence databases are sure to become available soon for use on microcomputers.

9.5 REVIEWS

Since much of the information in the sources described in the preceding sections of this chapter is bound to be at least two to three years old (much of it probably considerably older), the searcher may need to find more recent descriptions. Reviews are good sources for rapidly scanning a large body of primary literature. A recently published review may bring the coverage up to within a year or two of the present. (See Section 13.4.6 for more information on reviews.)

Review articles can be found in primary sources, such as journals or conference proceedings. Longer review articles are likely to appear in the secondary sources called **review serials**, for example, *Advances in X-Ray Analysis*. The **review articles** are written by experts in the subject area who scan and evaluate the most important literature which has appeared in a given period of time. If the review is published in a review serial, the time period is usually one year.

The authors of the review articles may take different approaches, depending on the intended audience. Some reviews are directed toward specialists in the field and are, therefore, written at an advanced level. Others tend to be written for the nonspecialist, where the goal is to present new techniques or advances which may find application in the reader's field. These review articles tend to be rather short in comparison to the specialist reviews, perhaps containing only a few dozen references to the key literature which has appeared in a period of time stretching over several years. The specialist reviews, on the other hand, can include hundreds of references in their bibliographies.

9.5.1 Special Review Issues of *Analytical Chemistry*

The June 15, 1989 issue of *Analytical Chemistry* is devoted to "*Application Reviews*," with topics such as air pollution, food, forensic science, industrial hygiene, particle size analysis, pesticides, petroleum, and water analysis, among others. Likewise, the June 15, 1988 issue of *Analytical Chemistry* covers "Fundamental Reviews," including thermal analysis, chemical sensors, Raman spectroscopy, thin-layer and paper chromatography, ion-selective electrodes, mass spectrometry, X-ray spectrometry, NMR spectrometry, etc. Continuing a pattern of covering theoretical topics one year, then applied topics in alternate years, the review articles have appeared in special issues of *Analytical Chemistry* for over 40 years. Authors of the reviews are required to survey the literature of their fields of specialization and select only the most significant material for inclusion in the review articles.

9.5.2. *Methods of Biochemical Analysis*

Since 1954 the review serial *Methods of Biochemical Analysis* [688] has covered the methodology and instrumentation used in biochemical analysis. Techniques for the determination of enzymes, vitamins, hormones, lipids, carbohydrates, proteins and their products, minerals, and antimetabolites are included. These range from chemical and physical techniques to those used in microbiology, and sometimes even animal assays. Each volume has an author and subject index for that volume and a cumulative author and subject index for it and all previous volumes. (This is typical of a review serial.) However, the subject indexes are generally limited to permutations of title words from the chapters of the individual volumes, such as "Nucleic Acids and Their Derivatives, Microbiological Assay of".

9.5.3 Other Review Serials

There are many other relevant review serials for constitutional chemistry, most of which are listed in Table 9.7.

TABLE 9.7
Analytical Chemistry Review Serials

Advances in Chromatography
Advances in Electrochemistry and Electrochemical Engineering
Advances in Magnetic Resonance
Advances in Mass Spectrometry
Advances in Spectroscopy (formerly, Advances in Infrared and Raman Spectroscopy)
Analytical Chemistry Application Reviews
Analytical Chemistry Fundamental Reviews
Annual Reports on Analytical Atomic Spectroscopy (superseded by Journal of Analytical Atomic Spectrometry)
Annual Reports on NMR Spectroscopy
Applied Spectroscopy Reviews
Contemporary Topics in Analytical and Clinical Chemistry
Chemtracts: Analytical/Physical/Inorganic Chemistry
Critical Reviews in Analytical Chemistry
Critical Reviews in Environmental Control
Critical Reviews in Toxicology
Electroanalytical Chemistry
Electrochemistry
Environmental Chemistry
High Performance Liquid Chromatography
Mass Spectrometry
Mass Spectrometry Advances
Mass Spectrometry Reviews
Methods of Biochemical Analysis
Modern Aspects of Electrochemistry
Molecular Spectroscopy
Molecular Structure by Diffraction Methods
Nuclear Magnetic Resonance
Photochemistry
Progress in Analytical Spectroscopy
Progress in Nuclear Magnetic Resonance Spectroscopy
Review of Scientific Instruments
Spectroscopic Properties of Inorganic and Organometallic Compounds
Vibrational Spectra and Structure

9.6 ABSTRACTING AND INDEXING JOURNALS

In addition to the abstracting and indexing journals discussed up to this point, such as *Chemical Abstracts*, *Science Citation Index*, and others devoted to particular primary literature formats, there are numerous abstracting and indexing services which cover analytical chemistry and related areas. These specialized services can be quite useful since they are more closely focused on the literature of relevance to constitutional chemistry. Examples include *Analytical Abstracts*, *Mass Spectrometry Bulletin*, and *Chromatography Abstracts*.

In order to identify appropriate abstracting or indexing services or databases to search, the database directories discussed in section 2.3 could be used. Alternatively, the searcher could consult a specialized directory or guide, such as *Abstracting and Indexing Services Directory* [532] (1982-83) or *Abstracts and Indexes in Science and Technology* [533] (1985).

The *Mass Spectrometry Bulletin* [299] is published by the Royal Society of Chemistry. The printed version covers the literature from 1966. From 1989, there is a PC version of the *Mass Spectrometry Bulletin* which is updated monthly with about 900 additional records. *Chromatography Abstracts* [1888] in its printed variants covers the literature of gas and liquid chromatography from 1951 and 1984 respectively. The publisher offers a CD-ROM version of the database, CD-CHROM.

A comprehensive abstracting journal which deals with all aspects of analytical chemistry is *Analytical Abstracts* [269] (1953-). From 1980, the database can be searched online. From 1984, abstracts are included in the records. The printed *Analytical Abstracts* covers well over a quarter million items and grows at the rate of over 1000 entries per month. By mid-1988, the database already contained over 100,000 records. Included are journal articles, technical reports, national standards, books, and conference proceedings. Printed cumulative author and subject indexes have been published for the period 1954-63, and there are annual cumulative author and subject indexes for later volumes.

The indexing of *Analytical Abstracts* is specifically tailored to the needs of the analytical chemist. There are three types of index entries. For the **analyte** (the substance which is to be identified, detected or separated in some manner) there are name entries for a specific chemical compound, as well as class names which refer to related groups of compounds. The **matrix** (the sample or medium in which the analyte is analyzed) also has both types of compound name index terms, and also includes terms for substances such as vegetable oil, washing powders, etc. A **concept** is everything which is not an analyte or matrix. Thus, reagents, techniques, and parameters such as pH fall into this category. CAS registry numbers are also searchable in *Analytical Abstracts* online.

There is a hierarchical arrangement of terms, using inverted phrases. Thus, one finds:

ANALYTE hydrocarbons, polycyclic aromatic
MATRIX waters, natural, sea-
CONCEPT chromatography, liquid, high-performance.

A neighbor or expand command can be profitably used to determine the order of the controlled-vocabulary index terms. This is a British database, so spelling differences must be taken into account (for example, "caesium" is used instead of "cesium"). There is a *Guide to Index Terms in Analytical Abstracts* [355] which can be obtained from the database producer, the Royal Society of Chemistry.

The abstract journal is divided into nine sections, as indicated in Table 9.8.

TABLE 9.8
Subject Sections of *Analytical Abstracts* on DIALOG

Online Code	Name
A	General Analytical Chemistry
B	Inorganic Chemistry
C	Organic Chemistry
D	Biochemistry
E	Pharmaceutical Chemistry
F	Food
G	Agriculture
H	Environmental Chemistry
J	Apparatus; Techniques

9.7 SUMMARY

Constitutional Chemistry, or as it has traditionally been called, analytical chemistry, has become very important in modern chemical research and development. Analytical techniques are involved in everything from quality control to structural analysis. The various methods used in the performance of those tasks are discussed in this chapter. Treatises, encyclopedias, and other large sets make the retrieval of analytical chemistry techniques much easier. Other smaller handbooks and compendia provide collections of the most important methods and techniques and can be frequently updated for standard methods.

In the area of spectral, crystallographic, and biochemical sequence collections, the chemist has many reference tools from which to choose. Increasingly these sources are available in computer-readable form, with options for access ranging from online searches of remote databases to CD-ROM databases on local microcomputers.

For more recent literature searches, reviews and abstracting or indexing services are the reference sources which should be used. The specialized reviews which appear each year in the journal *Analytical Chemistry* are very important, but there are many other potential sources of reviews in this area. Likewise, there is a range of specialized abstracting and indexing services to choose from. One of the best is *Analytical Abstracts*.

9.8 SELECTED READINGS

Ginsburg, Michelle. "Online Services," In *Nucleic Acid and Protein Sequence Analysis: A Practical Approach*; Bishop, M.J.; Rawlings, C.J., Eds.; IRL Press: Oxford, Washington, 1987; pp 115-146. [1338]

Heller, Stephen R. "The Chemical Information System and Spectral Databases." *J. Chem. Inf. Comput. Sci.* **1985**, *25*(3), 224-231.

Rumble, John R., Jr.; Lide, David R., Jr. "Chemical and Spectral Databases: A Look Into the Future." *J. Chem. Inf. Comput. Sci.* **1985**, *25*(3), 232-235.

Bergerhoff, G.; Hundt, R.; Sievers, R. "The Inorganic Crystal Structure Data Base." *J. Chem. Inf. Comput. Sci.* **1983**, *23*(2), 66-69.

Allen, Frank H.; Kennard, Olga. "The Cambridge Database of Molecular Structures." *Perspect. Comput.* **1983**, *3*(3), 28-43.

Jenkins, R.; Holomany, M. "'PC-PDF': A Search/Display System Utilizing the CD-ROM and the Complete Powder Diffraction File." *Powder Diffr.* **1987**, *2*(4), 215-219.

CHAPTER 10

SEARCHING WHICH INVOLVES CHEMICAL AND PHYSICAL PROPERTIES OF SUBSTANCES

10.1 INTRODUCTION

From the mid-19th century to the present there have been many efforts to collect in tables and handbooks the physical and chemical properties of substances. The compilers of such works assemble the most reliable data and arrange them in a rational manner for easy retrieval. Even far into the 20th Century, a second goal for some compilers was to comprehensively gather the best data for all relevant areas of science. The great German handbuchs had such worthy goals. However, as chemistry and other sciences entered a period of explosive growth during and after World War II, it became very difficult and ultimately impossible to be truly comprehensive when collecting and evaluating data.

Despite the difficulty of the task, data collecting and evaluating activities continue to grow in many areas of science. Chemistry is no exception, and reliable data compilations are very much in demand. Even common physical constants like the boiling point and melting point are important in identifying an unknown. However, finding a numeric value of a particular chemical or physical property for a substance, especially if it is not a common substance, can be a real treasure hunt. What handbook or data compilation among the hundreds in existence has flash points or provides azeotropic data? Unfortunately, there is no "superindex" which leads to all available data sources. However, there are ways to increase the chances of finding the needed information. Among these are the use of comprehensive indexes to numeric data compilations and searches of databases of numeric data. Another way is to take a broad look at the available data compilations by consulting a guide.

10.2 GUIDES TO THE LITERATURE

A guide to numeric data sources is a good place to start a search for numeric data. Arny's *The Search for Data in the Physical and Chemical Sciences* [1318] (1984) concisely surveys the processes of creation, compilation, and retrieval of data. Among the topics found in the first part of the book are critically evaluated data and data analysis centers (also called data centers, data compilation centers, or **information analysis centers**). In addition, the author presents an overview of "Selection Sources for Handbooks and Data Compilations" and "Data Journals and Related Publications". Also included are surveys of the National Bureau of Standards (now the National Institute of Standards and Technology) and methods of retrieving information from NBS data compilations. In the last part of the book, Arny has reviewed, analyzed, and indexed the NBS data compilations published before 1984.

The *CODATA Directory of Data Sources for Science and Technology* [389] (1977-85) gradually appeared in 12 parts in the CODATA *Bulletin*. Chapters dealing with crystallography (1977), chemical kinetics (1981), spectroscopy (1982), and chemical thermodynamics (1984) can be found there. CODATA, the Committee on Data for Science and Technology, was established under the auspices of the International Council of Scientific Unions in 1966 to promote and encourage on a worldwide basis the production and distribution of collections of critically evaluated numeric data. Originally, CODATA's efforts were concentrated on chemistry and physics, but its range of activities has been extended to the biosciences and geosciences in recent years.

A unique work which leads to relevant data compilations is *Handbooks and Tables in Science and Technology* [1145] (1983). The second edition is a comprehensive listing of nearly 4000 data compilations. Many of the entries are annotated, and access is provided by author/editor and subject indexes. Finally, the STN directory database NUMERIGUIDE [2058] provides information on the availability of numeric data in STN files. Initially, over 400 properties are included in NUMERIGUIDE. Search the Chemistry Reference Sources Database for **GUIDE AND INDX = "NUMERIC DATA"** or **GUIDE AND INDX = "PHYSICAL PROPERTIES"** to find additional sources which can lead to data compilations.

10.3 CRITICALLY EVALUATED DATA AND INFORMATION ANALYSIS CENTERS

The non-electrical physical constants are length, mass, and time. Several systems have been used to define the units in these categories, as shown in Table 10.1. The SI (International Metric or Système International d'Unités) system has steadily been gaining acceptance around the world. It is important to be aware of the units used in any data compilation which is consulted. To that end, *Quantities, Units, and Symbols in Physical Chemistry* [168] (1988) can be of great assistance.

TABLE 10.1
Selected Systems of Units for Non-Electrical Physical Constants

	Unit of		
System Name	Length	Mass	Time
Gaussian	millimeter	milligram	second
CGS	centimeter	gram	second
SI	meter	kilogram	second

If scientists are to be confident of the values found in various compilations of numeric data, there must be a reliable procedure by which to test the accuracy of the data. Such a mechanism is the **critical evaluation** process. By checking the data reported in the primary literature against theoretical values, by examining the experimental details, and in some cases even repeating the experiments, scientists who work in **information analysis centers** perform the critical evaluation function. They rate the data according to accepted definitions of critically evaluated data. These are:
- recommended values: those considered to be as reliable and as accurate as possible, given the state of the art of the equipment used to measure them
- provisional values: of lower accuracy than recommended values
- typical values: those which provide an indication of the behavior of a related group of substances with respect to the property
- selected values: the values which have been chosen in preference to another set of conflicting values as being the most likely to be accurate.

A listing of some centers which critically evaluate data can be found in the *Directory of Federally Supported Information Analysis Centers* [390] (1980).

10.4 LARGE DATA COMPILATIONS

There are three characteristics usually shared by large multi-volume sets of critically-evaluated data. First is a clear description of the physical theories on which the data are based. Next, the data sources in the primary literature are fully referenced, so that the user of the data compilation can verify its accuracy if necessary. Finally, there is the largest section containing the numeric data.

10.4.1 *Landolt-Börnstein Numerical Data and Functional Relationships*

The current edition of the *Landolt-Börnstein Zahlenwerte und Funktionen aus Physik, Chemie, Astronomie, Geophysik und Technik* [400] is the sixth edition, published between 1950 and 1980. By the 1960s it had become apparent to the compilers that the task of assembling in one mammoth set all of the relevant data for science and technology was no longer feasible. Thus, in 1961 there began to appear the New Series of *Landolt-Börnstein* [404] which concentrates on those areas most in need of critically evaluated data compilations.

Landolt-Börnstein covers such diverse subject areas as physics, chemistry, astronomy, crystallography, materials science, and technology. For years the users of *Landolt-Börnstein* had to familiarize themselves with the contents of the dozens of volumes into which the data had been classified in order to select the proper volume. There was no index covering the entire set. The sixth edition is organized in four broad volumes, each containing a number of parts, as follows:

I. Atomic and molecular physics
II. Properties of matter in its various states of aggregation
III. Astronomy and geophysics
IV. Basic techniques

On the other hand, the New Series is divided into:

Group I. Nuclear and particle physics
Group II. Atomic and molecular physics
Group III. Crystal and solid state physics
Group IV. Macroscopic and technical properties of matter
Group V. Geophysics and space research
Group VI. Astronomy, astrophysics, and space research
Group VII. Biophysics: Nucleic Acids.

The data are categorized into properties of the elements, optical, electrical, magnetic, mechanical, thermal, and kinetic properties. Substances for which data are collected are generally distinguished as vapors or gases, liquids, solids, and "atomistic" systems (the last covering elementary particles, nuclei, ions, atoms, molecules, and free radicals). Another class of substances includes materials such as minerals, rocks, ceramics, metallic and non-metallic materials, glasses, plastics, fibers, and biological materials, for example, wood and leather.

The publication of *Landolt-Börnstein* over such a long period of time has resulted in some inconsistencies due to changes in nomenclature. This has led to the same physical property appearing under different subject terms or even occurring in tables which have different headings in different volumes. Conversely, the same subject descriptor may have been used with different meanings. In order to alleviate the impact of such inconsistencies, the *Landolt-Börnstein* tables from the sixth edition and the New Series were supplemented by a comprehensive index. The *Comprehensive Index* [403] (1987) covers all volumes of the sixth edition and those volumes of the New Series published through 1985. Thus, 126 volumes of data published during a 35-year period can be accessed through one index.

There are two approaches found in the index, reflecting the two different methods used to codify the data over the years. In most cases, the physical property is listed first, qualified by the class of substance, as in:

reflectivity
–metals **II/8** 3, 17
— thin metallic films **II/8** 36.

In other cases, the main entry is a class of material, with the physical properties listed beneath, as in:

minerals
–density **III** 313 **NS V/1a** 67, 131

If the volume indexed itself contains an index of substances or a subject index, that fact is noted in the *Comprehensive Index*. In this manner, the user is alerted to the existence of the more specific index in that volume, for example:

reflectivity
–metals **II/8** 3, 17
.
.
—index of substances
——pure compounds **NS III/3** 571.

Unfortunately, not many of the separate volumes have their own index.

The *Comprehensive Index* is especially valuable to native speakers of English. Although the sixth edition of *Landolt-Börnstein* was published in German, the *Comprehensive Index* includes two separate alphabetical indexes, one in German, the other in English. Beginning in 1988, a cheaper paperback edition of the *Comprehensive Index* began to be published in an English-language version only. This version does not have as abundant use of keywords and cross references as does the original hardbound 1987 edition. It is planned to issue the paperback version every few years.

10.4.2 *Beilstein Handbook of Organic Chemistry*

The *Beilstein Handbook of Organic Chemistry* [669] [670] is certainly not limited to the presentation of physical data, but there is a considerable amount of data in *Beilstein*. Consequently, the physical data aspects of the work are discussed here, and the information dealing with the synthesis and reactions of organic compounds will be deferred until the next chapter.

10.4.2.1 *Coverage and Arrangement of Beilstein*

The *Beilstein Handbook of Organic Chemistry* is a thoroughly comprehensive multivolume set covering the entire field of organic chemistry. It is a marvelous example of the great German **handbuchs** of critically evaluated data compilations. All organic compounds reported in the world's chemical literature are covered in *Beilstein*. The coverage is so comprehensive that one can virtually assume that the compound was unknown if it was not found in a thorough search of *Beilstein* from the beginning to the present dates of coverage. Over 1.75 million compounds are described in *Beilstein*. Macromolecular compounds are excluded, as are naturally occurring compounds such as cellulose and proteins. The criteria for inclusion in *Beilstein* require that the substance be completely described and obtained in a pure enough form to allow a definite molecular formula and constitution to have been determined.

The contents of *Beilstein*, which is now in its 4th edition, are arranged by type of compound as shown in Table 10.2. Volumes 28 and 29, published in 1955-57,

TABLE 10.2
Arrangement of Compounds in the *Beilstein Handbook of Organic Chemistry*

Division	Volumes	Contents
1	1-4	Acyclic compounds (carbon chains)
2	5-16	Isocyclic compounds (carbon rings)
3	17-27	Heterocyclic compounds (at least one element other than carbon in the rings)
4	30-31	Natural products not assigned to divisions 1-3 (no longer covered)

constitute a General Subject Index (an index of compound names) and a General Formula Index, respectively, for the Basic and first two Supplementary Series. Beginning in 1990 the Beilstein Centennial Index will be published, replacing the older volumes 28 and 29 and extending the coverage through the third and fourth Supplementary Series.

The fourth edition of *Beilsteins Handbuch der Organischen Chemie* was completed in 1918 and covered the organic chemical literature up to the beginning of 1910. This is referred to as the Basic (or Main) Series (Hauptwerk). Only the Basic Series has volumes 30-31. All other volumes of the Basic Series have been supplemented by updates covering one or two decades. The current (fifth) supplement to the fourth edition brings the literature coverage to 1980. The Basic Series and Supplementary Series (the Ergänzungswerk volumes) and their coverage are summarized in Table 10.3.

The production of the work fell behind schedule during the period when the third and fourth Supplementary Series were being produced. Consequently, it was decided to publish a combined EIII/IV for volumes 17-27, covering heterocyclic compounds. Likewise, the publication of the fifth Supplementary Series, which is entirely in English, began with heterocyclic compounds. Information in *Beilstein* is classified in such a precise manner that a chemist who has thoroughly mastered the *Beilstein* system can go right to the place in the work where the compound of interest should be listed. Furthermore, the classification system has been maintained throughout the history of the 4th edition of the Basic Series and Supplementary Series. Every compound is assigned to a system number. Once the proper place has been found in any part of *Beilstein*, it is easy to move forward or backward in time to the same system number because it is always in the same numbered volume. Thus, the proper way to

TABLE 10.3
Time Periods Covered in the *Beilstein Handbook of Organic Chemistry*

Series	Abbreviation	Time Period
Basic Series	H	up to 1910
Supplementary Series I	EI	1910-1919
Supplementary Series II	EII	1920-1929
Supplementary Series III	EIII	1930-1949
Supplementary Series III/IV	EIII/IV	1930-1959
Supplementary Series IV	EIV	1950-1959
Supplementary Series V	EV	1960-1979

TABLE 10.4
Recommended Labeling of *Beilstein* Volumes

Volume Number:	1	1	1	1	1	1	2	2	2	2
Series Designator:	H	EI	EII	EIII	EIII	EIV...	H	EI	EII	EIII etc.
Part Number:				1	2	1				1

file *Beilstein* on the shelves, the way which leads to the most convenient access by the users, is to keep all of the Basic Series and Supplementary Series of a given volume number together on the shelf. This violates library cataloging rules which place the complete sets of supplements one after another, following the complete main work. A label placed at the top of the spine of each volume overcomes this difficulty (See Table 10.4).

10.4.2.2 Access to the Information in Beilstein

The presentation of the rules of the Beilstein system would require far more space than is available in this book. Nevertheless, the application of the rules is one avenue to the information in *Beilstein*. For those who wish to master the system, Weissbach's *The Beilstein Guide: A Manual for the Use of Beilsteins Handbuch der Organischen Chemie* [668] (1976) could be used, but a more readable guide is *How to Use Beilstein* [1826] (1987). There are also many helpful guides available at no cost from Springer Verlag, the publisher of *Beilstein*. Examples include *Contents of Beilstein: Representative Examples* [1832], and *Beilstein Dictionary: German-English* [1825] (1979). Search **GUIDE AND BEILSTEIN** in the Chemistry Reference Sources Database for other useful material.

Until the appearance of the 5th Supplementary Series, the *Beilstein Handbook of Organic Chemistry* was published in German. Thus, the language barrier, coupled with the sheer size of the work might discourage a potential user from seeking the answer in *Beilstein*. That would be unfortunate, because the wealth of information in *Beilstein* fully justifies the effort to use it. Listed in Table 10.5 are some of the more common abbreviations and terms for physical properties in Beilstein. Other abbreviations (Abkürzungen) and symbols are found at the front of each volume.

One way to access the information in *Beilstein* is through the indexes produced by the Beilstein Institute. As noted earlier, v. 28 is a General Subject Index (Sachsregister) which includes compound names in German. Volume 29 is a formula index in Hill system order. Both of these indexes cover the Basic Series (H) and the first two Supplementary Series (EI-EII), with coverage to be extended to EIII-EIV when the Beilstein Centennial Index is complete around 1994. Supplementary series through EIV already have indexes published for each volume as it is completed. Thus, a volume index for each basic volume and supplements EI-EIV is to be published, as well as a comprehensive index covering all volumes through EIV. The use of a good German-English/English-German chemical dictionary is advisable when approaching *Beilstein* through the subject indexes. Search the Chemistry Reference Sources Database for **GERMAN AND DICTIONARY** in order to find examples of such works. It is a good idea to first try the corresponding German trivial name of a chemical compound when searching the subject index (Sachsregister).

Another way to locate data on a compound in *Beilstein* is to use the microcomputer software package SANDRA [358]. SANDRA, the "Structure and Reference Analyzer" for *Beilstein*, permits freehand drawing of chemical structures with a mouse as well as insertion of pre-defined or user-defined fragments and stereochemical notations. The resulting chemical structure is then classified by the software into the Beilstein system. In this manner, the correct volume and part of Beilstein is identified for the user. Version 2.0 supports IBM graphics (and other high resolution graphics boards). Up to seventy non-hydrogen atoms can be entered for a molecule. An example found by using SANDRA is shown in Figure 10-1.

SANDRA lists the actual pages and system numbers for the Basic Series (H) Volume, but only a volume and part number for the later supplements. The user has to realize that all information about a given compound will be found in the same numbered volume of H, EI, EII, EIII, EIV, and EV. The indexes for the volume can lead to a more precise location.

TABLE 10.5
German Abbreviations and Terms for Physical Constants in *Beilsteins Handbuch der Organischen Chemie*

Abbreviction/Term	English Meaning
D	density, specific gravity
E	freezing point
F	melting point
Gew.-%	percent by weight
k(ks,kb)	ionization constant according to classical theory
K(Ks,Kb)	ionization constant according to Zwitterion theory
Kp	boiling point
Mol. Gew.	molecular weight
n	refractive index
Bildungs-konstante	equilibrium constant
Bildungs-wärme	heat of formation
Bindungs-energie	bond energy
Brechungs-index	index of refraction
Brenngeschwindigkeit	rate of combustion
Dampf-druck	vapor pressure
Dichte	density
Druck	pressure
Durchlässigkeit	permeability
Entzündungs-temperatur	ignition temperature
Erstarrungs-punkt	freezing point
Flamm-punkt	flash point
Flüssigkeit-Dampf-Gleichgewicht	liquid-vapor equilibrium
freie Enthalpie	free enthalpy (Gibbs energy)
Geschwindigkeit	velocity, rate
Gift-wirkung	toxic effect
Gitter-konstante	lattice constant
Gleichgewichts-konstante	equilibrium constant

SANDRA predicts the theoretical location of a compound in *Beilstein* even if the compound does not actually exist. Another microcomputer program which does the same thing in response to a series of questions is the "Beilstein Key". The BASIC program locates only the volume number for a compound of interest. The user must then consult the subject or formula indexes for that volume in order to find the precise location. On the other hand, SANDRA takes the user to within 40 or so pages of the correct location. Since *Beilstein* is a classified work, getting that close easily leads to the compound of interest.

A further unique feature which assists the user of *Beilstein* is the placement of coordinating page references in the top margin of the pages in the Supplementary Series. These are keyed to the volume and page numbers of the Basic Series where the same or similar compounds were discussed. These page references thus link the information in all series of a given volume. For example, the following is found at the top of page 351 of part 11 of the fifth supplement to volume 17:

Syst. No. 2479/H509; E III/IV 6278-6279

TABLE 10.5
German Abbreviations and Terms for Physical Constants in *Beilsteins Handbuch der Organischen Chemie*

Grundschwingungs-frequenz	fundamental vibrational frequency
Grund-zustand	ground state
Halbwertszeit	half-life
Hydrierungs-wärme	heat of hydrogenation
Ionisierungs-potential	ionization potential
Kern-quadrupol-kopplungs-konstante	nuclear quadrupole coupling constant
Kraft-konstante	force constant
Leitfähigkeit	conductivity
Löslichkeit	solubility
Lösungs-wärme	heat of solution
Mikrowellen-Spektrum	microwave spectrum
Mischungs-enthalpie	enthalpy of mixing
Normaldruck	standard pressure
Oberflächen-spannung	surface tension
Röntgen-Diagramm	X-ray pattern
Schemlz-punkt	melting point
Schmelz-wärme	heat of fusion
Schwingungs-relaxations-zeit	vibrational relaxation time
Siedepunkt	boiling point
Stabilitäts-konstante	stability constant
Trägheits-moment	moment of inertia
Überführungs-zahl	transport number
Übergangs-wahrscheinlichkeit	transition probability
Valenz-winkel	bond angle
Wärme-übergangs-koeffizient	heat transfer coefficient
Wellen-länge	wave length
Wellen-zahl	wave number
Zimmer-temperature	room temperature

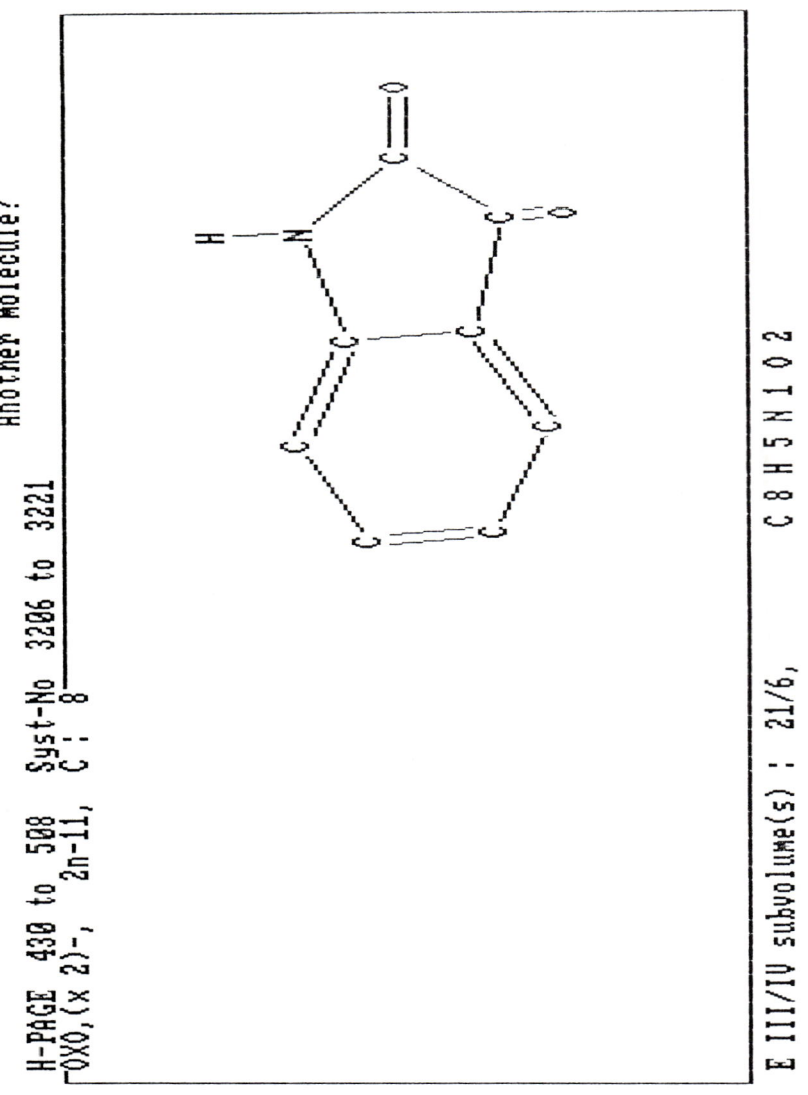

Figure 10-1
Use of SANDRA to Locate Isatin in the *Beilstein Handbook of Organic Chemistry*

This indicates that the volume 17 Hauptwerk has information on the compound on page 509 and that the Ergänzungswerk volume for volume 17 Supplement III/IV also has information on pages 6278-6279. Thus, if there have been any new developments concerning a given compound in the period of the Supplementary Series, the coordinating page numbers can help to access that information by moving forward through the supplements using the H page number.

Another method of locating information in *Beilstein* is to use the references listed in the physical constants sections of smaller one-volume handbooks such as the *CRC Handbook of Chemistry and Physics*, *Lange's Handbook of Chemistry*, or the *Aldrich Catalog Handbook of Fine Chemicals*. For example, the entry for isatin in the *CRC Handbook* has in the last column of information the designation "B21^4,4981". This means that isatin can be found in volume 21, EIV, on page 4981. In *Lange's Handbook of Chemistry*, the compound is listed as Indole-2,3-dione, with *Beilstein* designation "21,432". In the *Aldrich Catalog* it is also "**21**,432". Thus, the latter two handbooks have the same reference to volume 21 of the Basic Series, while the *CRC* shows a later reference to the fourth Supplementary Series to volume 21.

Once the proper place in *Beilstein* has been found, the following information about the compound is given:

- Molecular and structural formula
- History
- Occurrence (V. or Vorkommen)
- Formation (B. or Bildung)
- Preparation (Darst. or Darstellung)
- Physical properties
- Chemical reactions
- Applications
- Chemical behavior
- Addition compounds and salts
- Transformation products of unknown structure.

An example of a partial entry from the printed Beilstein is shown in Figure 10-2.

10.4.2.3 Beilstein Databases

The Beilstein database [1911] has been available on STN International since late 1988, with DIALOG loading the file in late 1989, and ORBIT planning to load it also. The first group of compounds included in the online file on STN was heterocycles, followed by acyclic compounds. It is estimated that when the database is complete in 1992, it will contain 3.5 to 4 million compounds, about 30 percent of which will not be in the CAS Registry System. CAS registry numbers are now included in the records of the vast majority of compounds in the Beilstein database on STN. Chemical and physical properties, keywords, and structures are both searchable and displayable.

The Beilstein database includes physical data for over 75 different properties. Since the values of some properties vary with pressure or temperature, additional data fields provide the capability to search a known value with the attendant conditions of the search. For example,

=>S 240/BP(P)760/BP.P

would find compounds in the database with boiling points of 240° C at 760 Torr pressure. Thus, over 140 data fields are searchable for physical properties. Table 10.6 indicates what a rich source of data the Beilstein database is.

---

432   HETERO: 1 N. — DIOXO-VERBINDUNGEN $C_nH_{2n-11}O_2N$ [Syst. No. 3206

2. *2.3 - Dioxo - indolin* bezw. ***2 - Oxy - 3 - oxo - indolenin*** $C_8H_5O_2N$, Formel I bezw. II, ***Isatin***. Literatur über Isatin: G. HELLER, Über Isatin, Isatyd, Dioxindol und

I. [structure: benzene ring fused with 5-membered ring containing NH, positions labeled 4,5,6,7,1; —CO 3 oder β; —CO 2 oder α]

II. [structure: benzene ring fused with 5-membered ring containing N, positions labeled 4,5,6,7,1; —CO 3 oder β; —C·OH 2 oder α]

Indophenin (AHRENS-HERZsche Sammlung chem. und chem.-techn. Vorträge, Neue Folge, Heft 5 [Stuttgart 1931]).

Vorkommen, Bildung und Darstellung.

V. Geringe Mengen Isatin wurden in einigen Indigo-Proben aus Java nachgewiesen (PERKIN, *Pr. chem. Soc.* **23** [1907], 30; vgl. P., THOMAS, *Soc.* **95**, 801). — B. Isatin, das Lactam der Isatinsäure (Bd. XIV, S. 648), bildet sich aus dieser beim Ansäuern ihrer alkal. Lösungen sowie beim Erwärmen ihrer wäßr. Lösung (ERDMANN, *J. pr.* [1] **24**, 14); daher sind nachstehend auch diejenigen Bildungsweisen angegeben, die primär zu isatinsauren Salzen führen. Isatin entsteht beim Kochen von 2-Nitro-phenylpropiolsäure mit wäßr. Alkalien oder Erdalkalien (BAEYER, *B.* **13**, 2259). Aus 2-Nitro-phenylglyoxylsäure durch Reduktion mit Ferrosulfat und Natronlauge und Ansäuern der erhaltenen Lösung (CLAISEN, SHADWELL, *B.* **12**, 353). Neben anderen Verbindungen beim Erhitzen von 2-Nitro-phenylbrenztraubensäure mit Natronlauge (REISSERT, *B.* **30**, 1038). In geringer Menge beim Erwärmen von Oxanilsäurethioamid (Bd. XII, S. 288) und von Thiooxanilid (Bd. XII, S. 289) mit konz. Schwefelsäure auf dem Wasserbad (REISSERT, *B.* **37**, 3724). Beim Erwärmen von Oxalsäure-bis-phenylimidchlorid (Bd. XII, S. 291) mit konz. Schwefelsäure auf dem Wasserbad (BAUER, *B.* **40**, 2653; vgl. D. R. P. 193633; *C.* **1908** I, 1001; *Frdl.* **9**, 518). Beim Erhitzen von N-Phenylglycin mit Alkalihydroxyden auf mehr als 200° unter beschränktem Zutritt von Luft (BASF, D. R. P. 105102; *Frdl.* **5**, 394). Durch Oxydation von Indoxyl (S. 69) mit Mangandioxyd in alkal. Lösung (BASF, D. R. P. 107719; *Frdl.* **5**, 397). Bei der Oxydation von Carbostyril (S. 77) mit Kaliumpermanganat in alkal. Lösung (FRIEDLAENDER, OSTERMAIER, *B.* **14**, 1921). Aus 2.4-Dioxo-3-oximino-1.2.3.4-tetrahydro-chinolin beim Kochen mit konz. Salzsäure (BAEYER, HOMOLKA, *B.* **16**, 2217). Aus Isatin-α-anil (S. 439) durch Erhitzen mit verd. Schwefelsäure (GEIGY & Co., D. R. P. 113979; *Frdl.* **6**, 583). Aus N-Oxy-dioxindol (Syst. No. 3239) beim Erhitzen für sich, beim Behandeln mit wäßr. Alkalien oder beim Kochen mit Acetanhydrid und Verseifen des entstandenen N-Acetyl-isatins durch siedendes Wasser (KALLE & Co., D. R. P. 184693, 184694; *C.* **1907** II, 198, 199; *Frdl.* **8**, 427; vgl. HELLER, *B.* **42**, 471, 474, 475). Bei der Oxydation von N-Oxy-indol-α-carbonsäure mit Chromessigsäure (REISSERT, *B.* **29**, 657). Durch Oxydation von Indoxylsäure (Syst. No. 3337) mit Salpetersäure oder mit Kaliumpermanganat, Kupfersulfat oder Kaliumferricyanid in alkal. Lösung (BASF, D. R. P. 107719). Durch Behandeln von 3-Amino-oxindol mit salpetriger Säure, Eisenchlorid oder Kupferchlorid (BAEYER, *B.* **11**, 1228). Bei der Oxydation von Indigo (Syst. No. 3599) mit Salpetersäure oder Chromsäure (ERDMANN, *J. pr.* [1] **24**, 10; LAURENT, *A. ch.* [3] **3**, 372). Durch Reduktion von Anthroxansäure (Syst. No. 4308) mit Ferrosulfat und verd. Ammoniak (SCHILLINGER, WLEÜGEL, *B.* **16**, 2224). Aus Isatogensäureäthylester beim Auflösen in konz. Schwefelsäure und nachfolgendem Verdünnen mit Wasser (BAEYER, *B.* **14**, 1742) oder (neben o.o'-Azobenzoesäure) beim Behandeln mit Soda-Lösung (B., *B.* **15**, 55).

Figure 10-2

Selected pages of the Isatin Listing in the *Beilstein* Basic Series and the Third/Fourth Supplementary Series

CHEMICAL AND PHYSICAL PROPERTIES 193

Darstellung aus Indigo: Man rührt 100 Tle. Indigo mit etwas verd. Natronlauge an und behandelt anfangs unter Kühlen und Rühren, zuletzt unter Erwärmen, mit 600 Tln. ca. 15%iger Salpetersäure, die ca. 60 Tle. Chromsäure enthält, bis das Reaktionsgemisch braun geworden ist, läßt erkalten, filtriert, löst den Rückstand in Natronlauge und fällt in der Hitze mit Säure (DIEZ & Co., D. R. P. 229815; *C.* **1911** I, 360; *Frdl.* **10**, 353). Ältere Verfahren zur Darstellung durch Oxydation von Indigo mit Salpetersäure oder Chromsäure: GERICKE, *J. pr.* [1] **95**, 177; KNOP, *J. pr.* [1] **97**, 86; v. SOMMARUGA, *A.* **190**, 369; FORRER, *B.* **17**, 976; KNAPE, *J. pr.* [2] **43**, 211; vgl. a. HOFMANN, *A.* **53**, 11; SCHÜTZENBERGER, *Bl.* [2] **4**, 170; KOLBE, *J. pr.* [2] **30**, 469 Anm.

Zur technischen Darstellung benutzt man die Verseifung von Isatin-α-anil mit verd. Schwefelsäure (s. o.) und die Behandlung von Oximinoessigsäureanilid mit konz. Schwefelsäure (SANDMEYER, *Helv.* **2** [1919], 234; GEIGY A.-G., D. R. P. 320647; *Frdl.* **13**, 450); vgl. darüber G. COHN in F. ULLMANNS Enzyklopädie der technischen Chemie, 2. Aufl., Bd. VI [Berlin-Wien 1930], S. 268.

Physikalische Eigenschaften.

Gelbrote Prismen. Monoklin prismatisch (BODEWIG, *J.* **1879**, 477; vgl. *Groth, Ch. Kr.* **5**, 564). F: 200—201° (BAEYER, OEKONOMIDES, *B.* **15**, 2094), 203,5° (HARTLEY, DOBBIE, *Soc.* **75**, 644). Verbrennungswärme bei konstantem Volumen: 867,4 kcal/Mol (D'ALADERN,

---

Syst. No. 3206] ISATIN 433

C. r. **116**, 1458). Leicht löslich in siedendem Alkohol, schwerer in Äther, löslich in siedendem Wasser mit rotbrauner Farbe, schwer löslich in kaltem Wasser (LAURENT, *A. ch.* [3] **3**, 374). Löslichkeit in Alkalien s. unten. Absorptionsspektrum in Alkohol: H., D., *Soc.* **75**, 647, 656; KORCZYŃSKI, MARCHLEWSKI, *Anz. Krakau. Akad.* **1902**, 248, 259; vgl. *B.* **35**, 4337. Elektrische Leitfähigkeit in Pyridin-Lösung: HANTZSCH, CALDWELL, *Ph. Ch.* **61**, 232.

Chemisches Verhalten.

Vgl. auch die Umsetzungen der Isatinsäure, Bd. XIV, S. 648. — Isatin gibt bei der Oxydation mit Chromtrioxyd in Eisessig Isatosäureanhydrid (Formel I; Syst. No. 4298) (KOLBE, *J. pr.* [2] **30**, 85, 469; vgl. ERDMANN, *B.* **32**, 2161). Liefert bei der Reduktion mit Natriumamalgam in alkal. Lösung Dioxindol (Formel II; Syst. No. 3239) (BAEYER, KNOP, *A.* **140**, 9; KN., *J. pr.* [1] **97**, 65), in schwefelsaurer Lösung Isatan (Formel III; Syst. No. 3636) (KN.,

I. [Formel] II. [Formel] III. [Formel]

J. pr. [1] **97**, 81; *J.* **1865**, 584; vgl. WAHL, HANSEN, *C. r.* **178** [1924], 394; HANSEN, *A. ch.* [10] **1** [1924], 132). Wird in alkoh. Lösung durch überschüssiges Schwefelammonium zu Isatan (ERDMANN, *J. pr.* [1] **24**, 15; vgl. W., HA.; HA.), durch geringere Mengen Schwefelammonium zu Isatyd (Formel IV; Syst. No. 3637) reduziert (LAURENT, *A. ch.* [3] **3**, 382;

IV. [Formel] V. [Formel] VI. [Formel]

vgl. KOHN, KLEIN, *M.* **33** [1912], 931; KOHN, OSTERSETZER, *M.* **37** [1916], 26; W., HA.; HA., *A. ch.* [10] **1**, 127; SUMPTER, *Am. Soc.* **54** [1932], 2917; STOLLÉ, MERKLE, *J. pr.* [2] **139** [1934], 329); Isatyd entsteht auch bei der Reduktion von Isatin mit Zink und sehr verd. Schwefelsäure (LAU., *A.* **72**, 285) oder mit Zinkstaub und Eisessig (BAEYER, *B.* **12**, 1309; HELLER, *B.* **37**, 943), während man bei der Reduktion mit Zinkstaub und siedender verdünnter

VII. [Formel] VIII. [Formel]

Salzsäure Dioxindol erhält (B., *B.* **12**, 1309). Durch Reduktion mit Zinn und konz. Salzsäure erhielt KNOP (*J. pr.* [1] **97**, 83) „Indiretin" (S. 436). Isatin gibt beim Chlorieren in wäßr.

Figure 10-2
Selected pages of the Isatin Listing in the *Beilstein* Basic Series and the Third/Fourth Supplementary Series

Suspension oder in Eisessig-Lösung 5-Chlor-isatin (LAU., *A. ch.* [3] **3**, 378; HOFMANN, *A.* **53**, 12 Anm.; MARCHLEWSKI, *B.* **29**, 1033; LIEBERMANN, KRAUSS, *B.* **40**, 2500; KALLE & Co., D. R. P. 206537; *C.* **1909** I, 1061; *Frdl.* **9**, 595). Bei der Einw. von Kaliumchlorat und Salzsäure entsteht Chloranil (Ho., *A.* **52**, 65). Gibt mit Bromwasser 5-Brom-isatin (Ho., *A.* **53**, 40; LIE., KR., *B.* **40**, 2501). Beim Erhitzen von Isatin mit Jodwasserstoffsäure (D: 1,4) auf 140—150° erhielt SCHÜTZENBERGER (*Bl.* [2] **4**, 171) ,,Isaton", ,,Isatochlorin" und ,,Isatopurpurin" (S. 436). Durch Einw. von Stickoxyden auf in Wasser suspendiertes Isatin entsteht Nitrosalicylsäure; beim Eindampfen des Reaktionsgemisches erhält man Pikrinsäure (Ho., *A.* **115**, 280, 282). BAEYER, KNOP (*A.* **140**, 4) erhielten bei der Einw. von Stickoxyden in Gegenwart von Alkohol Benzoesäure. Isatin liefert beim Nitrieren mit Kaliumnitrat und konz. Schwefelsäure in der Kälte 5-Nitro-isatin (BAEYER, *B.* **12**, 1312; LIEBERMANN, KRAUSS, *B.* **40**, 2501; vgl. RUPE, KERSTEN, *Helv.* **9** [1926], 579). Beim Erwärmen mit rauchender Schwefelsäure entsteht Isatin-sulfonsäure-(5) (Syst. No. 3381) (GEIGY & Co., D. R. P. 122233; *C.* **1901** II, 251; *Frdl.* **6**, 846; vgl. MARTINET, DORNIER, *C. r.* **172** [1921], 330). Isatin liefert mit Kaliumdisulfit-Lösung sowie mit schwefliger Säure und Äthylamin bezw. Anilin die entsprechenden Salze der isatinschwefligen Säure (S. 439) (LAURENT, *J. pr.* [1] **28**, 337; **35**, 112; SCHIFF, *A.* **144**, 49; HASLINGER, *B.* **41**, 1447). Isatin löst sich in konz. Kalilauge in der Kälte mit violettroter Farbe, die beim Erhitzen unter Bildung von isatinsaurem Kalium (vgl. Bd. XIV, S. 648) in Gelb übergeht (ERDMANN, *J. pr.* [1] **24**, 13; LAURENT, *A. ch.* [3] **3**, 376). Von verd. Soda-Lösung wird Isatin erst bei stärkerem Erwärmen unter Überführung in isatinsaures Natrium gelöst (BACK, Priv.-Mitt.). Beim Einleiten von Ammoniak-Gas in eine siedende absolut-alkoholische Lösung oder äther. Suspension von Isatin erhält man Imesatin (Formel V) (S. 440) (LAURENT, *A. ch.* [3] **3**, 484; *J. pr.* [1] **35**, 126; vgl. REISSERT, HOPPMANN, *B.* **57** [1924], 972, 976; vgl. a. v. SOMMARUGA, REICHARDT, *B.* **10**, 433); beim Überleiten von Ammoniak über ein schwach angefeuchtetes Isatin erhielt LAURENT (*J. pr.* [1] **35**, 122) außerdem ,,Isatimid", ,,Isatilim" und ,,Amisatim" (S. 436 u. 437). Bei der Einw. von alkoh. Ammoniak auf Isatin bei 100° unter Druck entstehen die Verbindungen $C_{16}H_{12}O_2N_4$ (S. 437), $C_{16}H_{14}O_3N_6$ (S. 437) und $C_{16}H_{11}O_2N_3$ (S. 437) (v. S., *A.* **190**, 371; **194**, 85; vgl. dazu R., H., *B.* **57** [1924], 976). Isatin gibt mit wäßrig-alkoholischem Ammoniak bei 40—50° Isamsäure (S. 442) (Formel VI), Isamid (S. 442) (Formel VII) und Imasatin (Formel VIII) (S. 442) (LAU., *A. ch.*

BEILSTEINs Handbuch. 4. Aufl. XXI.   28

E III/IV 21   Syst.-Nr. 3205—3206 / H 430—433   **4981**

B. Dioxo-Verbindungen

(Fortsetzung)

Dioxo-Verbindungen $C_nH_{2n-11}NO_2$

Dioxo-Verbindungen $C_8H_5NO_2$

1*H*-Indol-2,3-dion, Indolin-2,3-dion, Isatin $C_8H_5NO_2$, Formel I auf S. 4983 (H 432; E I 348; E II 327, 567).

Isatin liegt nach Ausweis des IR-Spektrums im festen Zustand, in der Schmelze und in Lösungen mit Dioxan, Pyridin, Aceton und Äthanol als Dion vor (*Schigorin, Ž. fiz. Chim.* **29** [1955] 1033, 1034, 1036; *C. A.* **1957** 138; s. a. *O'Sullivan, Sadler, Soc.* **1956** 2202, 2203, 2204).

B. Aus Oxalsäure-anilid-nitril beim Erwärmen mit Aluminiumchlorid in Trichlorbenzol oder beim Erhitzen in einer Schmelze von Aluminiumchlorid und Natriumchlorid auf

Figure 10-2
Selected pages of the Isatin Listing in the *Beilstein* Basic Series and the Third/Fourth Supplementary Series

110° (*I.G. Farbenind.*, D.R. P. 541 924 [1928]; Frdl. **18** 643). Beim Erwärmen von 1-Acetylindolin-3-on mit Eisen(III)-chlorid in wss. Salzsäure (*Spencer*, J. Soc. chem. Ind. **50** [1931] 63 T). Beim Erhitzen von 3-Hydroxy-indolin-2-on mit Ammoniumnitrat in Essigsäure (*Klein*, Am. Soc. **63** [1941] 1474).

Atomabstände und Bindungswinkel (aus dem Röntgen-Diagramm ermittelt): *Goldschmidt, Llewellyn*, Acta cryst. **3** [1950] 294, 298.

Rote Kristalle (nach Sublimation bei 180°/1 Torr); F: 205° (*Piozzi, Favini*, R.A.L. [8] **18** [1955] 647, 651). Monoklin; Raumgruppe $P2_1/c$ $(=C_{2h}^5)$; aus dem Röntgen-Diagramm ermittelte Dimensionen der Elementarzelle: a = 6,19 Å; b = 14,46 Å; c = 7,17 Å; β = 94,82°; n = 4 (*Goldschmidt, Llewellyn*, Acta cryst. **3** [1950] 294, 295). Dichte der Kristalle: 1,51 (*Cox et al.*, Pr. roy. Soc. [A] **157** [1936] 399, 401). Verbrennungsenthalpie bei 15°: *Stern, Klebs*, A. **504** [1933] 287, 296. Brechungsindices der Kristalle: *Cox et al.* IR-Spektrum (3500 — 3000 cm⁻¹) der Kristalle, der Schmelze und des Dampfes (bei 260°): *Schigorin*, Ž. fiz. Chim. **29** [1955] 1033, 1035; C. A. **1957** 138. IR-Spektrum (Nujol; 2000 — 700 cm⁻¹): *Bergmann*, Am. Soc. **77** [1955] 1549. IR-Banden (KBr) im Bereich von 3445 cm⁻¹ bis 660 cm⁻¹: *O'Sullivan, Sadler*, Soc. **1956** 2202, 2203; von 3420 cm⁻¹ bis 1615 cm⁻¹: *Sadler et al.*, Soc. **1959** 667, 669; von 1620 cm⁻¹ bis 1295 cm⁻¹: *O'Sullivan, Sadler*, J. org. Chem. **22** [1957] 283, 285. IR-Banden von Lösungen in 1,1,2,2-Tetrachlor-äthan im Bereich von 3420 cm⁻¹ bis 1610 cm⁻¹: *Sad. et al.*; in Chloroform von 3450 cm⁻¹ bis 1320 cm⁻¹: *O'Su., Sa.*, Soc. **1956** 2203; in Dioxan, in Pyridin, in Aceton und in Äthanol von 3210 cm⁻¹ bis 1625 cm⁻¹: *Sch.*, l. c. S. 1034. CO-Valenzschwingungsbanden (CHCl₃): 1755 cm⁻¹ und 1740 cm⁻¹ (*O'Sullivan, Sadler*, Soc. **1957** 2839). Absorptionsspektrum von Lösungen in Äthanol (200 — 500 nm): *Dabrowski, Marchlewski*, Bl. [4] **53** [1933] 946, 948; *Ault et al.*, Soc. **1935** 1653, 1654; *Mangini, Passerini*, G. **85** [1955] 840, 842, 858; s. a. *Jones et al.*, J. Assoc. agric. Chemists **38** [1955] 949, 958; in Methanol (220 — 325 nm): *Julian, Printy*, Am. Soc. **75** [1953] 5301, 5303; in Wasser (220 — 550 nm): *Ault et al.*; s. a. *Jo. et al.*; in konz. Schwefelsäure (200 — 500 nm): *Ma., Pa.*; in wss. Salzsäure (220 nm bis 400 nm): *Jo. et al.*; in Natriumäthylat enthaltendem Äthanol (220 — 600 nm): *Ault et al.*; von gepufferten äthanol. Lösungen vom pH 2,7, pH 9,7 und pH 12,8 (200 — 400 nm): *Dobrinškaja, Neĭman*, Izv. Akad. S.S.S.R. Ser. fiz. **14** [1950] 520, 523; C. A. **1951** 3240. Absorptionsmaxima: 292 nm und 406 — 410 nm [Dioxan] bzw. 208 nm, 242 nm und 302 — 304 nm [W.] (*Ma., Pa.*), 419 nm [DMF], 550 nm [geringe Mengen wss. Tetraäthylammonium-hydroxid enthaltendes DMF] bzw. 513 nm [geringe Mengen wss. Tetraäthylammonium-hydroxid enthaltendes A.] (*Sawicki et al.*, Anal. Chem. **31** [1959] 2063). Redoxpotential in wss. Äthanol (polarographisch ermittelt): *Cassebaum*, Z. El. Ch. **62** [1958] 426, **58** [1954] 515, 518; *Langenbeck et al.*, J. pr. [4] **4** [1957] 136, 138. Polarographie: *Sumpter et al.*, J. org. Chem. **14** [1949] 713, 714, **16** [1951] 1777, 1778; *Korschunow et al.*, Zavod. Labor. **15** [1949] 1287; C. A. **1950** 3845; *Korschunow, Schtschen-*

Figure 10-2
Selected pages of the Isatin Listing in the *Beilstein* Basic Series and the Third/Fourth Supplementary Series

The default display on STN includes the substance identification information (the IDE display format) and the display fields which include the hit terms. Other display options on STN are:

- HIT   displays only the fields which contain the search terms
- FA   the field availability option, which lists all of the display fields with a tally of the number of times they occur
- OCC   displays a table of fields which contain the hit terms with the number of occurrences of each
- TRIAL   displays molecular formula, field availability.

TABLE 10.6
Physical Property Data Searchable and/or Displayable in the Beilstein Database

Property	Selected Example
Electrochemical Behavior	Redox Potential (RDXP)
Electrical Data	Dielectric Constant (DIC)
Magnetic Data	Magnetic Susceptibility (MSUS)
Multi-Component System Data	Solubility (SLB)
Mechanical Properties	Density (DEN)
Optical Data	Optical Rotatory Power (ORP)
State of Aggregation	Melting Point (MP)
Structure and Energy Parameters	Ionization Potential (IP)
Spectral Data	Infrared Spectrum (IRS)
Thermodynamic Data	Heat of Fusion (HFUS)
Transport Phenomena	Thermal Conductivity (TCND)

There are also "super" display options on STN to display all physical data (PHY), all chemical data (CHE), and all general data (GEN). The IND format shows the controlled-vocabulary terms used in a record.

The Beilstein database can be searched with substructure, textual, or numeric search options. Thus, the techniques for searching chemical compounds which were described in chapters 7-8 are also available in the Beilstein database. In most cases, the search proceeds exactly as it would in the Registry File. Thus, field labels such as CN (Chemical Name), MF (Molecular Formula), PG (Periodic Group), and FW (Formula Weight) can all be employed with the appropriate search statement. Note, however, that chemical name segments can be searched in Beilstein with the label CNS, whereas the Registry File performs such searches only with unlabeled terms in the Basic Index. A further option in Beilstein is the possibility of including (or excluding) a compound in an answer set if it has a particular element symbol in the Molecular Formula, for example, =>S N/ELS

Searching for the existence of a physical property in the record of a substance can be done with controlled terminology. This implies that index terms indicate the presence of a property in a primary source without the actual data being present in the Beilstein record. On the other hand, the abbreviations for properties searched with the Field Availability (FA) label will limit the answer set to those records where specific numeric data are listed. In order to encompass both types of searches (either the presence of actual data or an indication of its existence) the search phrase can be labeled with the /PH (Property Hierarchy) label, for example, =>S BP/PH

Figure 10-3 indicates the powerful techniques available for searching the Beilstein database. In the search, we are looking for oximes with one halogen atom (designated as an X) and melting points between 200° and 225° C.

Also available for lease is the Beilstein Inhouse System, a subset of the full database which offers structures and substructures, citations to the literature, and the Beilstein registry number. The Inhouse System thus lacks data for preparation and methods and physical and chemical properties. However, the Beilstein registry number can be used to link the Inhouse system to the full Beilstein database offered by the online search vendors. Unlike the CAS Registry System, the Beilstein Registry Con-

CHEMICAL AND PHYSICAL PROPERTIES 197

```
=> s 1/x and oxime?/cns and 200-225/mp
        217240    1/X
          4684    OXIME?/CNS
        143666    200-225/MP
L19         22    1/X AND OXIME?/CNS AND 200-225/MP
=> d
L19    ANSWER 1 OF 22
BRN    385107 Beilstein
MF     ***C21 H14 Cl N O2***
CN     ***6-chloro-2,3-diphenyl-chromen-4-one oxime***
       6-Chlor-2,3-diphenyl-chromen-4-on-oxim
FW     347.80
SO     2-17-00-00420
LN     18173
```

```
                :C.          .O                                    C:
              :    .       .    .                               .    :
           C:           .C.              .C.............C.        :C
           .            :                :              :         .
           .            :                :       . C :            .
           C:           .C.              .C....C.       : CC:.    :C
              .       :                :    :        .    .  .   .
                :C.          .C.       :    :        .        .C:
  .                                    :    :        .
  Cl                                       C.        :C
                                   NOH          . C:
```

Melting Point:			
Value (MP) (Cel)	Solv.(MP.SOL)	Ref.	Note
206.00 - 206.50	methanol	2	1

Reference(s):
2. Wittig, Liebigs Ann.Chem. 446 <1926>, 190, CODEN: LACHDL

Note(s):
1. Handbook Data

FIGURE 10-3
Use of the Beilstein Database for Numeric Data Searching.

nection Table allows structure and substructure searches with stereochemical specifications through the use of **ROSDAL** (Representation of Organic Structure Descriptions Arranged Linearly) structure codes. ROSDAL translates organic structures into an alphanumeric string of characters. Both the Inhouse database and the online file include structures which have been abstracted for the printed *Beilstein Handbook of Organic Chemistry*, but not yet published.

10.4.3 *Gmelin Handbook of Inorganic Chemistry*

The *Gmelin Handbuch der Anorganischen Chemie* [797] is now in its eighth edition. *Gmelin* is perhaps the oldest continuing data compilation, having begun publication in 1817. *Gmelin* is to inorganic and organometallic chemistry what *Beilstein* is to organic chemistry. Both sets contain the most nearly comprehensive collection of information in existence for their respective areas of coverage. Since 1982 *Gmelin* has been published entirely in English under the title *Gmelin Handbook of Inorganic Chemistry*. The set covers every element in existence, including carbon. The Gmelin Institute is also working to produce an online version of *Gmelin*, called GMELIN ONLINE Data System, or GOLD.

10.4.3.1 Coverage and Arrangement of Gmelin

The *Gmelin Handbook of Inorganic Chemistry* encompasses the entire body of published knowledge in the field of inorganic chemistry from its beginnings in the mid-18th Century to the present. On the verso of the title page of each volume is printed the date up to which (bis) the coverage of that volume extends. Supplements (the *Ergänzungsband* volumes) to the earlier volumes of the eighth edition (which began in 1922) have been issued periodically. In addition, there is a new Supplement Series (the *Ergänzungswerk* volumes) which covers important topics and is not limited to a single element and its compounds.

Gmelin was, of course, originally published in German, but over the years, more and more English-language chapters and sections began to appear. English had been used in the tables of contents and subsection headings for quite some time prior to 1982, when the switch to English was made for the entire contents of future volumes.

Information is arranged in *Gmelin* by means of a classification scheme based on the chemical elements. Each element (plus ammonium) or group of elements has been assigned a system number (See Table 10.7). Compounds are assigned to the system which has the highest number. (This is called the **principle of last position**.) The lowest system number is 1, "Rare Gases," and the highest is 71, "Transuranium Elements". Closely related elements, such as the rare earths, the rare gases, and the transuranium elements, share Gmelin system numbers. For many years, libraries shelved *Gmelin* in system number order. However, the appearance of the New Supplement Series complicated matters. Finally, in 1978 the New Supplement Series was integrated into the Main Series filing order by placing all volumes in alphabetical order by the element symbol.

The principle of last position separates salts from the corresponding acids in their placement in *Gmelin*. Nevertheless, once the proper volume is found, the information is presented in a standard order:

- Discussions of the element
- Its binary compounds with lower-numbered elements
- Compounds consisting of more than two elements.

Within a given volume, the compounds are arranged from the lowest to the highest of the remaining system numbers in the compounds. Accordingly, Cu_3P would precede Cu-As alloys, which in turn would precede Cu_3BiS_3 in the Copper volume, because Cu is system number 60, P is 16, As is 17, and Bi is 19. For a given compound, one can find such data as occurrence and methods of preparation, physical properties, and chemical properties (including toxicology).

10.4.3.2 Access to the Information in Gmelin

One means of finding the proper place in the *Gmelin Handbook of Inorganic Chemistry* is to consult the periodic table which is published in later volumes on the inside front cover. Listed there are the system numbers for each element. A second option within the volumes is the list of elements printed on the inside back cover. By applying the *Gmelin* principle of last position, the appropriate volume can then be selected.

Another approach to *Gmelin* is to use the Formula Index [799]. The original 12-volume Formula Index covered elements, compounds, and ions which had appeared

TABLE 10.7
The *Gmelin Handbook of Inorganic Chemistry* System Numbers

System No.	Element	System No.	Element
1	Rare gases	36	Gallium
2	Hydrogen	37	Indium
3	Oxygen	38	Thallium and isotopes
4	Nitrogen	39	Rare Earths
5	Fluorine	40	Actinium and isotopes
6	Chlorine	41	Titanium
7	Bromine	42	Zirconium
8	Iodine	43	Hafnium
8a	Astatine	44	Thorium and isotopes
9	Sulfur	45	Germanium
10	Selenium	46	Tin
11	Tellurium	47	Lead and isotopes
12	Polonium and isotopes	48	Vanadium
13	Boron	49	Niobium
14	Carbon	50	Tantalum
15	Silicon	51	Protactinium and isotopes
16	Phosphorus	52	Chromium
17	Arsenic	53	Molybdenum
18	Antimony	54	Tungsten
19	Bismuth and isotopes	55	Uranium and isotopes
20	Lithium	56	Manganese
21	Sodium	57	Nickel
22	Potassium	58	Cobalt
23	Ammonium	59	Iron
24	Rubidium	60	Copper
25	Cesium	61	Silver
25a	Francium	62	Gold
26	Beryllium	63	Ruthenium
27	Magnesium	64	Rhodium
28	Calcium	65	Palladium
29	Strontium	66	Osmium
30	Barium	67	Iridium
31	Radium and isotopes	68	Platinum
32	Zinc	69	Technetium
33	Cadmium	70	Rhenium
34	Mercury	71	Transuranium ele⋅
35	Aluminum		

10.4.3.2   Access to the Information in Gmelin

One means of finding the proper place in the *Gmelin Handbook of Inorganic Chemistry* is to consult the periodic table which is published in later volumes on the inside front cover. Listed there are the system numbers for each element. A second option within the volumes is the list of elements printed on the inside back cover. By applying the *Gmelin* principle of last position, the appropriate volume can then be selected.

Another approach to *Gmelin* is to use the Formula Index [799]. The original 12-volume Formula Index covered elements, compounds, and ions which had appeared

TABLE 10.7
The *Gmelin Handbook of Inorganic Chemistry* System Numbers

System No.	Element	System No.	Element
1	Rare gases	36	Gallium
2	Hydrogen	37	Indium
3	Oxygen	38	Thallium and isotopes
4	Nitrogen	39	Rare Earths
5	Fluorine	40	Actinium and isotopes
6	Chlorine	41	Titanium
7	Bromine	42	Zirconium
8	Iodine	43	Hafnium
8a	Astatine	44	Thorium and isotopes
9	Sulfur	45	Germanium
10	Selenium	46	Tin
11	Tellurium	47	Lead and isotopes
12	Polonium and isotopes	48	Vanadium
13	Boron	49	Niobium
14	Carbon	50	Tantalum
15	Silicon	51	Protactinium and isotopes
16	Phosphorus	52	Chromium
17	Arsenic	53	Molybdenum
18	Antimony	54	Tungsten
19	Bismuth and isotopes	55	Uranium and isotopes
20	Lithium	56	Manganese
21	Sodium	57	Nickel
22	Potassium	58	Cobalt
23	Ammonium	59	Iron
24	Rubidium	60	Copper
25	Cesium	61	Silver
25a	Francium	62	Gold
26	Beryllium	63	Ruthenium
27	Magnesium	64	Rhodium
28	Calcium	65	Palladium
29	Strontium	66	Osmium
30	Barium	67	Iridium
31	Radium and isotopes	68	Platinum
32	Zinc	69	Technetium
33	Cadmium	70	Rhenium
34	Mercury	71	Transuranium elements
35	Aluminum		

in *Gmelin* by the end of 1974. An 8-volume supplement, which brought the coverage up to 1979, was published in 1986. Supplements continue to appear for the period 1980-87. In addition to the printed Formula Index and its supplements, the user of *Gmelin* now has available the Gmelin Formula Index database [612]. The database is actually more than an online copy of the printed Formula Index, since it incorporates the "Complete Catalog" of published *Gmelin* volumes. The "Catalog" includes bibliographic and other information such as the language of publication and the last year of literature coverage of the volume. In addition, the entry contains an abstract which describes the hundreds of compounds contained in the volume. An enhancement planned for the GFI database is the addition of CAS registry numbers.

Table 10.8 shows the searchable fields in File GFI, STN's version of the database. It is also available on DIALOG.

10.5 SMALLER COLLECTED WORKS AND HANDBOOKS

There are hundreds of handbooks published in science and technology. Since most of the handbooks are vying for the same market, they cover essentially the same compounds and the same properties for those compounds. The data in the handbooks are most often arranged by the names of the compounds. This may lead to complications due to different names being used for the same substance in different handbooks (as we saw earlier with isatin) or even a change of names for the compound from one edition of a handbook to another. A few printed handbooks are arranged by the values of the physical properties themselves, thus allowing a researcher to use a value measured in the laboratory as a way of identifying an unknown. Some of those are discussed in section 10.5.2.

10.5.1 Important Smaller Collected Works and One-Volume Printed Handbooks

There are many competing data compilations smaller than *Gmelin*, *Beilstein*, or *Landolt-Börnstein*. Most of them have one thing in common: they copy their data from larger, critically-evaluated sets. For organic compounds, this is often *Beilstein*, as was shown earlier to be the case with the *Lange's*, *CRC*, and *Aldrich* handbooks. The most successful of the smaller compilations or handbooks have gone through many editions. Since an attempt is made to keep their size to a few volumes or less, some earlier data must inevitably be dropped as important new data are added to the next edition.

Perhaps the best known one-volume handbook in the world is the *CRC Handbook of Chemistry and Physics* [179]. A new edition of the *CRC* is now published each year. Thus, the 1988-1989 volume was the 69th edition, billed as the 75-year anniversary edition. Obviously, not all sections of a handbook containing nearly 2500 pages could be thoroughly revised each year, but the publication of such frequent editions of the work does allow selected sections to be revised annually. The information in the *CRC Handbook of Chemistry and Physics* is divided into Mathematical tables, Elements and inorganic compounds, Organic Compounds, General Chemical

TABLE 10.8
Search Examples from STN's GFI (Gmelin Formula Index) Database

SEARCH FIELD NAME	SEARCH CODES	SEARCH EXAMPLES	DISPLAY CODES
ACCESSION NUMBER			AN
BASIC INDEX: (1)	NONE	S CRYSTALLIZATION	AB,CI,CT,ELN,MF,TI
ABSTRACT TEXT (2)	(or	S ISOTOPE(W)SEPARATION	
COMPOUND CLASS IDENTIFIER	/BI)	S GLASSES	
CONTROLLED TERMS		S CHROMIUM	
ELEMENT NAME (2)		S SI	
MOLECULAR FORMULA		S LINB2O3	
TITLE OF HANDBOOK (2)		S TRANSURANIUM(W) ELEMENTS	
FILE SEGMENT	/FS	S CAT/FS	not displayed
UPDATE CODE (3)	/UP	S 870220/UP	not displayed
INDEX FIELDS			
ATOMIC COUNT (4)	/ATC	S 5-7/ATC	not displayed
CHARGE (4)	/CHA	S 2-3/CHA	not displayed
COMPOUND CLASS IDENTIFIER	/CI	S SOLUTIONS/CI	CI
CITATION TITLE	/CIT	S AF?/CIT	CIT
CONTROLLED TERMS	/CT	S MELTING POINT/CT	CT
ELEMENT COUNT	/ELC	S 2/ELC	not displayed
ELEMENT SYMBOL	/ELS	S NI/ELS	not displayed
FORMULA WEIGHT (4)	/FW	S 323,235-324,172/FW	FW
LINEARIZED STRUCTURE FORMULA	/LSF	S ALBR3/LSF	LSF
MOLECULAR FORMULA (6)	/MF	S AG4 CS0.5 I5 K0.5/MF	MF
NUMBER OF COMPONENTS	/NC	S 3/NC	not displayed
PERIODIC GROUP	/PG	S A1/PG AND A7/PG	not displayed
SYSTEM COMPONENT FORMULA (5)	/SCF	S NA B O2 - H2 O/SCF	SCF
ELEMENT COUNT using (4) standard element symbols	Element symbols	S AU=1 and 5/BE	not displayed
COMPLETE CATALOG FIELDS: (2)			
ABSTRACT			AB
CITATION TITLE	/CIT	S H?/CIT	CIT
ELEMENT NAME	/ELN	S ARSENIC/ELN	ELN
ISBN (International Standard Book Number)	/ISN	S 3-540-93530-4/ISN	ISN
LANGUAGE	/LA	S GERMAN/LA	LA
LITERATURE CLOSING YEAR	/LY	S 1972-1978/LY	LY
PUBLICATION YEAR	/PY	S 1981-1983/PY	PY
RELATED MOLECULAR FORMULA (5) (Abstract Formula)	/RMF	S D2 O/RMF	not displayed
TITLE OF HANDBOOK	/TI	S TECHNOLOGY/TI	TI

(1) With the exception of /MF, all fields in the /BI are searchable with the (L) and (W) operators.
(2) All entries within one specific Handbook are searchable with the (L) operator.
(3) Entry Date /ED is alias name for /UP.
(4) The content of this field may be an integer or a decimal number.
(5) Elements in the fields SCF and RMF have to be entered with a blank in-between for clear distinction between elements and compounds.
(6) Elements in the MF field can be entered with spaces between element symbols for clear distinction between elements and compounds.
(7) All DISPLAY field codes can be used as DISPLAY and PRINT formats.

section, General Physical Constants section, Miscellaneous, and the Index, which is a subject index.

The sections devoted to chemical compounds or elements in the *CRC Handbook* include many physical constants. In the organic section are also found biochemical substances such as steroids and amino acids, as well as some natural products. The

"General Chemical" section covers a large number of physical constants for both inorganic and organic compounds. In the "Miscellaneous" section are published references to other collections of critically evaluated data, plus the major symbols, units, and terminology recommended by IUPAC. In 1987, CRC Press published the first Student Edition of the *CRC Handbook of Chemistry and Physics* [1697].

Lange's Handbook of Chemistry [178] is currently in its 13th edition (1985). In addition to the normal constants found in such a handbook (for example, density, melting point, and boiling point), *Lange's* has sections on atomic and molecular structure, analytical chemistry, electrochemistry, spectroscopy, and thermodynamic properties. The "Miscellaneous" section includes a good deal of information on thermocouples.

Tables of Physical and Chemical Constants and Some Mathematical Functions [401] (1986) was originally published in 1911 by Kaye and Laby. Throughout its long history, the data have been selected so as to be of value not only to specialists, but to scientists working in a variety of fields. The set of mathematical tables which had appeared in previous editions was dropped from the 15th edition, since the editors felt that pocket calculators now satisfy most needs with more precision than is possible in a printed handbook. Nevertheless, the handbook does include mathematical functions and a section on statistical methods for the treatment of experimental data.

The basic division of data in the Kaye and Laby *Tables* is between "Chemistry" and "Atomic and Nuclear Physics". In the chemistry section, one finds the usual properties of inorganic and organic compounds, plus properties of solutions, properties of chemical bonds, and electrochemical sections. The "Atomic and Nuclear Physics" section contains, among other things, data on free electrons and ions in gases and electrons in atoms. The *Tables* have less thermodynamic data than either the *CRC* or *Lange's Handbooks*.

A more specialized one-volume handbook is Dean's *Handbook of Organic Chemistry* [1458] (1987). Over 4000 organic compounds are listed with properties such as density, refractive index, melting point, boiling point, flash point, and solubility in various solvents. A considerable amount of UV-visible and IR spectral data is included. In addition, there is a section dealing with electrolytes, electromotive force, and chemical equilibrium, plus solvent data and azeotropic data.

Much more comprehensive in the scope of organic chemicals covered is the *Dictionary of Organic Compounds* [667] (1982), which is really a handbook. The *Dictionary* contains data on more than 150,000 compounds, and increases in size as the basic seven-volume set is supplemented each year. In addition to providing structural diagrams and physical and chemical properties, references to the original literature for such things as synthesis, structure, and spectra have been included. Furthermore, data and information on derivatives are often given. The CAS registry number is also listed.

The compounds are arranged in the *Dictionary* by common names, but a Name Index, Molecular Formula Index, and CAS Registry Number Index are provided. The sample entry in Figure 10-4 indicates the type of information found in the *Dictionary of Organic Chemistry*.

Isatin I-00855
1H-Indole-2,3-dione, 9CI
[91-56-5]

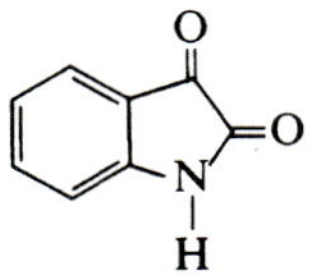

$C_8H_5NO_2$ M 147
Intermediate for indigoid dyestuffs. Orange cryst. Spar. sol. H_2O. Mp 203.5°. Subl.
▷NL7873000.
2-Oxime: Mp 198-200°.
3-Oxime: Yellow cryst. Mp 214°.
3-Semicarbazone: Mp 266° dec.
N-Ac: Mp 141°.
N-Me: see *1-Methylisatin, M-02179*
1-Methoxycarbonyl: Mp 170° dec.
1-Ethoxycarbonyl: Yellow cryst. (pet. ether). Mp 117°.

Hantzsch, A., *Ber.*, 1921, **54**, 1242
Org. Synth., 1925, **5**, 71
Popp, F.D., *Adv. Heterocycl. Chem.*, 1975, **18**, 1 (*rev*)
Gassmann, P.G. *et al*, *J. Org. Chem.*, 1977, **42**, 1344 (*synth*)

Figure 10-4
Isatin Entry in the *Dictionary of Organic Compounds*

A number of sets derived from or based on the concept of the *Dictionary of Organic Compounds* have now appeared. These include the *Dictionary of Organometallic Compounds* [712] (1984), the *Dictionary of Organophosphorous Compounds* [1654] (1987), the *Dictionary of Alkaloids* [1929] (1989), and the series of "Chapman and Hall Sourcebooks," such as *Carbohydrates* (1987). The full *Dictionary* is also available online as Heilbron, so named for one of the original compilers, Heilbron and Bunbury.

For thermodynamic data, a major one-volume handbook is the *JANAF Thermochemical Tables* [788] (1986). The third edition of the tables utilizes SI units and follows international standards for notation. Both name and formula indexes are provided. The *JANAF Thermochemical Tables* are also available on computer tape from the National Institute of Standards and Technology and as File JANAF on STN International. In addition to normal searching parameters, the JANAF file can be searched by a range of thermodynamic data values.

The *Merck Index* [177] (1989) has data on important organic and inorganic chemicals and drugs which are commercially available. Coverage was broadened in the 10th edition to include more information on biochemistry, pharmacology, toxicology, metabolism, and even agriculture and the environment. With the publication of the 11th edition in 1989, *Merck* celebrated its 100th anniversary. The user encounters in the *Merck Index* the term **monograph** to refer to the individual entries in the handbook, a use of that term which is common in the pharmaceutical area. Over 10,000 monographs which include some 8,000 stereochemically exact structures and line formulas are found in the 11th edition of the *Merck Index*. The type of information given for each substance is comparable to that found in the *Dictionary of Organic Compounds*. As in the *DOC*, the entries in *Merck* are arranged by common name, but CAS names, tradenames, and generic names are also given, resulting in over 62,000 such names in the 11th edition.

Some other smaller data compilations are *Perry's Chemical Engineers' Handbook* [1024] (1984), *Data for Biochemical Research* [882] (1986), and the *CRC Handbook of Biochemistry and Molecular Biology* [861] (1975-76).

10.5.2 Printed Compilations Which Lead From a Known Value of a Property to a Compound

There are a few printed works which are arranged in such a manner that the name of a compound is not required to use the work. Instead, the researcher uses physical or chemical data measured in the laboratory to assist in identifying an unknown. There are several indexes of this kind for spectral data. Far fewer of them exist for other types of data.

The *CRC Handbook of Data on Organic Compounds* [1460] (1989) covers over 25,000 organic compounds. Although the primary table is in alphabetical order by CAS Index Name, there is a molecular formula index and two special indexes, the melting point and boiling point indexes. In addition, compounds included in the *Handbook* are referenced to various spectral data compilations for IR, UV, NMR, and mass spectra. *Beilstein* references are also given. As with the *Dictionary of Organic Compounds*, yearly supplements are planned for this work, which is now online on STN. An earlier CRC Press work of this type is the *CRC Handbook of Tables for Organic Compound Identification* [1473] (1967).

The *Aldrich Microfiche Library of Chemical Indices* [1857] (1985) claims to be the largest single printed collection of chemicals in alphabetical and molecular formula sequences. Aldrich Chemical Company, a well-known supplier of chemicals, publishes two catalogs, both of which include considerable data on the compounds. One is the *Aldrich Catalog/Handbook of Fine Chemicals* [1037] which lists chemicals intended for use primarily in laboratory applications and as components in chemical specialties. The *Aldrich Library of Rare Chemicals* [1038] covers chemicals available only in small quantities, including research products and intermediates. Both of the catalogs are outstanding examples of **trade literature**. Trade literature is generally free, and while it has as its main purpose the marketing of the products or services offered by a given company, there is often a wealth of technical data included.

In the *Aldrich Microfiche Library of Chemical Indices*, the molecular formula index is similar in its arrangement to standard Hill formula indexes. However, inorganic chemicals precede carbon-containing compounds, and any inorganic compound with hydrogen always has the H listed first. Physical properties are listed in ascending order for formula weight, melting point, refractive index, density, boiling point, and wavelength of IR absorption maximum. In addition, the flash point is listed and references are given to the third edition of the *Aldrich Library of Infrared Spectra* [1252] (1981) and the second edition of the *Aldrich Library of NMR Spectra* [1256] (1983). Within each index, a secondary ordering is made by another property. For example, the wavelength of the IR maxima determines the order when several compounds are found with the same melting point.

Utermark and Schicke's *Melting Point Tables of Organic Compounds* [800] (1963) is another standard work with melting points arranged in ascending order. It covers approximately 3200 compounds with melting points ranging from -189.9° to 500° C.

10.6 INDEXES TO NUMERIC DATA

With so many different data compilations available, one might assume that a number of comprehensive indexes to physical and chemical properties might exist. Let us distinguish at this point between a **comprehensive index to numeric data compilations**, which is an index covering one or more sets of secondary data compilations and an **index to primary numeric data** as reported in primary journal articles and other primary literature. The latter type of index includes specialized secondary indexing journals and other works devoted exclusively to covering the literature which contains significant physical properties. It also encompasses the treatment of numeric data in the traditional subject- or mission-oriented abstracting and indexing services.

10.6.1 Comprehensive Indexes to Numeric Data Compilations

Unfortunately, there are not many comprehensive indexes to numeric data compilations in printed form. Recall that the *Comprehensive Index to Landolt-Börnstein* did not appear until 1987. There were a few attempts prior to that time to produce comprehensive indexes to more than one data set. In 1962, the *Consolidated Index of Selected Property Values: Physical Chemistry and Thermodynamics* [488] was published. It covers six different compilations of critically evaluated numerical property values. These are:

Selected Values of Properties of Hydrocarbons and Related Compounds [1930]
Selected Values of Properties of Chemical Compounds [1931]
Selected Values of Chemical Thermodynamic Properties [491]
Thermodynamic Properties of the Elements [540]
Contributions to the Data on Theoretical Metallurgy [1932]
Selected Values for the Thermodynamic Properties of Metals and Alloys [1933].

The *Comprehensive Index of API44-TRC Selected Data on Thermodynamics and Spectroscopy* [1266] (1974) provides rapid access to the data in the serial publications produced by the American Petroleum Institute's Research Project 44 and the Thermodynamics Research Center (TRC) Data Project. Included are physical properties such as boiling point, refractive index, viscosity, density, and second virial coefficients of gases, among others. Also P-V-T data, properties of the real gas, and vapor pressures are indexed, as are thermodynamic properties for 9679 compounds. Infrared, Ultraviolet, Raman, mass, and NMR spectra are also indicated. There are Wiswesser Line Notation Index entries included with the references in the Formula Index to this work. In 1990, the TRC introduced a PC version of a comprehensive index to its databases for chemistry and engineering [1934].

In 1977, CRC Press published the *CRC Composite Index for CRC Handbooks* [1430]. Nearly 60 volumes of data published in over two dozen CRC handbooks are covered in the massive index. Everything from the *CRC Handbook of Biochemistry and Molecular Biology* to the *CRC Handbook of Tables of Functions for Applied Optics* is found here. The volume is particularly useful in institutions which have their library resources scattered in many buildings. A quick check of the *CRC Composite Index* can save many steps in a fruitless search in such cases. A new edition is scheduled for publication in late 1990.

A major guide to publications of the National Institute of Standards and Technology (formerly, NBS) is *Standard Reference Data Publications: 1964-1984* [381] (1985). Included are author, materials, and properties indexes to the many NBS publications. It was noted earlier that Arny includes an "Index to Selected NBS Data Compilations" in her *The Search for Data in the Physical and Chemical Sciences* [1318] (1984).

A source which provides access to solubility data is the set of *Cumulative Indexes* to the volumes published in the IUPAC Solubility Data Series [806]. Touted as the most authoritative and comprehensive compilation of critically evaluated solubility data, the series is expected to stretch to 100 volumes when complete.

10.6.2 Secondary Indexes to Data in the Primary Literature

A well-known index to thermodynamic data is the annual *Bulletin of Chemical Thermodynamics* [1148]. A unique feature of the *Bulletin* is the inclusion of not only published reports in the primary literature, but also research which is current, but not yet published. The unpublished results are summarized in the Reports section. In the Bibliography section are papers which have been published in a given year. The contents of this section are currently divided among organic substances, organic systems (mixtures), inorganic (and metal-organic) substances, and biochemical and macromolecular subsections. The Substance-Property Index is arranged by classes of compounds and delineates the property measured for each substance with coded references to items in the Bibliography section. Whenever possible, the data are indexed from the original publications.

The *Thermophysical Properties Research Literature Retrieval Guide, 1900-1980* [1118] (1982) covers mostly primary literature for 44,338 substances, including elements; inorganic compounds; organic compounds and polymeric materials; alloys, intermetallic compounds, and cements; oxide mixtures and minerals; mixtures and solutions; coatings; systems, composites, foods, and animal and vegetable products. The Materials Directory in each volume indicates which of the 13 thermophysical properties have been covered. These are thermal conductivity, thermal diffusivity, specific heat, viscosity, emittance, reflectance, absorptance, transmittance, absorptance-to-emittance ratio, Prandtl Number, thermal linear expansion coefficient, thermal volumetric expansion coefficient, and thermal radiative properties. The Search Parameters section of each volume allows searching by physical state, subject, language, temperature, and year. An author index is also provided in each volume.

The *Journal of Physical and Chemical Reference Data* [1874] (1972-) publishes critically evaluated data in the areas of atomic and molecular properties, chemical kinetics parameters, colloid and surface properties, and thermodynamic and transport properties. This is often an overlooked source of data since the *Journal* may be filed with the primary journals in a library. It really belongs with the reference collection. In addition to the articles which appear in the *Journal* itself, a number of valuable data compilations have been issued as supplements. In fact, the latest edition of the *JANAF Thermochemical Tables* was published as one of the supplements. Periodically there are indexes produced for the articles which appear in the *Journal of Physical and Chemical Reference Data*. By the end of 1989, the indexes listed in Table 10.9 had appeared. The most recent Property Index, Author Index, and Index to Selected Classes of Materials for v.11-16 (1982-1987) appeared in v.16 no.4 (1987) of the *Journal*.

10.7 OTHER PHYSICAL PROPERTY AND RELATED DATABASES

The integration of a battery of physical property databases into a vendor's arsenal of available resources seems like an attractive and reasonable course of action. However, there are attributes of numeric databases which have impeded this from happening. Numeric data are not as constant as the data which make up bibliographic databases. Once a correct bibliographic reference is entered into a database, it never changes. However, numeric databases are in constant need of revision of *existing* records as better, more reliable data are produced. Furthermore, there is very little standardization of the manner in which numeric data are entered in a database. Consequently, designing search software which will work across widely different data elements in different numeric databases is a real challenge. On the other hand, for a bibliographic database, an author is always an author, a title is always a title, etc. The field designations among bibliographic databases are, therefore, quite similar, even though they may be produced by many different organizations.

Some of the printed handbooks mentioned earlier in this chapter now have online counterparts. For example, the *Dictionary of Organic Compounds* and the *Dictionary of Organometallic Compounds* [667] can be searched on DIALOG as the file HEILBRON (file 303). With over 150,000 organometallic and organic substances or natural

TABLE 10.9
Indexes to the *Journal of Physical and Chemical Reference Data*

Title	Available As:
Property Index and Author Index to Volumes 1-5	Reprint 89
Property, Materials, and Author Indexes to...Vol. 1-10	Reprint 194

products in the database, there is a great deal of numeric and other non-bibliographic data which can be retrieved. For example, it is possible to search the file for information on organometallic compounds which contain cobalt and have melting points in the range of 100°-120° C. All of the numeric fields can be searched for a range of data in HEILBRON. These include boiling point, freezing point, melting point, molecular weight, optical rotation, dissociation constant, and relative density. Other properties which can be searched are the physical state, solubility, and hazard/toxicity data. Of course, combinations of different properties are also possible. For example, a search for a yellow crystalline substance with a melting point of 128° C would be entered as:

?SELECT PS = (YELLOW(w)CRYST)

followed by:

?SELECT S1 AND MP = 128

Structural images are available in the Heilbron file when the DialogLink telecommunications software is employed in the search.

The *Merck Index* is also available online through several vendors, including Questel, Chemical Information Systems, Inc., and BRS. As with HEILBRON, this database can be searched by physical or chemical property. Unlike HEILBRON, however, *Merck* lumps all physical data into a single field. Thus, abbreviations such as mp or bp must be included in the actual search strategy.

For a period of time, the most frequently used tables from the CRC handbooks were available in SuperHandbook [1762]. The database was only accessible through the vendor SuperSearch, Inc., an affiliate of the CRC company. As with the other databases of this type, it was possible to retrieve an unknown substance by entering actual data or to search by entering ranges of relevant data. Over a dozen of the printed CRC handbooks served as sources for the data in SuperHandbook. Now, HODOC, the database for the *CRC Handbook of Data on Organic Compounds*, is online on STN and includes the most commonly used physical data.

Numerica is another vendor which has mounted physical property databases. Such databases as the Log P and Related Parameters database [1860], the Physical Property Data Service [1862], and the TRC Thermophysical Property Datafile I: Vapor Pressure [1863] are to be found there.

Other numeric databases or databases which extensively index numeric properties are available through the vendors of primarily bibliographic databases. DIPPR [1767] is a compilation of physical property data for several hundred chemicals important in industry. (A printed version of the database is *Data Compilation Tables of Properties of Pure Compounds* [1035] (1985).) DIPPR is a project of the American Institute of Chemical Engineers Design Institute for Physical Property Data. There are 26 single-value property constants and 13 temperature-dependent properties for the substances. Exact values or ranges of values may be input for properties such as melting point, boiling point, refractive index, and vapor pressure. There is also a considerable amount of thermodynamic data such as enthalpy of combustion, Gibbs energy of formation, and enthalpy of formation. In addition, transport data such as conductivity and viscosity are found in the file. DIPPR is available on STN International. The tape can also be obtained from the National Institute of Standards and Technology. There are a number of other data compilations available in machine-readable form from NIST. Among them are the JANAF Thermochemical Tables Database, the NIST/NIH/EPA/MSDC Mass Spectral Database, and the NIST Chemical Thermodynamics Database. As noted earlier, JANAF is now available on STN International.

EMIS, the Electronic Materials Information Service [1867], is produced by INSPEC, the information division of the Institution of Electrical Engineers in Great Britain. In EMIS, extensive properties of electronic materials are included. Sample and measurement details are found in the Materials Properties File which has approximately 10,000 records for many solid-state materials. The Materials Supply File lists suppliers for materials used in the production, development, and research in solid-state devices. INSPEC is also the publisher of *Physics Abstracts* [531], a component of the online database INSPEC. From January 1987, the indexing of INSPEC has been considerably enhanced for both chemical substances and numeric data. For example, on DIALOG it is possible to search INSPEC for a document indexed with a wavelength of 0.0000106m as follows:

?**Select NI = WAVELENGTH(S)MT = 1.06E-05**

where MT stands for meter, and E-05 indicates the placement of the decimal point.

The Materials Property Data Network promises to be an authoritative source of data. The pilot program has been operated by Stanford University, and the production version is to be loaded on STN International in 1990.

Finally, it should be noted that many of the databases available through Chemical Information Systems, Inc. are full-text or numeric databases. These include OHM/TADS [975]; MSSS, the Mass Spectral Search System [1603]; IRSS, the Infrared Search System [1604]; CRYST, for crystallographic data [1607]; Carbon-13 Nuclear Magnetic Resonance Spectral Search System [1610]; THERMO, for Thermodynamic Property Values [1613]; and Baker, a database of materials safety data sheets [1771].

10.8 MISCELLANEOUS SOURCES

Since numeric data are often thought of as the purview of physical chemists, some of the more important reference sources for physical and theoretical chemists will be noted here. There are a number of relevant treatises for this area, among them *An Advanced Treatise on Physical Chemistry* [492] (1949-54), *Physical Chemistry: An Advanced Treatise* [1493] (1967-74), *Comprehensive Chemical Kinetics* [1504] (1969-), *Modern Theoretical Chemistry* [1873] (1976-), and the series of books published as *The International Encyclopedia of Physical Chemistry and Chemical Physics* [1643] (1960-). Emiliani's *Dictionary of the Physical Sciences: Terms—Formulas—Data* [154] (1985) is an interesting work. Although the bulk of the *Dictionary* defines various terms, about one-third is devoted to the kinds of tables of data found in one-volume handbooks.

As for review serials, those listed in Table 10.10 are the most important ones for this area:

TABLE 10.10
Physical Chemistry Review Serials

Advances in Chemical Physics
Advances in Quantum Chemistry
Annual Reports on the Progress of Chemistry
  Section C: Physical Chemistry
Annual Review of Physical Chemistry
Chemtracts: Analytical/Physical/and Inorganic Chemistry
International Reviews in Physical Chemistry
Review of Modern Physics
Theoretical Chemistry

10.9 SUMMARY

Locating a chemical or physical property of a substance can be a real challenge. Even the most nearly comprehensive of the available printed sources or numeric databases contain data on just a fraction of the existing compounds. Therefore, it is necessary to know the limitations of the numeric data collections and to use such aids as comprehensive indexes and guides to the literature in order to locate potential sources of data. Numeric data are in many cases not as stable as bibliographic data. When better equipment and more accurate techniques of measurement are utilized to compute an earlier value, it may be necessary to modify the existing value in a database. Consequently, it is important to find the most recent reliable source which is available. The process of critical evaluation of numeric data has greatly enhanced the reliability of physical or chemical data compilations. A new era in searching for physical property data has been ushered in with the appearance of numeric databases.

10.10 SELECTED READINGS

Touloukian, Y. S. "Designations of Critically Evaluated Numerical Property Data." *Thermophys. Electron. Newslett.* **1980**, *9*(4), 1-2, 6.

Sunkel, J.; Hoffmann, E.; Luckenbach, R. "Straightforward Procedure for Locating Chemical Compounds in the Beilstein Handbook." *J. Chem. Educ.* **1981**, *58*(12), 982-986.

Luckenbach, Reiner; Sunkel, Josef. "Problem Solving with the *Beilstein Handbook*." *J. Chem. Inf. Comput. Sci.* **1989**, *29*(4), 271-278.

CHAPTER 11

SEARCHING FOR THE SYNTHESIS OR REACTIONS OF COMPOUNDS (REACTION CHEMISTRY)

11.1 INTRODUCTION

In previous chapters, we learned techniques for determining whether a compound is known (chapters 7-8) and what its properties are (chapters 9-10). In this chapter, attention is focused on the broad area of reaction chemistry. There are many aspects of potential interest in a chemical reaction search. The starting materials, catalysts, reaction sites, mechanism of the reaction, experimental conditions, yields, products, by-products, and changes in the bonding of the substances involved in the reactions— all of these may be of interest to the synthetic chemist. Perhaps the most difficult of these to describe in an information retrieval system is the **reaction mechanism**:

> ...a detailed description of a particular reactant to product path, together with information pertaining to intermediates, transition states, stereochemistry, the rate limiting step, electronic excitation and transfer, and the presence of any loose or intimate electron ion pairs.[1]

One approach to the indexing of reaction mechanisms is to ascribe a name to a particular type of reaction. This might be an eponym formed from the name of the chemist(s) who discovered the synthetic method, such as the Friedel-Crafts reaction. It might also depict what is taking place in the reaction, for example, hydrohalogenation.

Unfortunately, there are many reactions which cannot easily be described with words. Consequently, indexers have often turned to reaction diagrams or codes in order to classify reactions.

Large traditional printed abstracting or indexing journals have until recently not been very successful in their treatment of reaction chemistry. A good information retrieval system for chemical reactions requires the capability to search by structure of the starting material, the product, and perhaps even the **substrate**, the minimal structure that contains the reaction site. It is often difficult to find information on reactions or transformation of a particular functional group. Therefore, reaction chemists must frequently resort to seeking information on simple model compounds which contain the functional groups of interest. Furthermore, chemists are often interested in structural features which undergo *no* change in a reaction. Compounding the difficulty of constructing a good reaction chemistry database is the need to index on the experimental conditions, such as the pressure and temperature or pH requirements of a reaction.

In a sense, the reaction chemist is akin to the engineer when approaching the literature. Both have a specific problem to be solved and a set of conditions within which the solution to the problem must be fitted. Each might have in mind at least a partial solution before beginning the search of the relevant literature. Whether the chemist is interested in the preparation of a particular compound (or group of related compounds) or has a type of reaction as the main area of interest, reaction chemistry searching is a real challenge, involving perhaps as much art as science.

Synthetic organic chemists have a tendency to classify reaction information by functional groups. On the other hand, the inorganic chemist may have to contend with compounds which cannot be described by stoichiometric formulas or those which may have a variable or unknown water content. Inorganic chemists tend to deal with related classes of compounds, such as those formed with the alkaline earth metals, when searching the reaction chemistry literature. Therefore, the various reference tools and databases for reaction chemistry have generally developed along quite different paths, with distinct works available for organic and for inorganic chemistry.

11.2 GUIDES TO THE LITERATURE

Loewenthal's *Guide for the Perplexed Organic Experimentalist* [295] (1978) is full of helpful hints and suggestions for methods and techniques. Appropriately, the author begins with a chapter entitled "On Searching the Literature". Some of the other chapters deal directly with reactions and preparative organic chemistry, for example, "On Carrying Out Small-Scale Reactions," "On Catalytic Hydrogenation," and "Some Detailed Reaction Examples". Loewenthal clearly states his opinion of abstracting practices, the publishing habits of other authors, and the utility of certain types of primary literature. By and large, the advice given is very practical and helpful. For example, he notes that patent literature should not be ignored, for there are whole groups of compounds whose reaction chemistry is found only in the patent literature.

There are several somewhat dated, but still useful guides which have appeared as chapters or articles in the serial literature. See, for example, "The Literature of

Heterocyclic Chemistry'' [1878] (1966) [1879] (1979), "The Organometallic Chemistry of the Main Group Elements—A Guide to the Literature'' [1880] (1975), "Retrieval and Use of the Literature of Inorganic Chemistry'' [1881] (1972), and "An Introduction to the Literature of Organic Chemistry'' [1882] (1968). More recent is a chapter in March's *Advanced Organic Chemistry* [1932] (1985) entitled "The Literature of Organic Chemistry''.

11.3 SOME TRADITIONAL SOURCES OF INFORMATION

There are a number of printed abstracts, indexes, treatises, handbooks, etc. (some of which were encountered in earlier chapters) with significant coverage of reaction chemistry.

11.3.1 Beilstein and Gmelin

Recall that the basic division of information in the *Beilstein Handbook of Organic Chemistry* is among acyclic (v.1-4), isocyclic (v.5-16), and heterocyclic (v.17-27) compounds. *Beilstein* concentrates on the physical and chemical properties of a substance, but includes references to its formation (*Bildung*) or preparation (*Darstellung*). Table 11.1 may help to spot useful references in the *Beilstein* H-EIV volumes, which are published entirely in German.

The problem with using the printed *Beilstein* for synthetic chemical searches is that the information in the set is classified into the system numbers of specific compounds. Often a synthetic chemist is interested in finding a number of procedures for making related types of compounds from certain starting materials. With *Beilstein*, each compound of potential interest would have to be searched independently. *Beilstein* is quite useful for finding a synthetic method of making a known compound. However, there are no entries in the subject index which delineate the type of reaction used to create the compound or anything else about the reactions involved. Furthermore, the user of *Beilstein* must always go to the original literature for the details of the synthesis. For methods of accessing the information in *Beilstein*, see section 10.4.2.2.

In the early years, all reported preparations of a substance were included in *Beilstein*. Now, the yield, ease of handling of the starting materials, and number of steps involved in the preparation all influence the selection. Reactions which are judged to be new or are considered interesting are also included in the Beilstein database. The Beilstein database overcomes much of the difficulty of using the printed *Beilstein* for synthesis or reaction data. In the first place, substructure searching and other techniques can be used to gather structurally similar substances into a single set. Secondly, the indexing allows several approaches of interest to reaction chemists. One can even find in the database information on the isolation of substances from natural products.

On STN International, the following preparative information is searchable in conjunction with compound names:

- PRE.SM   Starting material(s)
- PRE.EDT Educt(s) (starting material(s))

- PRE.RGT   Reagent(s) (catalysts, solvents, inorganic reagents)
- PRE.BPRO   By-products
- PRE.YD   Yield

Likewise, reaction indexing permits searching for:

- REA.PRO   Reaction product
- REA.RP   Reaction partner (solvent, catalyst)
- REA.RGT   Reagents

Finally, a substance name can be labeled /INP to determine if it was a natural product from which certain compounds were isolated as in => **S DIGITALIS/INP** Similar

TABLE 11.1
Selected German Words Relevant to Reaction Chemistry

German	English
Abbau	degradation
Alkylierung	alkylation
Ausbeute (an)	yield (of)
Ausgangsmaterial	starting material
Ausgangsverbindung	starting compound
ausgehen(d) von	to start(ing) from
behandeln	to treat
beim Behandeln	on treating
nach Behandlung	after treatment
bereiten	to prepare
beständig	stable
bilden	form
bildet sich	is formed
Bildung	formation
Darstellung	preparation
Derivat	derivative
einbringen	to introduce
einleiten	to feed, pass into
Einschlussverbindung	inclusion compound, clathrate
eintragen	to introduce
enthalten(d)	to contain(ing)
ergeben	to result
Ersatz	substitution, replacement
Folge	sequence
formulieren	to formulate
geben	to give
gibt	gives
Gemisch	mixture
Geschwindigkeitskonstante	rate constant
Gewinnung	isolation, extraction
Hauptprodukt	main product
herstellen	to prepare, manufacture
Herstellung	preparation, manufacture
hervorgehen (aus)	to originate (from), result (from)
hindern	to inhibit, prevent

search techniques are possible with the DIALOG and ORBIT implementations of the database.

Let us assume that a substructure search on the isatin structure had been performed in the Beilstein database on STN and assigned set number L2. Then a strategy to find all preparation information for those substances would be:

=> S L2 AND PRE/FA

Note that the field availability label (FA) is used as a search delimiter in this case. It can also be used to define a display option to determine the number of items in a Beilstein database record which satisfy the search. This is an important step because many substances have hundreds of answers associated with a particular descriptor (PRE or REA).

TABLE 11.1
Selected German Words Relevant to Reaction Chemistry

Katalysator	catalyst
künstlich	artificial
liefern	to give, deliver
Nebenprodukt	by-product
Präparat	preparation
Quelle	source
Raumtemperatur	room temperature
Reagenz	reagent
Reinheit	purity
Reaktionsfolge	reaction sequence
Rohprodukt	crude product
Säure	acid
Salz	salt
Schluss	conclusion, end
Seitenkette	side chain
Sprengstoff	explosive
Stoff	matter, substance
Summenformel	stoichiometric formula
überführen (in)	to transform into
Überführung	transformation
übergehen	to be converted into
Umsetzung	reaction
umwandeln	to transform, convert
Umwandlung	transformation
ungesättigt	unsaturated
unrein	impure
unvollständig	incompletely
Verbindung	compound
vereinigen	to combine, join
Verfahren	procedure
Verseifung	saponification
versetzen	to add
Versuch	experiment
vollständig	complete(ly)
zersetzen (sich)	to decompose
Zwischenstufe	intermediate

Figure 11-1 shows three examples of reaction searching in the Beilstein database. In the first, 143 substances were found which have been isolated from various species of *Digitalis*. The second search resulted in 139 methods of preparing methyl isocou-

```
=> s digitalis/inp

L20    143 DIGITALIS/INP
=> d

L20    ANSWER 1 OF 143

BRN    1799681 Beilstein
MF     C26 H52 O2
CN     hexacosanoic acid
       Hexacosansaeure
FW     396.70
SO     3-02-00-01089; 4-02-00-01310; 2-02-00-00380
LN     1291
RN     506-46-7

HO2C(CH2)24Me

Isolation from Natural Product:
INP    in den Blaettern von Digitalis purpurea
       Reference(s):
       1. Schwarz, J.Am.Pharm.Assoc. 21<1932>856,994,997, CODEN: JPHAA3 Centralblatt: 1933 I
          3331
       Note(s):
       2. Handbook Data
```

```
=> s (methyl or dimethyl or trimethyl or tetramethyl)(s) (isocumar? or isocoumar?)
          311722 METHYL
           95737 DIMETHYL
           21053 TRIMETHYL
            9277 TETRAMETHYL
             209 ISOCOUMAR?
             395 ISOCUMAR?
L1           150 (METHYL OR DIMETHYL OR TRIMETHYL OR TETRAMETHYL)(S) (ISOCU-
                 MAR? OR ISOCOUMAR?)

=> d 1,4,14 hit

L1     ANSWER 1 OF 150

SY     ***3-Methyl-6-chlor-isocumarin***

L1     ANSWER 4 OF 150

SY     ***5-Methyl-8-hydroxy-3,4-dihydro-3R-methyl-isocumarin (5-Methyl-mellein)***

L1     ANSWER 14 OF 150
```

FIGURE 11-1
Use of the Bellstein Database in a Reaction Chemistry Search.

marins. The last finds a reaction in which a phthalic acid anhydride is derived from a reaction in which mesitylene is a reaction partner.

Coverage of reaction chemistry in the *Gmelin Handbook of Inorganic Chemistry*

SY   ***4-Carboxy-8-methoxy-3-methyl-isocumarin***

=> s l1 and (pre/fa or inp/fa or rea/fa)
     1526007 PRE/FA
       19693 INP/FA
     178464 REA/FA

L2   139 L1 AND (PRE/FA OR INP/FA OR REA/FA)

=> d hit

L2   ANSWER 1 OF 139

SY   ***3-Methyl-6-chlor-isocumarin***

Preparation:
PRE
     Reference(s):
     1. Korte et al., J.Org.Chem., 42, <1977>, 1329,1335, CODEN: JOCEAH
PRE
     Reference(s):
     1. Korte et al., J.Org.Chem., 42, <1977>, 1329,1334, CODEN: JOCEAH

=> s (phthalic(s)anhydride)/cns and mesitylene/rea.rp

      377 PHTHALIC/CNS
     4589 ANHYDRIDE/CNS
      239 (PHTHALIC(S)ANHYDRIDE)/CNS
       15 MESITYLENE/REA.RP
L8   1 (PHTHALIC(S)ANHYDRIDE)/CNS AND MESITYLENE/REA.RP

=> d hit
L8   ANSWER 1 OF 1
CN   ***3-chloro-phthalic acid-anhydride***
     3-Chlor-phthalsaeure-anhydrid

     Chemical Reaction:
     REA
       Part.: mesitylene, aluminium chloride, 1,1,2,2-tetrachloro-ethane
       Prod.: 2-chloro-6-<2,4,6-trimethyl-benzoyl>-benzoic acid,
          3-chloro-2-<2,4,6-trimethyl-benzoyl>-benzoic acid
       Reference(s):
       1. Newman, Scheurer, J.Amer.Chem.Soc. 78 <1956> 5004, 5005, CODEN: JACSAT
       Note(s):
       2. Handbook Data

FIGURE 11-1
Use of the Bellstein Database in a Reaction Chemistry Search.

is comparatively sparse. The formation of the compounds is generally listed near the beginning of the section, with some general references to the usual methods of preparation. Complete references to the primary literature allow the chemist to obtain further preparative details.

11.3.2 Methoden der Organischen Chemie (Houben-Weyl)

One of the most celebrated works in the area of reaction chemistry is the *Houben-Weyl Methoden der Organischen Chemie* [674] (1952-). The fourth edition of *Houben-Weyl* is projected to be complete in the early 1990s. Unlike other great German compilations of this type, *Houben-Weyl* has continued to be published in German. This should be viewed by those who do not read German not as an impediment, but rather as an incentive to learn the minimal vocabulary needed to understand the work. *Houben-Weyl* is much more than just a treatise on the reaction chemistry of organic compounds. Indeed, there are volumes on general laboratory practice, analytical methods, and physical and chemical methods. Nevertheless, the main thrust of the work is the preparation and synthesis of classes of organic, organometallic, macromolecular, and even biochemical compounds and specific functional groups. Experimental conditions are given in great detail in *Houben-Weyl*, so that it is relatively easy for a knowledgeable synthetic organic chemist to reproduce a given synthesis.

There are detailed tables of contents for each volume and very useful indexes as well. In 1987, two comprehensive index volumes appeared for the *Houben-Weyl Methoden der Organischen Chemie*. Volume 16, part 1 is the *Register der Arbeitsvorschriften* (Index of Experimental Procedures). This includes an alphabetic index, arranged by the name of the parent compound, which covers the preparation of classes of substances. A formula index of individual compounds leads to the place in *Houben-Weyl* where sufficient experimental details are given to allow the synthesis of the compound without further reference to the primary literature. The formula index is

TABLE 11.2
Contents of *Inorganic Reactions and Methods*

Vol.	Year	Title
1-2	1986-	The Formation of Bonds to Hydrogen, 1-4
3-4	1989-	The Formation of Bonds to Halogens, 1-2
5-6		The Formation of Bonds to O, S, Se, Te, Po, 1-2
7-8	1988-	The Formation of Bonds to N, P, As, Sb, Bi, 1-2
9-12	1988-	The Formation of Bonds to C, Si, Ge, Sn, Pb, 1-4
13		The Formation of Bonds to B, Al, Ga, In, Ti, and to Elements of Group I and II
14		The Formation of Bonds to Transition Metals, Oxidative Additions and Reductive Eliminations, Insertion Reactions and Their Reverse
15	1986	Electron Transfer and Electrochemical Reactions; Photochemical and Other Energized Reactions
16		Reactions Catalyzed by Inorganic Compounds
17		Oligomerizations and Polymerizations, The Formation of Intercalation Compounds
18		The Formation of Ceramics

called the *CH-Register der Herstellungsvorschriften*. It is in Hill formula order, except for the final two pages which contain *all* formulas which have no carbon atoms in them. The index covers over 80,000 synthetic procedures for about 70,000 compounds (approximately seven percent of the compounds mentioned in *Houben-Weyl*). Volume 16, part 2 is the *Register der Stoffklassen* (Index of Substance Classes). Both of the index volumes cover volumes 1-15 of the basic set and volumes E1-5 and E11 of the *Erweiterungs* and *Folgebande* volumes. A cumulative index for the entire set will be issued when the fourth edition of *Houben-Weyl* is finally completed.

11.3.3 English-Language Treatises Devoted Largely to Reaction Chemistry

A treatise, though somewhat difficult for a novice to use, organizes the material in a logical manner by utilizing a classified subject approach. There are a number of extremely useful treatises for reaction chemists.

11.3.3.1 *Inorganic Reactions and Methods*

The major chemistry publisher VCH (formerly, Verlag-Chemie) is currently producing *Inorganic Reactions and Methods* [764] (1986-). When complete, the treatise is expected to contain the volumes listed in Table 11.2. Most of the volumes of *Inorganic Reactions and Methods* deal with ways to form bonds to inorganic elements, with the remainder of the volumes devoted to methods of effecting various types of reactions or methods of characterizing the compounds. Both the classical chemistry of the elements and modern aspects of the topic, such as organometallic, polymer, and solid state chemistry, are included.

Inorganic Reactions and Methods is indexed via author, compound, and subject indexes. The compound index has the empirical formula interfiled with a permuted formula. This groups together all compounds which contain a given number of a specific element. Index terms which serve to define the context in which the compound is discussed are tied to the formulas. Names for compound classes are found in the subject index, which also covers subjects like methods, techniques, reaction types, apparatus, etc.

11.3.3.2 *Pergamon's Comprehensive Treatises*

Pergamon Press has performed a real service to reaction chemists by publishing several very important treatises in the decades of the seventies and eighties. Each of them has a title like *Comprehensive [XXXXX] Chemistry*. Although their cost may prohibit smaller libraries from acquiring all of them, it is important to know of their existence. Consequently, they are all briefly discussed below.

Comprehensive Inorganic Chemistry [652] (1973) was conceived with the idea that it should be of use to a wide range of users, not just chemists. Consequently, a consistent arrangement of the material within the chapters on the elements was followed. The contents of the work are divided into the volumes listed in Table 11.3.

TABLE 11.3
Contents of *Comprehensive Inorganic Chemistry*

Vol.	Contents
1	H, Noble Gases, Group IA, Group IIA, Group IIIB, C, and Si
2	Ge, Sn, Pb, Group VB, Group VIB, Group VIIB
3	Group IB, Group IIB, Group IIIA, Group IVA, Group VA, Group VIA, Group VIIA, Group VIII
4	Lanthanides, Transition Metal Compounds
5	Actinides, Master Index

Comprehensive Organic Chemistry [666] (1979) is subtitled "The Synthesis and Reactions of Organic Compounds". Contents are divided as shown in Table 11.4.

TABLE 11.4
Contents of *Comprehensive Organic Chemistry*

Vol.	Contents
1	Stereochemistry, Hydrocarbons, Halo Compounds, Oxygen Compounds
2	Nitrogen Compounds, Carboxylic Acids, Phosphorous Compounds
3	Sulphur, Selenium, Silicon, Boron, Organometallic Compounds
4	Heterocyclic Compounds
5	Biological Compounds
6	Author, Formula, Subject, Reagent, Reaction Indexes

The editors of *Comprehensive Organic Chemistry* decided that theoretical organic chemistry would not be given a special section in the work since they believed that subject to be better treated in monographs. The mechanisms of the reactions are stressed throughout the work. All important classes of organic compounds are discussed in detail, including both synthetic organic and biosynthetic compounds. Of particular note is the index volume. The reaction index covers both named reactions and types of reactions. One section lists references in the treatise itself which deal with the reaction. Another section of the reaction index is for reagents which can be used in the reaction, and a third section contains other literature references, especially reviews, in which the reaction has been extensively discussed. Over 2500 organic and inorganic compounds are found in the reagents index.

Comprehensive Heterocyclic Chemistry [708] (1984) is subtitled "The Structure, Reactions, Synthesis, and Uses of Heterocyclic Compounds". This large class of ring compounds containing at least one non-carbon atom in the ring is very important, especially in areas related to biochemistry. Heterocycles include such compounds as nucleic acids, most vitamins, coenzymes, biopigments, sugars, hormones, antibiotics, and many synthetic pharmaceutical compounds. Many other industrial products are heterocycles. The contents of *Comprehensive Heterocyclic Chemistry* are arranged as shown in Table 11.5.

The chapters in the volumes devoted to rings are arranged in a uniform basic outline. Initially, the characteristics of the class of compounds found in a given major subdivision are treated in general chapters which compare the structure, reactivity, and synthesis. Then, each of the more specific chapters has a structure section (covering

TABLE 11.5
Contents of *Comprehensive Heterocyclic Chemistry*

Vol.	Contents
1	Introduction, Nomenclature, Review Literature, Biological Aspects, Applied Aspects, Less-Common Heteroatoms
2	Six-Membered Rings with One Nitrogen Atom
3	Six-Membered Rings with Oxygen, Sulfur, or Two or More Nitrogen Atoms
4	Five-Membered Rings with One Oxygen, Sulfur, or Nitrogen Atom
5	Five-Membered Rings with Two or More Nitrogen Atoms
6	Five-Membered Rings with Two or More Oxygen, Sulfur, or Nitrogen Atoms
7	Small and Large Rings
8	Subject Index, Author Index, Ring Index, Physical Data Index.

theory, molecular dimensions, spectra, ring strain, aromaticity, shape, thermodynamic aspects, and tautomerism). The reactivity of the compounds is discussed next for both the rings and substituents, including reactions with electrophiles, nucleophiles, free radicals, and electrocyclic reactions. Finally, there is a standard treatment in the synthesis sections.

The ring index is arranged in the same manner as other ring indexes previously discussed. For example, looking in the 5,6 section under C_4N-C_6 leads to many references on 1*H*-Indole. There is also a section of the index volume to *Comprehensive Heterocyclic Chemistry* for spectroscopic and physical property data. Purchasers of the set also received the *Handbook of Heterocyclic Chemistry* [709] (1985), an abridged version of the larger work.

Comprehensive Organometallic Chemistry [710] (1982) puts particular emphasis on industrial applications, but includes much valuable information of interest to chemists involved in pure research with organometallic compounds. All of the major types of organometallic compounds are treated in 8 volumes. The first six volumes of the set deal with the organic chemistry of both main group and transition elements. Volumes 7 and 8 concentrate on the use of organometallic compounds in synthesis and catalysis. A final index volume contains subject, formula, and author indexes, as well as an "Index of Structures by Diffraction Methods" and an "Index of Review Articles and Books".

Coordination compounds are those which result as the product of association of a Brønsted base with a Lewis acid. In *Comprehensive Coordination Chemistry* [1617] (1987), an arbitrary distinction is made between coordination compounds and organometallic compounds with metal-carbon bonds. Species in which the number of metal-carbon bonds is at least half the coordination number are left to the sister publication, *Comprehensive Organometallic Chemistry*. Coordination compounds of both main group and transition elements are found in the later work. (Chapters on phosphorus and technetium were not included at the time of publication, but were to be issued as later supplements.) Coordination compounds are used in such diverse environments as the petrochemical, pharmaceutical, and plastics industries. They are important compounds for medicine, mineral extraction, industrial catalysis and many other areas.

Comprehensive Coordination Chemistry is subtitled "The Synthesis, Reactions, Properties, and Applications of Coordination Compounds." Contents are arranged as

TABLE 11.6
Contents of *Comprehensive Coordination Chemistry*

Vol.	Contents
1	Theory and Background
2	Ligands
3	Main Group and Early Transition Elements
4	Middle Transition Elements
5	Late Transition Elements
6	Applications
7	Indexes

shown in Table 11.6. The introductory volume covers diverse topics, including history, nomenclature, mechanism of reactions, and analysis. Volume 2 treats the properties of ligands and is arranged by a system based on the nature of the binding atom. As with *Comprehensive Organometallic Chemistry*, particular emphasis is given to industrial applications, with the final volume of this treatise devoted entirely to that topic.

The index to *Comprehensive Coordination Chemistry* is more than just an index to the set itself. It also contains an "Index of Review Articles and Specialist Texts" published from 1945 to early 1986. All of the reviews are in English or are translations into English. A classified index provides access to the reviews, which are broken down into 22 categories roughly corresponding to the arrangement of material in the treatise. Representative categories are "Complexes According to Ligands," "Bonding in Complexes," and "Metal-Containing Enzymes". There is, of course, a cumulative subject index and a cumulative formula index which form the bulk of the index volume. In addition, each separate volume has its own subject and formula indexes. This was not provided in any of the earlier treatises published by Pergamon.

Comprehensive Polymer Science [1988] (1988), like the other Pergamon treatises, covers the synthesis, application, reactions, and applications of polymers. The coverage of the volumes is shown in Table 11.7. A cumulative subject index is included in volume 7, and each volume has its own subject index. The nomenclature chapter in volume 1 is designed to provide a rational basis for the discussion of polymers.

The latest addition to the Pergamon sets is *Comprehensive Medicinal Chemistry* [1931] (1990). The chapters in Volume 1 deal with the historical perspectives, targets of biologically active materials, bioactive materials, and socio-economic factors in

TABLE 11.7
Contents of *Comprehensive Polymer Science*

Vol.	Contents
1	Polymer Characterization
2	Polymer Properties
3	Chain Polymerization I
4	Chain Polymerization II
5	Step Polymerization
6	Polymer Reactions
7	Specialty Polymers & Polymer Processing

drug development. Volume 2 covers enzymes, agents acting on oxygenases, electron transport systems and pyridoxal-dependent systems, agents acting on metabolic processes, hydrolases and peptidases, cell walls, and nucleic acids. Membranes, membrane receptors and second messenger pathways, neurotransmitter and autocoid receptors, peptidergic receptors, drugs acting on ion channels and membranes, lymphokines and cytokines, and intracellular receptors are found in Volume 3. Volume 4 discusses drug design and molecular modeling, quantitative descriptions of the physicochemical properties of drug molecules, quantitative descriptions of biological activity and drug transport, molecular graphics and drug design, quantitative structure-activity relationships, and pattern recognition and other statistical methods for drug design. Volume 5 covers analytical methodology, principles of pharmacokinetics and metabolism, and chemistry and pharmacy in drug development. Volume 6 contains a cumulative subject index as well as a drug compendium covering over 5000 compounds.

11.3.3.3 Other Treatises

Other publishers have been active in the production of chemistry treatises and advanced textbooks. Some of the following may be of use to the reaction chemist:

> The Chemistry of Functional Groups series [1935]
> *Advanced Organic Chemistry: Reactions, Mechanisms, and Structure* [1995] (1985)
> *Comprehensive Organic Transformations* [1996] (1989)
> *Advanced Inorganic Chemistry* [1997] (1988)
> *Mellor's A Comprehensive Treatise on Inorganic and Theoretical Chemistry* [464] (1922-37) and supplements [463] (1956-)
> *Chemistry of the Elements* [748] (1984)
> *Comprehensive Inorganic Chemistry* [1849] (1953-61)
> *Rodd's Chemistry of Carbon Compounds* [673] (1964-)
> *Comprehensive Biochemistry* [716] (1962-)
> *New Comprehensive Biochemistry* [1614] (1981-)
> *Comparative Biochemistry: A Comprehensive Guide* [859] (1964)
> *Biotechnology* [1642] (1981-).

11.4 ABSTRACTING AND INDEXING SERVICES

Several abstracting or indexing services cover the literature of reaction chemistry, some dating back several decades. In addition to the more general treatment given by *Chemical Abstracts*, there are specialized services which have developed elaborate methods for coding information about reactions.

11.4.1 Searching *Chemical Abstracts* and CAS ONLINE for the Preparation or Reactions of a Compound

In order to receive an entry in the *CA* Chemical Substance Index, there must be something new reported about a known compound, or it must be a new compound never before reported in the literature. In choosing terms for both the Chemical Substance Index and the General Subject Index, CAS elects to use the most specific term possible and may not tie the document to a more general indexing term for the class of compound, class of reaction, etc. Thus, if a particular stereoisomer of a compound is the subject of an article, but there is nothing new said about the basic compound (of unspecified stereochemistry), only the registry number and *CA* Index Name for that stereoisomer will be used in the indexing. (A substance is assumed to be in the form in which it most commonly occurs or is used unless the author states otherwise.) Likewise, if an article deals with a technique like bromination, the indexer makes no attempt to tie it to the more general term, such as halogenation, in the General Subject Index. The more general entries are used for both substances and techniques only when three or more individual members of a given class are described in the document. Thus, it must always be kept in mind when searching for information on a compound in *Chemical Abstracts* that the Chemical Abstracts Service indexing policies may exclude certain information in an abstracted work or may scatter the indexing for closely related compounds among several different registry numbers.

The latest (1987) Index Guide to *Chemical Abstracts* states in paragraph 14A:

> New information about a known substance may take the form of a new source or preparative method, a new reaction in which it takes part, newly reported kinetic or mechanism studies, chemical or physical properties or methods of detection, or a new use or biological effect.

Paragraph 14B deals with ''Entries for chemical reaction studies''. For preparative studies, the CAS indexing policy allows entries for the reactants, intermediates, and products. The reactant is always qualified by a text modification phrase such as ''reaction of, with....''. Novelty again is the criterion used to assess whether entries will be made for reagents. Intermediates are indexed only if isolated and characterized. Entries are made in the General Subject Index for named reactions only if there is general information about the reaction which has been stressed by the author. Chemical Abstracts Service has never made any effort to cover in *CA* routine solvents or catalysts.

In searching the printed *Chemical Abstracts* for the preparation of a compound, there are several approaches which could be taken. If the volume and accession number is listed directly after the name of the compound in the Chemical Substance Index with *no* text modification, one can assume that the preparation of the substance was the main point of discussion in the abstracted primary document.

For the approximately 600 so-called qualified substances with voluminous entries in *CA* (such as carbon monoxide, benzene, etc.), one of the seven standard subdivisions into which the abstracted items are classified *in the Chemical Substance Index* is ''Preparation''. Chemical Abstracts Service includes under that rubric synthesis, manufacture, incidental formation (other than biochemical), recovery, and separation and purification. Another standard subdivision used in *CA* with the qualified substances is ''Reactions''. This term covers chemical changes leading to new products, as well as nuclear interactions (other than simple scattering), corrosion, neutralization, enolization, isomerization, and tautomerism.

Functional derivatives, such as salts, esters, hydrazones, etc., follow the entry for the basic compound (the Heading Parent) in the Chemical Substance Index. These may serve as starting materials for the desired compound if no preparative method is found for the compound of interest.

One way of searching *CA* online for reaction chemistry is to link the reactant and product in the same record before introducing terms for the type of reaction. As noted above, in the printed *CA*, one often finds entries of the form "PRODUCT, preparation from REACTANT" or "REACTANT, reaction of, with REAGENT". In either case, the second-named substance is likely to be indicated by a common name. Words like "manufacture", "preparation" (and the abbreviation "prepn"), "reaction," and others must then be combined to assure effective recall.

If a registry number is found in an online search, the location of a primary work involving the preparation of the substance is quite easy. For example, in the STN International CA File, one simply appends a "P" to the registry number and searches as:

=> S 91-56-5P

or

=> S L#/P

where the L# refers to an answer set formed in the Registry File. The "P" suffix in the online files of *Chemical Abstracts* is no guarantee that details of the method of preparation are given, only that the preparation is reported. Another useful online technique is to use proximity operators. Index terms like:

alcohols, preparation
alcohols, primary, preparation

could be found in the search by inputting:

=> S alcohols(2w)preparation

or

=> S alcohols(2w)prepn

In summary, CAS ONLINE is most effective for retrieving information on the preparation of specific compounds via registry numbers which have the suffix "P". As noted earlier, the technique can be very useful when extended to a group of related substructures which are represented by an L# in the Registry File answer set. By crossing over into the CA File, all of the registry numbers in the set which have the "P" appended in the index terms will be pulled.

11.4.2 CASREACT

Also developed by Chemical Abstracts Service is the database CASREACT [1780], which provides much better access to the literature of reaction chemistry than does

Chemical Abstracts. Based on 106 selected organic chemistry journals, the records for 1985 to the present are accessible primarily through CAS registry numbers labeled as reactants, products, reagents, catalysts, or solvents. The answer set includes chemical structural and reaction diagrams, with the added feature of highlighted reacting bonds. Each record covers all examples of reactions found in the article, with all steps of a given reaction included.

The CAS registry number is the main search avenue in CASREACT, so searches which start in the Registry File and switch to CASREACT should be quite common. A special screen has been created for the Registry File to facilitate searches of small structures which would otherwise not run to completion in a structure search. It is screen 2082, which designates structures found in the CASREACT database. You can also use the dictionary search term CASREACT/LC to limit structure searches in the Registry File. For example, a structure search which resulted in an answer set, L2, could be searched as follows:

=> **S L2 AND CASREACT/LC**

This would restrict the answer set to just those substances in the CASREACT database.

Besides registry numbers, the only other subject search terms which are permitted come from the reaction information fields and include:

- m/NS   Number of steps
- XX/YD   Percentage yield
- NONE/YDT Yield data
- /NTE   Note field for special reactions conditions.

Thus, the /NS and /YD fields are searchable with numbers. The search statement 1/NS is useful for searches with solvents, reagents, or catalysts as one of the search terms, since it requires them to be in the same reaction step as the rest of the query. For the percentage yield, it is possible to input a range of values as part of the search statement, for example, 95-100/YD. The adjacency operator (A) must be used in such searches. When the paper did not specify a yield value (or it was indexed prior to October 1986, the date CAS began to index yield information), the article is indexed with NONE/YDT. For a comprehensive search on yield data that statement should be part of the search. The /NTE field contains words like "temp," "pressure," "safety," "photochem," etc.

In general, the proximity operators are used in searching CASREACT. It is possible to conduct structure searches in the Registry file, then use the L#'s to represent the reactants, products, etc. in the CASREACT database, as in:

=> **S L2/RCT (L) L4/PRO**

Codes for the various possibilities are:

- /RCT for reactants
- /PRO for products
- /CAT for catalysts
- /RGT for reagents
- /SOL for solvents.

=> s 91-56-5

L1   1 91-56-5

=> d

L1   ANSWER 1 OF 1

RX(7) OF 527   ***T***   ===>   UNKWN-U...

```
                    H
        : C .     . N .      : O
    C:       . C .      . C :
     .         :           .
     .         :           .
     .         :           .
    C:       . C . . . . . . . . C:
     :  .                   :
         C                  : O
```

T

UNKWN

U

RX(7)   RCT T   ***91-56-5***
            PRO UNKWN-U
            SOL 67-68-5 DMSO, 7727-37-9 N2
            RGT 1310-58-3 KOH

CA108(25):221504t Electrophilic substitution in indoles. Part 15. The reaction between methylenediindoles and p-nitrobenzenediazonium fluoroborate. Jackson, Anthony H.; Prasitpan, Noojaree; Shannon, Patrick V. R.; Tinker, Alan C. (Dep. Chem., Univ. Coll., Cardiff CF1 1XL, UK). J. Chem. Soc., Perkin Trans. 1 (11), 2543-51 (Eng) 1987. CODEN: JCPRB4. ISSN: 0300-922X.

FIGURE 11-2
Sample Record from the CASREACT Database

The answer set actually consists of CA File records, supplemented by reaction information. Thus, a great deal of information, such as abstract text, index terms, keywords, titles, etc. is displayable, but not searchable in CASREACT. The additional information consists of a reaction map with alphabetic identifiers. Finally, the alphabetic identifiers are linked to their roles in the reactions and to the registry numbers for the substances depicted in the reaction diagram, as shown in Figure 11-2.

Another interesting feature of the CASREACT database is the link which has been provided to the CJO (Chemical Journals Online) database. Once the records have been included in the answer set in CASREACT, the corresponding text record in CJO can be found by selecting the crossover key field (/CK). This special field consists of the CODEN abbreviation for the journal title linked to its volume, issue, and starting page for the article. An example is:

=> S TELEAY-28-1-55/CK

11.4.3   *Index Chemicus* and *Current Chemical Reactions*

ISI's *Index Chemicus* [1150] is one of the best examples of unnecessary title changes in all of the scientific literature. When it began in 1960, the service was simply called *Index Chemicus*. In 1970, the title changed to *Current Abstracts of Chemistry and Index Chemicus*, then in 1977, to *Current Abstracts of Chemistry*. Finally in 1987, the original title was rediscovered, so it is now once again *Index Chemicus*! None of the variant titles adequately reflects the subject matter of *Index Chemicus*. It is a weekly abstracting service for new organic compounds synthesized or isolated and reported in the primary journal literature. At the present time, over 110 of the main chemistry and pharmaceutical journals are scanned for relevant compounds. ISI estimates that the 3500 new organic compounds entered into the database each week represent 90 percent of all important new organic compounds which appear in the journal literature. The product currently abstracts isolated compounds, unisolated compounds, and even theoretical intermediates. (In contrast, CAS insists on complete characterization of intermediates before assigning *Chemical Abstracts* registry numbers.) By 1987, over 4,000,000 substances, tied to over 300,000 source articles, had entered the database.

A characteristic of *Index Chemicus* from the very beginning is the liberal use of reaction diagrams in conjunction with a printed abstract. In the reaction diagram, new compounds are identified by a line beneath the number assigned to the structure. Figure 11-3 shows a typical abstract.

The black box in the upper right-hand corner of each abstract in *Index Chemicus* reveals several things about the article in a concise coded format. A star indicates that labeled compounds were found in the article. From 1968, a coded abbreviation tells what analytical techniques were used to identify the substances. Some abbreviations used for analytical techniques (of which 24 are now employed) are: CC, ESR, TLC, IR, GC, and UV. Other pieces of information in the black box alert the user to an explosive reaction (1978-), the presence of a new or improved synthetic method (1968-), biological activity (tested or predicted, from 1985), and even the amount of experimental detail (full, partial, or none, from 1983).

Compounds may be found in *Index Chemicus* by molecular formula, name (subject), and labeled compound indexes. There is also a biological activity index and a list of ''Publications Indexed'' in each weekly issue. By consulting the list, the searcher can monitor favorite journals.

The indexes to *Index Chemicus* are cumulated four times a year, the last forming the annual index for a volume. Each cumulated index has another special molecular formula index, called the Rotaform (Rotated Molecular Formula) Index. Compounds which contain only carbon, hydrogen, nitrogen, or oxygen are excluded from the Rotaform Index, but all other compounds are listed under each element other than hydrogen. This facilitates the location of compounds having a specific number of a

413253 SYNTHESIS OF SOME 5-MEMBERED AND 6-MEMBERED OXASTEROIDS OF CHOLESTANE SERIES BY RING CONTRACTION AND THE MASS-SPECTROMETRIC FRAGMENTATIONS OF OXASTEROIDS.

EXPTL | FULL | CC | IR | MS | NMR | TLC

SUGINOME H, YAMADA S.
HOKKAIDO UNIV, FAC ENGN, DEPT CHEM PROC ENGN, SAPPORO, JAPAN.
BULL CHEM SOC JPN 60(7),2453-61(1987).

Synthesis of several new five- and six-membered oxasteroids of the cholestane series by the ring contraction of those oxasteroids whose oxygen-containing ring is larger by one member has been achieved. Ring contraction utilizes a series of reactions recently developed by us for the transformation of cyclic ketones into cyclic ethers; it involves a regioselective β-scission of the alkoxyl radical generated by irradiation of the lactol hypoiodite derived from the starting oxasteroid to give the iodoformate followed by its cyclization. The present work demonstrates the versatility of our new method for the synthesis of cyclic ethers and extends the scope of its application. The mass spectral fragmentation of some oxasteroids synthesized in the present and previous work is discussed.

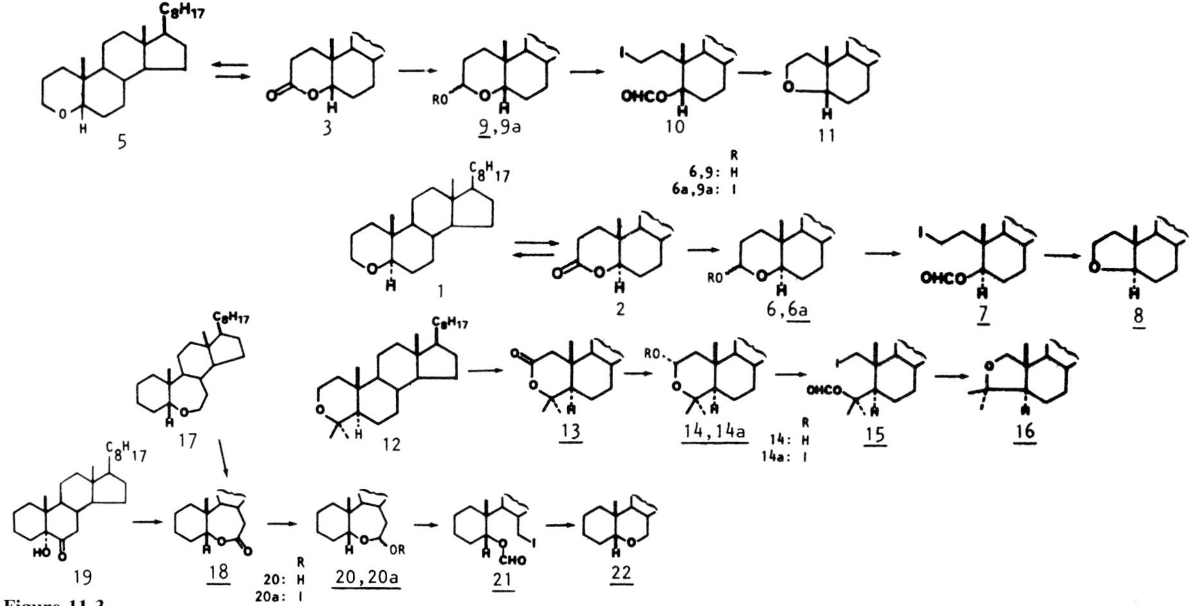

Figure 11-3
Typical Abstract from *Index Chemicus*

INDEX CHEMICUS VOLUME 109, ISSUE 1268, 1988

certain element in the molecular formula. A special index for *Index Chemicus* from 1966 onward is the *Chemical Substructure Index* [192]. The *CSI* is a Wiswesser Line Notation Index which allows the searcher to manually locate compounds containing a specific substructure. The WLNs are rotated to form an alphabetical sequence which groups together related structures. The *Chemical Substructure Index* is only available in microform at this time, although printed versions were produced in the 1970s.

Cumulated indexes to *Index Chemicus* are available in a 22-year microform cumulation covering 1960-1981 for each year in which the indexes originally appeared. The full text of the abstracts themselves have also been cumulated in microform for the same period.

At the present time, there is no online version available for *Index Chemicus*. However, for a few years in the mid-1980s, the database was searchable on Questel. Although the Questel DARC substructure search system could be used for starting material, intermediates, and products, the venture was apparently not a commercial success, so it was discontinued by 1988. Neither the author summary nor the reaction diagrams were available online.

Index Chemicus has since 1960 reported new reactions or new syntheses, even for older known compounds. This includes new ways of using old reagents which result in higher yields or use faster or easier reactions, as well as the first laboratory or total synthesis of a natural product. In 1979, ISI launched another abstracting service, *Current Chemical Reactions* [1149], to provide the actual details of all new reactions. *Index Chemicus* continues to indicate the presence of a new reaction in its entries, but refers the user to *Current Chemical Reactions* for more information. Each month *CCR* covers nearly 450 articles taken from over 125 primary organic chemistry and pharmaceutical journals. Even review articles are within the scope of the service. These are taken not only from the primary source journals for *Index Chemicus*, but also from additional journals, review serials, and books.

Entries in *Current Chemical Reactions* are quite detailed, as Figure 11-4 illustrates. Access to the information in *Current Chemical Reactions* is via four indexes: subject, corporate, author, and journal indexes. The subject index (Index to New Synthetic Methods) utilizes permuted words and phrases which indicate the name or type of reaction. These are taken from the summary and from the text of the article.

A new development in the ISI product line for reaction chemistry is the availability of databases for use on a personal computer. The sixteen topics for which the databases from *Index Chemicus* are available are shown in Table 11.8. Each database from *Index Chemicus* has about 200 representative compounds for heterocyclic compounds, organoboron compounds, etc.

There are also databases which are taken from *Current Chemical Reactions*. The *Current Chemical Reactions* personal databases cover the topics shown in Table 11.9. Each of the *CCR* topics contains 100 reactions, describing in each record the reaction type, catalysts, solvents, reagents, experimental conditions, yields, etc. Quarterly updates are available for certain of the ISI personal databases. Both the *Index Chemicus* and *Current Chemical Reactions* databases can be searched with the chemical database management software ChemBase [73] or ChemSmart [79].

■ J ORGANOMETAL CHEM 328(3),1987 — J 3531

040392 ALPHA-ACETYLENIC AND ALPHA-ETHYLENIC TRIETHYL ORTHOCARBOXYLATES

REACTION OF ACETYLENIC AND VINYLIC ORGANO-LITHIUM REAGENTS WITH TRIETHOXYCARBENIUM TETRAFLUOROBORATE - PREPARATION OF ALPHA-ACETYLENIC AND ALPHA-ETHYLENIC TRIETHYL ORTHOCARBOXYLATES.
PICOTIN G. MIGINIAC P.
UNIV POITIERS, CHIM ORGANOMET LAB, CNRS, F-86022 POITIERS, FRANCE.
(P MIGINIAC, UNIV POITIERS, F-86022 POITIERS, FRANCE).
J ORGANOMETAL CHEM 328(3),249-54(1987).

Acetylenic and vinylic lithium derivatives react with triethoxycarbenium tetrafluoroborate to give α-unsaturated triethyl orthocarboxylates.

$$RC\equiv CLi + C(OEt)_3^+ BF_4^- \xrightarrow[RT]{Et_2O} RC\equiv CC(OEt)_3$$
R = Bu (65%)*

$$RCH=C(R')Li + C(OEt)_3^+ BF_4^- \xrightarrow[RT]{Et_2O} RCH=C(R')C(OEt)_3$$
R ~ R' = Me (81%)*

*Nine other examples (42-59%)
*Five examples (46-80%)

040393 ALPHA-HYDROXYCARBOXYLIC ACIDS

AN APPROACH TO THE STEREOSELECTIVE SYNTHESIS OF ALPHA-HYDROXYCARBOXYLIC ACIDS.
DAVIES S G, WILLS M.
DYSON PERRINS LAB, S PARKS RD, OXFORD OX1 3QY, ENGLAND.
J ORGANOMETAL CHEM 328(3),C 29-33(1987).

The complex [(η^5-C$_5$H$_5$)Fe(CO)(PPh$_3$)COCH$_2$OCH$_2$Ph] undergoes, via the corresponding enolate, highly stereoselective alkylation, deuteration and aldol reactions, can be hydrogenated to the corresponding hydroxyacetyl complex, and gives the ester PhCH$_2$OOCCH$_2$OCH$_2$Ph on oxidative decomplexation in the presence of benzyl alcohol.

040394 ASYMMETRIC ADDITION

DIASTEREOSELECTIVE ADDITION OF ORGANOCUPRATES TO UNSATURATED CHIRAL IMIDES.
POURCELOT G, AUBOUET J, CASPAR A, CRESSON P.
ECOLE NATL SUPER CHIM, RECH CHIM ORGAN LAB, F-75231 PARIS 05, FRANCE.
J ORGANOMETAL CHEM 328(3),C 43-5(1987). IN FRENCH.

Some organocuprates add to unsaturated imides synthesized from chiral imidazolidones with very high diastereoselectivity.

*When the chiral imidazolidone (R = Ph) was reacted with the magnesium cuprate (R^1 = Me), the (R)-(-)-3-Ph-butanoic acid was obtained, (80% overall;82% e.e.).

Figure 11-4
Typical Entry in *Current Chemical Reactions*

TABLE 11.8
Index Chemicus Personal Databases Available for Microcomputers

Amino Acids/Peptides	Macrocyclic Compounds
Antibiotics	Natural Products and Derivatives
Antihypertensives/Hypotensives	Nucleosides/Nucleotides
Antiinflammatory Agents	Organoboron Compounds
DNA, RNA, Protein Synthesis Inhibitors	Organometallic/ Organosilicon Compounds
Herbicidal Compounds	Prostaglandins
Heterocycles	Steroids/Antifertility Compounds
Introduction to Organic Chemistry	Unisolated Intermediates

Another machine-readable option for ISI reaction chemistry products utilizes larger computers. It is the Current Chemical Reactions In-House Database, which started in 1987 with over 5000 new synthetic methods. The file is expected to grow to over 50,000 new synthetic methods by the end of the century. Included in the database are reactions selected by chemists at nearly two dozen chemical and pharmaceutical companies in the United States and Europe. The software to search the database is REACCS [251] (discussed in section 11.8.1.). Essentially all of the access points available in the printed *CCR* plus substructure searching capabilities are employed in the In-House *CCR*.

TABLE 11.9
Current Chemical Reactions Personal Databases Available for Microcomputers

General and Name Reactions	New Reagents
Asymmetric Synthesis	Protecting Groups
Synthesis of Natural Products and Derivatives	Catalysis

11.4.4 The *Journal of Synthetic Methods* and *Theilheimer*

The *Journal of Synthetic Methods* [590] (1975-) is a British publication with an online counterpart, the Chemical Reactions Documentation Service database (CRDS). Approximately 6000 reactions are summarized annually, a large proportion of them taken from patent specifications. About half of the total are completely new reactions and are given full abstracts throughout the year. The other half, judged to be modifications or improvements of known reactions, are listed as supplementary references in the *Journal*'s June and December issues. Approximately 150 journals are scanned for relevant citations. In addition, other works are taken from the organic chemistry sections of *Chemical Abstracts*. These are supplemented through the publisher's (Derwent's) extensive coverage of the patent literature. Reactions are selected for their practical synthetic value. Hence, novel or unusual reactions or new synthetic methods are included. Also selected are new publications containing known reactions, but which involve new reagents or better synthetic methods. Furthermore, extensions or applications of known reactions are sometimes selected if in the editor's judgment, they are interesting. Likewise, the application of organometallic compounds is an area not

NC⇑N   75353D

H   Iron/acetic acid/silica
R
         Indoles
         from o-nitrotoluenes
         via β-amino-o-nitrostyrenes
         Modified Leimgruber-Batcho synthesis

cf. Synth.Meth. 29, 469. A 2:3 mixture of 2-nitrotoluene and tripiperidino-
methane stirred in vacuo (water aspirator) at 120° for 5.5 hrs. ⟶
crude 2-nitro-β-piperidinostyrene, in 5:3 toluene/acetic acid added to a
well-stirred mixture of Fe and silica gel in the same solvent at room
temp., and refluxed under argon for 1 hr. ⟶ indole. Y 82% (overall
yield). The method is suitable for preparing 2,3-unsubst. indoles
having halogen or alkoxy groups in the benzene ring. F.e. (62-94%) s.

         M. Kawase, A.K. Sinhababu, R.T. Borchardt,
         J.Heterocyc.Chem. 24, No.6, 1499-1501 (1987).

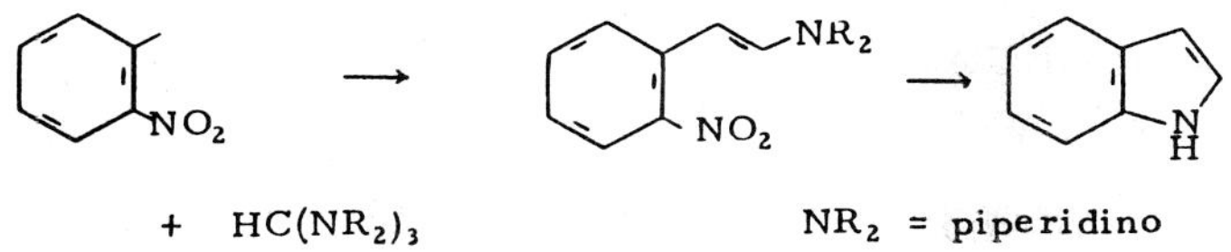

         + HC(NR$_2$)$_3$ NR$_2$ = piperidino

NC⇑Hal   75354D

H   Sodium hydride

         1-Acyl-3-indolones
         from o-acylamino-α-chloroketones

A soln. of startg. chloroketone in tetrahydrofuran added dropwise to a
suspension of NaH in the same solvent with cooling, and acidified when t.l.c.
indicated completion of reaction ⟶ 1-acetyl-5-methyl-3(2H)-indolone. Y
71%. F.e. (37%, 74-79%) s.

         M. Nimtz, G. Häfelinger,
         Ann.Chem. 1987, No.9, 765-70.

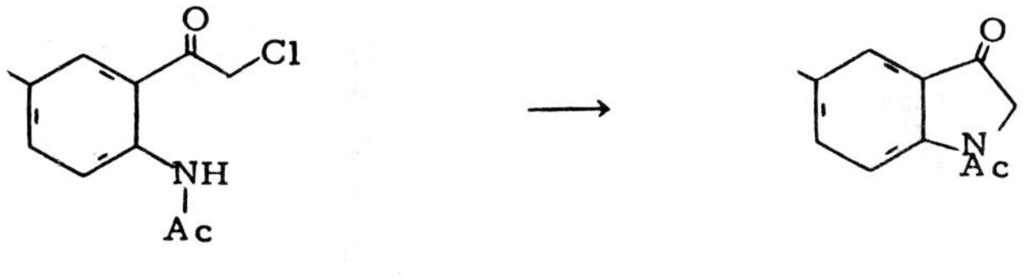

Figure 11-5
Typical Entry from the *Journal of Synthetic Methods*

TABLE 11.10
Thematic Group Codes Used in the *Journal of Synthetic Methods*

Code	Meaning
A	Aromatic substitution and exchange. Aromatization and dearomatization. Aromatic ring opening (in addition to I). Aromatic ring closure (in addition to J).
B	Biochemical reaction
C	Carbohydrate chemistry
E	Electrochemical reaction
G	Heterocyclic ring opening
H	Heterocyclic ring closure
I	Isocyclic ring opening
J	Isocyclic ring closure
K	Ring contraction
L	Ring expansion
N	Non-cyclic reactions
O	Oxidation
P	Peptide and amino acid chemistry
Q	Patent
R	Reduction
S	Steroid chemistry
U	Supplementary reference
V	Irradiation
W	Rearrangement
X	Stereochemistry
Y	Selective and preferential reactions
Z	Protective group chemistry

neglected in the *Journal of Synthetic Methods*. Figure 11-5 shows a typical entry. The capital letters in the left-hand column designate thematic groups, coded according to the scheme in Table 11.10.

There is a subject index in each issue of the *Journal of Synthetic Methods*. Starting materials, reagents, reactions, and products are all covered in the single index. The functional groups which influence the course of the reaction are given full indexing.

The online file, CRDS [590], includes not only the *Journal of Synthetic Methods* entries, but also the references which have appeared in Theilheimer's *Synthetic Methods of Organic Chemistry* v.1-30, 1942-1974. (The printed *Theilheimer* indexing service [684] is still being published.) The database can be searched by controlled-vocabulary index terms for ring systems, functional groups, reagents, reaction types, reaction conditions, etc. A second method of searching is based on Derwent's Ring Code and a code devised by Theilheimer to indicate the bonds broken or formed, the reagents, and the conditions of the reaction. The reaction codes are shown in the Table 11.11. Theilheimer elected to consider the nature of the bonds as the basis for classifying reactions, ignoring the reaction mechanisms in the process.

A useful guide is *Getting the Best Out of Theilheimer's Synthetic Methods of Organic Chemistry* [1883] (1984). The relationship between the *Journal of Synthetic Methods* and *Theilheimer* is explicitly defined in the guide by the statement that the abstracts and supplementary data for *Theilheimer* are now derived from the *Journal*. Items are selected for inclusion if they meet any of the following criteria:

- novel organic reactions and methods of synthetic interest
- novel reagents
- significant modifications of known reactions
- interesting and unusual applications of known reactions
- reviews, especially if devoted to synthetic aspects of organic chemistry
- other items dealing with the preparation of reagents and special apparatus for conducting reactions.

Only general reaction types are selected for inclusion. Most of the information on heterocyclic chemistry is now excluded from the scope of *Theilheimer*'s coverage since the editors believe that information to be adequately covered elsewhere. Likewise, the vast majority of organometallic chemistry does not fall within its scope. The general principle followed in selecting material for *Theilheimer* is that the reaction must have a high yield and give full experimental procedures. Of special importance for selection is the inclusion in the work of a typical experimental procedure. No patents are referenced in *Theilheimer*, however.

TABLE 11.11
Reaction Codes in *Theilheimer's Synthetic Methods of Organic Chemistry*

Symbol	Meaning	Symbol	Meaning
⚡	Electrolysis	⊘	Ring expansion
↑↑↑	Irradiation	C	Ring opening
O	Ring closure	⊖	Ring hydrogenation
⊘	Ring contraction	⇑	Addition
↷	Rearrangement	⇓	Elimination
		↓↑	Exchange

The *Theilheimer* classification scheme distributes the entries according to the bonds which are broken or formed during the reactions. Subject indexes are found in each volume and are cumulated for each set of five volumes. Between five-year cumulations, the subject indexes are cumulated in the second and third of the volumes published in that period. *Classes* of compounds are the main entries in the subject indexes. Within broader classes, such as glycols, more specific groups of compounds can be found, either as starting materials or products. Name reactions and names for types of reactions are generally avoided. A formula index for complex functional groups was introduced with volume 24. For example, NOSiC refers the searcher to entries such as Siloximes and Silylisocyanates. Note that the element symbol for carbon always comes *last* in this index.

Theilheimer makes a special effort to update earlier references through the Supplementary Reference Indexes. Up to 1200 improved or modified reactions which had been cited previously are included in each volume. A final feature of *Theilheimer* is the inclusion of "Trends in Synthetic Organic Chemistry" in the introduction to each annual volume. These highlight the 50-60 most important developments in synthetic organic chemistry during the year of coverage.

11.4.5 Other Abstracting Services with Reaction Diagrams

SYNFORM [1876] (1983-) is devoted to total syntheses of selected natural products. The references are identified through searches of the literature as covered in *Chemical Abstracts*, (1967-). For research published as journal articles, *SYNFORM* records the details of the synthesis. Such information as the starting materials, reagents, reaction conditions, by-products, end products, yields, and stereochemistry are reported. Other types of publications are only referenced to the *CA* abstract number.

ChemInform [1875] (1987-) covers over 200 journals which contain material of interest to synthetic chemists. This is a weekly service which aims for rapid publication of new results (within two months of the appearance in the journal literature). *ChemInform* is mostly for organic chemists, but some attention is paid to preparative inorganic chemistry and other areas of interest to synthetic chemists. Each issue has a special section devoted to new review articles.

The Royal Society of Chemistry publishes *Methods in Organic Synthesis* [1158] (1984-). Such topics as functional group changes, new reagents, asymmetric synthesis, protective groups, carbon-carbon bond-forming reactions, and biological methods are found in the monthly issues. The reaction diagrams include the yields and other appropriate information. There are five indexes in each issue of *Methods in Organic Synthesis*: author, reactant, reagent, products, and reaction indexes. The indexes are cumulated annually.

Annual Reports in Organic Syntheses [1884] (1970-) attempts to promptly publish the results of its survey just after the close of the period being indexed. It is actually an abstracting service which covers about 50 of the most important journals relevant to synthetic organic chemists. The entries consist mostly of structures and reaction diagrams with few comments. There is no subject index, but a detailed table of contents to some extent compensates for this lack. The author index lists only the senior or first author.

11.5 ANNUAL SURVEYS AND REVIEWS

With such a huge volume of literature devoted to reaction chemistry, it is essential that a broad look at the field be taken periodically. The annual surveys or reviews provide such coverage. Already noted above was the "Trends in Synthetic Organic Chemistry" articles which appear each year in *Theilheimer's Synthetic Methods of Organic Chemistry*. Other important surveys are discussed in this section.

11.5.1 *Organic Syntheses*

Organic Syntheses [682] is one of the oldest and best documented annual surveys of the literature, having begun in 1921. The procedures described in the annual volumes are collected periodically (and revised if necessary) to provide easier access in the collective volumes. To date, six of the collective volumes have been published. Special index volumes are also available, including a cumulative index to Collective Volumes

I-V. Nine separate indexes are found in the Cumulative Indices volume. Another index to the set is Sugasawa's *Reaction Index of Organic Syntheses* [683] (1967).

The procedures in *Organic Syntheses* are published with enough detail that consultation of the original primary literature is generally unnecessary. In recent years there has been a tendency to emphasize model compounds and procedures which illustrate important types of reactions.

11.5.2 *Organic Reactions* and Other Reviews of Organic Synthesis

The series *Organic Reactions* [672] (1942-) concentrates on name reactions and other well-defined reactions. An attempt is made to accurately portray from the preparative viewpoint the scope and usefulness of the various processes. Unlike the procedures in *Organic Syntheses*, however, they have not been carefully tested in two or more laboratories. Of interest is the tabular surveys in each chapter. These relate the molecular environment to the appropriateness of using a particular reaction. Compounds which are listed in the tables are not repeated in the indexes. The latest volume contains a cumulative list of all chapters in earlier volumes. Likewise, there are cumulative author and chapter/topic indexes covering all volumes published.

Organic Reaction Mechanisms [1381] (1965-) covers the specific time period of one year in each annual volume. Excluded or restricted in coverage are photochemical reactions, biosynthesis, electrochemistry, organometallic chemistry, surface chemistry, and heterogeneous catalysis.

Natural Product Reports [1885] (1984-) is published bimonthly. Subtitled "A journal of current developments in bio-organic chemistry," the reviews cover the general chemistry and biosynthesis of alkaloids, terpenoids, steroids, fatty acids, and O-heterocyclic, aliphatic, aromatic, and alicyclic natural products. Some attention is also given to techniques for the separation and spectroscopic identification of natural products, as well as some more general methods.

11.5.3 Other Surveys and Reviews

Although *Inorganic Syntheses* [653] began publication in 1939, there has been no attempt to provide separate collective volumes or collective indexes as with *Organic Syntheses*. However, indexes at the back of each volume cover several previous volumes in groups of five each. There is a subject index and a formula index in each volume.

Biochemical Preparations [872] (1949-71) eventually ceased publication due to the existence of competing sources like *Methods in Enzymology* and the journals *Analytical Biochemistry* and *Preparative Biochemistry*. Nevertheless, some quite useful methods were summarized over the course of its lifetime.

Table 11.12 lists some additional review serials which survey fields of interest to reaction chemists.

TABLE 11.12
Selected Review Serials for Reaction Chemistry

Advances in Catalysis
Advances in Heterocyclic Chemistry
Advances in Inorganic Chemistry and Radiochemistry
Advances in Organic Chemistry
Advances in Organometallic Chemistry
Advances in Polymer Science (Fortschritte der Hochpolymeren-Forschung)
Aliphatic and Related Natural Products
Alkaloids
Amino Acids, Peptides and Proteins
Annual Reports in Organic Synthesis
Annual Reports on the Progress of Chemistry:
   A. Inorganic Chemistry
   B. Organic Chemistry
Annual Review of Biochemistry
Biosynthesis
Catalysis
Catalysis Reviews
Chemtracts: Biochemistry and Molecular Biology
Chemtracts: Macromolecular Chemistry
Chemtracts: Organic Chemistry
Coordination Chemistry Reviews
Critical Reviews in Biochemistry and Molecular Biology
Critical Reviews in Biotechnology
General and Synthetic Methods
Heterocyclic Chemistry
Inorganic Reaction Mechanisms
Macromolecular Chemistry
Macromolecular Reviews
Mechanisms of Inorganic and Organometallic Reactions
Organic Reaction Mechanisms
Organic Reactions
Organometallic Chemistry
Organometallic Chemistry Reviews
Progress in Heterocyclic Chemistry
Progress in Inorganic Chemistry
Progress in Lipid Research
Progress in Reaction Kinetics
Progress in the Chemistry of Organic Natural Products
Terpenoids and Steroids

11.6 COMPENDIA OF SYNTHETIC METHODS AND IMPORTANT SERIES

The *Compendium of Organic Synthetic Methods* [678] (1971-) includes several thousand examples of published methods for the preparation of monofunctional and difunctional compounds, classified according to the reacting functional group of the starting material and the functional group which is formed. There is an index of monofunctional groups which relates the preparation of one functional group from another. The difunctional compounds are those formed from the groups acetylene, carboxylic acid, alcohol, aldehyde, amide, amine, ester, ether, epoxide, halide, ketone, nitrite, and olefin. Reactions are classified in this section by the two functional groups of the product.

The second edition of *Organic Functional Group Preparations* [924] (1983-86) (formerly *Synthetic Organic Chemistry*) is a source of reliable preparative procedures for the most common functional groups. Organized by functional group (allenes, enamines, ketones, nitrites, etc.), the book follows a basic outline according to reaction types (condensation, elimination, oxidation and reduction, etc.).

Volume 1 of *Formation of C-C Bonds* [681] (1973-75) is devoted to "Introduction of a Functional Carbon Atom," and volume 2 to "Introduction of a Carbon Chain or an Aromatic Ring". Percent yields and references are given for all entries.

Fieser and Fieser's Reagents for Organic Synthesis [677] (1967-) is undoubtedly the most popular work of its type in all of chemistry. Not only does it present a wealth of information on reagents for use in organic syntheses, it also provides many references to the literature where additional information can be found. The authors of the work have drawn upon *Organic Syntheses* for many of the references. The work is especially useful for planning a synthesis and for answering questions which arise along the way. *Reagents for Organic Synthesis* presents in addition to the structural formula and physical constants such information as the preferred methods of preparation or purification, suppliers, uses, and excellent reaction diagrams. Volumes include an "Index of Reagents according to Types," which essentially indexes reactions and methods. There is also an author and a subject index.

Preparative Inorganic Reactions [656] (1964-71), *Macromolecular Syntheses* [680] (1963-), *Organometallic Syntheses* [1378] (1965-86), Pizey's *Synthetic Reagents* [1467] (1974-), the *Bibliography of Electro-Organic Synthesis, 1801-1975* [1094] (1980), and the two monograph series "Heterocyclic Compounds" and "The Chemistry of Functional Groups" are examples of other important works. The last-named (very expensive) series now has a useful means of accessing the individual titles in *Saul Patai's Guide to the Chemistry of Functional Groups Series* [1938] (1989).

A final category of reference material which should be mentioned in this section is a **formulary**, a kind of "cookbook" for making many useful products. The best known and largest formulary is Bennett's *Chemical Formulary* [1020] (1933-). Whether it is a formulation to strip wax from floors or to lacquer a violin, *The Chemical Formulary* will usually give directions for making several preparations which satisfy your need. There is now a cumulative index covering v.1-25.

11.7 NAME REACTIONS

Over the years various methods of preparing compounds have been named for the people who devised the synthetic techniques. Thus, reactions such as the Fischer indole synthesis, Claisen rearrangement, Wolff-Kishner reduction, Beckmann rearrangement, and the Friedel-Crafts reaction are well known. When important new reactions are first reported, they may become quite famous in a short period of time. Since the reactions were very well known at that time, authors tended to refer to the techniques by the eponyms, without further definition. However, the reader of a decades-old article may be quite perplexed by a reference to a name reaction which is no longer in fashion. To assist in finding the description of the synthetic method, a number of name reaction dictionaries have appeared. Among the reference books available are:

> *Name Reactions and Reagents in Organic Synthesis* [1936] (1988)
> *Name Reactions in Organic Chemistry* [496] (1961)
> *Named and Miscellaneous Reactions in Practical Organic Chemistry* [676] (1967)
> *The Vocabulary of Organic Chemistry* [692] (1980)
> *Organic Name Reactions* [695] (1964)
> *Name Index of Organic Reactions* [699] (1960).

The *Merck Index* also has a good section on organic name reactions.

11.8 COMPUTER-BASED REACTION SEARCHING SYSTEMS FOR IN-HOUSE USE

In recent years, more and more attention has been given to applying computers to the storage and retrieval of reaction information. The most important commercially available systems are discussed in this section. All of those must be used in conjunction with larger stand-alone computers. (At present, the only system for reaction chemistry which is available through an online database vendor is CASREACT, discussed in section 11.4.2.)

11.8.1 REACCS

REACCS (the Reaction Access System) [251] became available in the early 1980s after testing at the Eastman Kodak Company. The system was developed by Molecular Design, Ltd. REACCS portrays all chemically significant information about the substances involved in a reaction, whether reactants, reagents, catalysts, solvents, or products. The powerful structure searching capabilities of MDL's MACCS (Molecular Access System) are part of the search package in REACCS, including the capability to search on the stereochemistry of the molecule. Additionally, reacting centers can be specified as part of the search strategy.

In version 6.0 of REACCS, a number of databases are included, among them about 5000 entries from *Organic Syntheses*, *Theilheimer's Synthetic Methods of Chemistry* (1946-1980), REACCS-JSM, entries from the *Journal of Synthetic Methods* (1982-), the *Fine Chemicals Directory* [1937] (a REACCS version of ChemQuest), and the Current Literature File (a database produced by Molecular Design, Ltd. from data submitted by various synthetic organic chemistry research groups which abstract the current literature, 1984-). The Current Literature File stresses reactions which are contingent on stereochemical control. New databases from MDL include CHIRAS, for asymmetric syntheses and METALYSIS for synthetic applications of organometallic methods. The number of reactions in the MDL databases is rapidly approaching 100,000. (See also section 11.4.3. for Index Chemicus In-House Database.)

The REACCS system uses very efficient data storage techniques. Data for both molecules and reactions are stored only once, but allowance is made for variations in the reactions. Search commands for REACCS are presented as menus from which the user selects the appropriate option with a light pen or mouse. These devices can also be used to draw a molecule. The main menu contains commands such as FIND, VIEW, BUILD, DELETE, EXIT, etc. Other menus are used to DRAW the molecule or reaction.

REACCS is typically used by chemists in pharmaceutical or other chemical industries to build in-house databases, since both reaction data and physicochemical data on the molecules can be entered. Typically, an in-house system might contain several tens of thousands of molecules and reactions. A structure can be drawn, the structure search run, and results viewed at a terminal or printed out. Synthetic chemists can use REACCS to draw a complete reaction and look for an exact match. Molecules with specific substructures can be defined as reactants or products. The bonds which are to be reacting centers can also be specified. The chemist can then designate textual data such as the type, volume, and physical properties of a solvent or the pressure and temperature requirements. For example, the system could be instructed to search for all reactions which occur at temperatures higher than 150° C or those which use Xylene as a solvent.

REACCS is certainly the most widely-used reaction searching system in industry. Molecular Design Ltd. has assembled the largest collection of searchable reactions at the present time. However, REACCS is by no means the only reaction search system which has been developed in recent years, as the following sections show.

11.8.2 ORAC

Another computer-based reaction searching system is ORAC (Organic Reactions Accessed by Computer) [254]. Like REACCS, ORAC draws heavily on menus and graphics to enter both the data and search query. However, this approach is not slavishly followed. The developers of ORAC were careful to include text-searching capabilities in the system. Hence, words which describe what is happening in the reaction with respect to the mechanism, selectivity, or types of reagents used are searchable as keywords.

A maximum of three individual compounds can be entered as reactants and two compounds as products. At the data input stage, the chemist can enter reaction search keys, reagent keys, and solvent keys, plus bibliographic information for the source document. A thesaurus can be constructed to control the vocabulary used as search keys.

Separate databases can be created with ORAC, and a single search query can be run simultaneously on all of them. The Boolean logical operators "AND," "OR," "AND NOT" are used to combine graphic and textual searching. Author and journal names can be searched in addition to the textual search terms assigned to the reaction type, reagents, solvents, etc. Both exact-match and substructure searches are possible, with graphic input of the structures of interest. Version 7.4 of ORAC allows tautomer searching, the entry of up to 256 non-hydrogen atoms in a structure search, and salt searches.

ORAC comes with its own database of reactions. In 1989, this included over 50,000 reactions, several thousand of which are for heterocyclic compounds. The criterion for selection is the novelty of the reaction or a novel use of an existing reaction. ORAC is available in the United States from Chemical Design, Inc., Mahwah, N. J.

11.8.3 SYNLIB

The SYNLIB (Synthesis Library) software [1768] was developed by chemists at Smith, Kline and French Laboratories and Columbia University. It is available in the United States from Distributed Chemical Graphics, Inc., Meadowbrook, PA. Like the previously discussed systems, SYNLIB is designed to use the graphics capabilities of today's modern computers to depict chemical structures of interest to the reaction chemist. The goal of the database is to include *representative* chemical reactions. Over 125 research groups at academic institutions throughout the world furnish the data for the reaction library, which now numbers over 50,000 reactions. The software can also be used to create and search proprietary databases.

As with other systems, data entry for compounds can be done with a light pen, mouse, or datatablet. Common structures can be selected from pre-drawn structures on a menu. Additionally, the chemist can create and store other frequently used structures on templates for later recall. The most common elements in organic chemistry (C, O, N, S, and the halogens) are pre-defined, but the user may install any element in the periodic table through a special input routine.

The user of SYNLIB can specify the reacting center and its environment. Thus, both the atoms and bonds which are to be modified and the neighboring atoms which have an influence on the reaction can be defined. The compound search mode allows for both broad, generic searches or exact matches for a given structure. Constraints on the reaction conditions, types of bonds formed or broken, stereochemistry, or minimum yield can also be specified. Textual input of search keys provides an additional dimension to the reaction search using SYNLIB. Thus, such things as the yield, solvents, reagents, etc. can be searched.

It remains to be seen what impact the CASREACT database will have on the market for the in-house reaction systems discussed in this section. It is likely that the capability to create a local database will keep them very much in demand for some years to come.

11.9 SUMMARY

Reaction chemists have been rather slow to adopt computer-based search techniques in comparison to analytical or physical chemists. This is perhaps because the reaction chemists have not had to rely on a computer as an essential component of their everyday work. The design of a synthesis is a complex process, drawing on many sources. In addition to traditional sources of information like *Beilstein*, *Gmelin*, and *Chemical Abstracts*, reaction chemists can use a number of specialized treatises, abstracting services, reviews and surveys, and compendia in their quest for information. Furthermore, computer-based assistance is now available in forms ranging from microcomputer databases to large databases on in-house standalone systems, and finally to reaction databases on the commercial vendors' systems.

11.10 SELECTED READINGS

Gund, Peter; Hoff, Dale R. "Computer-Assisted Synthesis Design." In *Components of Scientific Instruments and Applications of Computers to Chemical Research*; 2nd ed.; Rossiter, Bryant W.; Hamilton, John F., Eds.; Physical Methods of Chemistry; Wiley: New York, 1986; Vol. 1, Chapter 9, pp 775-805.

Modern Approaches to Chemical Reaction Searching; Willett, Peter, Ed.; Gower Publishing Company Limited: Aldershot, Hants, England, 1986. [1505]

French, Stephen E. "Our Reaction Access System." *Chemtech* **1987**, *17*(2), 106-111.

Mills, John E. "Reaction Search Strategies using REACCS." *Am. Lab.* **1988**, *20*(2), 154-159.

Buntrock, Robert E. "Chemical Reaction Searching: CASREACT and REACCS." *Database* **1988**, *11*(6), 124-127

REFERENCES

[1] "Reaction Indexing," In *Communication, Storage, and Retrieval of Chemical Information*; Ash, Janet E., Ed.; E. Horwood: Chichester; Halsted Press: New York, 1985; Chapter 8, p 208.

CHAPTER 12

SEARCHING FOR CHEMICAL SAFETY OR TOXICOLOGY INFORMATION

12.1 INTRODUCTION

Each year there are stories in the news media of chemical laboratory or plant accidents, chemical spills, negative results of past disposal methods, etc. Every chemist can contribute to improving the image of chemistry by engaging in the safest possible practices. Safe practices in the handling or use of chemicals will prevent accidents or injuries. However, if an emergency occurs, procedures for dealing with it must be easily accessible, especially when toxic substances are involved. **Toxicity** is the ability of a chemical substance to cause injury once it reaches a susceptible site in or on the body. **Toxicology** is the study of chemical, physical, and biological hazards to biological systems and the ecosphere.

There have been many reference works and databases developed in the last few decades to assist in dealing with chemical safety or toxicology questions. Some were generated as a result of legislative action aimed at regulating the chemical industry. In compliance with these regulations, the chemical manufacturing industries have produced very useful trade literature about the potential hazards of their products. Under current "right-to-know" rules, workers must be warned by their employers of the hazards posed by chemicals to which they are exposed. This may involve labeling of the substances or the provision of **material safety data sheets**. Such sources provide complete, accurate, and current information on substances which are **carcinogenic** (cancer-causing), **mutagenic** (causing biological mutations), or **teratogenic** (causing

TABLE 12.1
Toxic Chemical Information Needs

1. IDENTIFICATION OF CHEMICAL
    Name
    Molecular formula
    Structural formula
    CAS registry number
    NIOSH number
    RTECS number
2. ECONOMIC DATA
    Production volume
    Marketing volume
    Import/Export data
3. USE OF CHEMICALS
    Use
    Use pattern
4. DETECTION IN THE ENVIRONMENT
    Air
    Soil
    Sediment
    Water
    Biosphere
    Food
    Drinking water
    Sewage sludge
5. PHYSICAL-CHEMICAL PROPERTIES
    Molecular weight
    Melting point
    Boiling point
    Flash point
    Hydrolysis
    Vapor pressure
    Density
    Surface tension
    Solubility in organic solvents
    Partition coefficients (K_{ow}, K_{oc}, etc.)
    Henry constant (H)
6. DEGRADATION/ACCUMULATION
    Biodegradation
    Photodegradation
    Bioaccumulation
    Bioconcentration
7. ECOTOXICITY
    Daphnia toxicity
    Fish toxicity
    Algae toxicity
    Plant toxicity
    Bacteria toxicity
8. ACUTE MAMMALIAN TOXICITY
    Oral toxicity
    Dermal toxicity
    Inhalative toxicity
    Skin sensation
    Skin irritation
    Eye irritation
9. SUBACUTE, SUBCHRONIC, CHRONIC TOXICITY (MAMMALS)
    Subacute toxicity
    Subchronic toxicity
    Chronic toxicity
10. GENOTOXICITY
    Cancerogenicity
    Mutagenicity
    Reproduction toxicity

birth defects). In addition, other symptoms or problems, such as reproductive disorders or behavioral problems caused by chemical substances must be disclosed. Table 12.1 shows the variety of information needs which may arise in this area.

Some useful printed guides to relevant information sources are Wexler's *Information Resources in Toxicology* [48] (1987), Webster's *Toxic and Hazardous Materials: A Sourcebook and Guide to Information Sources* [1590] (1987), and *Some Publicly Available Sources of Computerized Information on Environmental Health and*

Toxicology [1772] (1988). Finding environmental data is often quite difficult because there are so many diverse sources. In order to keep track of widely scattered environmental data, the National Environmental Data Referral Service (NEDRES) was created. The NEDRES database [1773] is available online. Providing descriptions of environmental data files, of published data sources, of data file documentation references, and of organizations that make environmental data available, the database is analogous to the broader service offered by the Library of Congress National Referral Center. Information on accessing the service can be obtained by calling the NEDRES Program Office at 202-634-7722. The American Chemical Society has a Health and Safety Referral Service designed to locate appropriate resources, whether books, journals, films, or individuals at government agencies. The number is 202-872-4515. However, for emergencies, ACS recommends that organizations such as Poison Control Centers or the Chemical Manufacturers Association's Chemical Transportation Emergency Center (800-424-9300) should be called.

12.2 MAJOR COLLECTIONS OF DATABASES CENTERED ON ENVIRONMENTAL OR TOXICOLOGICAL CONCERNS

For the last several decades, the public's increasing concern with questions of environmental pollution and toxicology has been paralleled by increasingly sophisticated computer techniques for information handling. Thus, there are some vendors of online information which have attempted to gather onto their systems a battery of databases which can help in answering environmental and toxicological questions. Chief among them are the National Library of Medicine and Chemical Information Systems, Inc.

12.2.1 NLM's Toxicology Information Program and the TOXicology Data NETwork (TOXNET)

The National Library of Medicine maintains two online search systems, one chiefly devoted to bibliographic databases, and the other, TOXNET, consisting primarily of factual databases. On July 1, 1985, the National Library of Medicine unveiled the TOXicology Data NETwork (TOXNET). TOXNET is a collection of databases oriented toward toxic problems associated with chemical substances. NLM has produced very good overviews of the available databases in the *TIP Files Demo Disk* [1939] and *CHEMLEARN, Microcomputer-Based Training for CHEMLINE* [1842].

NLM's Toxicology Information Program files include:

- CHEMLINE [593], NLM's online chemical dictionary (See section 7.4.2.), covering over 800,000 substances
- TOXLINE [219], a bibliographic file with over 800,000 citations from other secondary sources (See Table 12.2)
- TOXLIT [1940], a bibliographic file with over 1,600,000 citations drawn from other secondary sources (See Table 12.2)
- RTECS [211], the Registry of Toxic Effects of Chemical Substances, a database of approximately 100,000 potentially toxic substances

- HSDB [1776], the Hazardous Substances Data Bank, a database of about 4200 toxic and potentially toxic substances
- CCRIS [1609], the Chemical Carcinogenesis Research Information System, a database with test results on carcinogenicity, tumor promotion, and mutagenicity
- TRI [1941], the Toxic Chemical Release Inventory, a database which tracks industrial emissions of toxic chemicals into the environment
- DIRLINE [2020] and DBIR (Directory of Biotechnology Information Resources) [1974], directory files with information on organizations and resources which can be consulted for toxicological/environmental or biotechnology/molecular biology questions.

The major bibliographic files in the NLM Toxicology Information Program are TOXLINE and TOXLIT. These databases are actually formed from others, both governmental and private, as shown in Table 12.2. Royalty fees must be paid by the National Library of Medicine to Chemical Abstracts Service, the American Society of Hospital Pharmacists, or BioSciences Information Service respectively if citations are retrieved from any of the first three sources in Table 12.2. TOXLINE is now available as a CD-ROM product from SilverPlatter Information, Inc.

The TOXNET files include RTECS, HSDB, CCRIS, and TRI. RTECS [211], the *Registry of Toxic Effects of Chemical Substances*, contains information on research data for over 90,000 substances. Included are both acute and chronic toxic effects, such as data on skin or eye irritation, carcinogenicity, mutagenicity, and reproductive studies. There is comparatively less data on the individual substances in RTECS than is found in HSDB. (See Section 12.3.1 for more information on RTECS.)

TABLE 12.2
Secondary Sources From Which NLM's TOXLINE and TOXLIT Database Citations Are Drawn

Title (Database)	Coverage
Chemical-Biological Activities (CA)	1965-
International Pharmaceutical Abstracts (IPA)	1970-
Toxicological Aspects of Environmental Health (BIOSIS)	1970-
Pesticides Abstracts (PESTAB)	
Environmental Mutagen Information Center File (EMIC)	
Environmental Teratology Information Center File (ETIC)	1950-
Toxicology Research Projects (CRISP)	
Toxicology Document and Data Depository (NTIS)	
Toxicity Bibliography (TOXBIB)	1965-
Epidemiology Information System (EPIDEM)	
International Labour Office (CIS)	
Hazardous Materials Technical Center (HMTC)	
National Institute of Occupational Safety and Health Technical Information Center (NIOSHTIC)	
Poisonous Plants Bibliography (PPBIB)	
Aneuploidy File (ANEUPL)	
Toxic Substances Control Act Test Submissions (TSCATS)	

The Hazardous Substances Data Bank (HSDB) [1776] is more restricted in the number of substances covered, having detailed data on just over 4200 chemicals produced in large quantities. HSDB supplanted the old Toxicology Data Bank at the end of 1986. It includes all of the data found in that file, plus more environmental fate and exposure data, as well as more information on standards and regulations for the substances, more safety information, and more information on monitoring and analytical methods. Unlike RTECS, HSDB is a peer-reviewed and edited database. The information on an individual substance is exhaustive, and pages of printout can be obtained for a search on a single compound. Examples of the types of information which might be retrieved from a search of HSDB are listed in Table 12.3. Data for HSDB are obtained from a variety of sources, including government documents and reports, standard textbooks and monographs, and the primary journal literature.

CCRIS [1609], the Chemical Carcinogenesis Research Information System database, is also an evaluated database. Produced by the National Cancer Institute, CCRIS has information on carcinogenicity, tumor promotion, and mutagenicity for over 1200 chemical substances.

TRI, the Toxic Chemical Release Inventory database [1941], contains information on routine or accidental toxic chemical releases to the air, water, and land. The names and addresses of the facilities, plants, and factories, and the amounts of chemicals released into the atmosphere or transferred to waste sites are included in the records.

12.2.2 The Chemical Information System

For over a decade prior to 1985, a significant program of the U.S. federal government was devoted to environmentally related problems of chemical substances. That program, the Chemical Information System (CIS), was primarily the responsibility of the U.S. National Institutes of Health and the Environmental Protection Agency. As many as two dozen databases were included in the CIS, and all were bound together by the Structure and Nomenclature Search System (SANSS) [222], the CIS's online chemical

TABLE 12.3
Data Fields in the National Library of Medicine's Hazardous Substances Data Bank

Name	Category Label
Substance identification number	ID
Manufacturing/use information	MANF
Chemical and physical properties	CPP
Safety and handling	SAFE
Toxicity and biomedical effects	TOXB
Pharmacology	PHCY
Environmental fate/exposure potential	ENVS
Exposure standards and regulations	EXSR
Monitoring and analysis methods	MAM

dictionary (See section 7.4.3.). In November 1984, the CIS was "privatized," and eventually made available through only one private vendor, Chemical Information Systems, Inc., in a form which closely resembles the original structure. The new Chemical Information System has continued to grow and now includes more databases than ever.

There are five main areas covered by the Chemical Information System:

- chemical and physical properties databases (including spectroscopic and diffraction data, as well as some thermodynamic data)
- regulatory databases
- toxicological databases
- environmental databases
- mathematical analysis and modeling databases.

All but the last category of database can be accessed through searches of the SANSS file. CIS, Inc. has developed an "Index to the CIS," a work which lists topics covered by the various components of the CIS. For an overview of the databases available in the CIS as of January 1, 1990, search **CIS** in the Chemistry Reference Sources Database (CRSD).

The major CIS databases in the toxicology area are:

- BAKER [1771]
- MALLIN [1942]
- Suspect Chemicals Sourcebook [1775]
- Registry of Toxic Effects of Chemical Substances (RTECS) [211]
- Chemical Carcinogenesis Research Information System (CCRIS) [1609]
- Clinical Toxicology of Commercial Products (CTCP) [940]
- Chemical Evaluation Search and Retrieval System (CESARS) [1943]
- GENETOX [1944].

GENETOX contains mutagenicity data on more than 3000 biological systems, whereas BAKER and MALLIN provide material safety data sheets for chemical substances. Note that RTECS and CCRIS are also available through the National Library of Medicine. Another source in the online CIS system is MERCK, the *Merck Index* [177], which includes some information on the toxicity of the approximately 10,000 substances it covers. (Some of the CIS databases are discussed in more detail in Section 12.3.1.)

There are also a number of databases on the CIS which are of interest primarily for environmental questions, including emergency responses. Some of these are:

- Oil and Hazardous Materials Technical Assistance Data System (OHMTADS) [975]
- Plant and Production Data (TSCAPP) [1601]
- Scientific Parameters for Health and the Environment: Retrieval and Estimation (SPHERE) [1945]
- Chemical Hazard Response Information System (CHRIS) [1534]

- Database of Off-Site Waste Management (DOWM) [1946]
- PHYTOTOX [1947]
- Toxic Substances Control Act Test Submissions (TSCATS) [1948]
- Hazardous Chemicals Information and Disposal Guide (HAZINF) [1949]
- Information System for Hazardous Organics in Water (ISHOW) [1605].

SPHERE includes the ENVIROFATE database, as well as the DERMAL file for effects on skin and the ACQUIRE database for aquatic effects. DOWM has information on 365 disposal facilities searchable by particular types of wastes, geographic region, types of treatment used, etc. PHYTOTOX includes about 70,000 records for toxic chemical substances applied to plants. ISHOW covers more than 5000 chemicals and lists data such as partition coefficient, acid dissociation constant, solubility in water, and vapor pressure. (Some of the other databases listed above are discussed in Section 12.3.1.)

12.2.3 Other Vendors of Environmental and Toxicology Databases

All of the major vendors of online databases include some files of interest to those doing research in environmental and toxicological chemistry. Some databases, like RTECS, are found on almost all of them. Other databases, however, are accessible only through vendors of specialized databases like Technical Database Services, Inc.'s Numerica.

Numerica serves as a vendor of several numeric databases produced by Syracuse Research Corporation, a major information provider in the environmental area. They have developed a chemical dictionary, SYNDEX [1950], to help find databases on the system which include the compound. SYNDEX can be searched by name or CAS registry number. Software and databases of interest include:

- AQUIRE (aquatic toxicity test results) [1611]
- Carcinogenicity Predictor [1951]
- Environmental Fate Database (including BIOLOG, BIODEG, DATALOG, and CHEMFATE) [1864]
- CHEMEST [1865]
- Log P and Related Parameters Database [1860].

A product of the Commission of the European Communities Joint Research Centre is the Environmental Chemicals Data and Information Network (ECDIN) [1952]. The ECDIN database is designed to enable rapid retrieval of reliable information on chemicals of environmental significance. If a chemical is produced in Europe in quantities greater than 500 kilograms per year, it will be included in ECDIN. Also included are data on compounds known to be highly toxic even if produced in quantities under 500 kg., as well as data on selected toxic natural products. ECDIN includes information on about 30,000 substances, such as odor and taste threshold concentrations, analytical methods, occupational safety and health considerations, and toxicity data.

12.3 SOURCES OF INFORMATION ON CHEMICALS KNOWN OR THOUGHT TO BE TOXIC OR HAZARDOUS

It is not possible to know with certainty that chemicals with which we come into contact or are introduced into the environment will have no harmful effect. Letters to the editors of chemical news magazines frequently tell of unexpected explosions or other negative results of dealing with chemicals. Nevertheless, a tremendous body of information sources has been produced, especially for the most common substances encountered in industry.

As with any specialized area of science or technology, there is a specialized jargon used in the fields of toxicology and environmental studies. The *Dictionary of Toxicology* [1953] should help in defining terms in the field such as:

Threshold Limit Values (TLVs) - the highest levels of exposure allowed for humans; the level at which no harmful effect is noted

LD_{50} - the median lethal dose; the estimate of the amount of a chemical (expressed in mg/kg of body weight) required to be ingested in order to kill 50 percent of a population of test animals.

12.3.1 Sources Available as Databases

The U.S. Coast Guard has developed the Chemical Hazards Response Information System (CHRIS) [1534]. In early 1989, the CHRIS database contained 1,156 records. CHRIS is also available as a four-volume printed set consisting of:

v.1. A Condensed Guide to Chemical Hazards
v.2. Hazardous Chemical Data
v.3. Hazards Assessment Handbook
v.4. Response Methods Handbook.

The Environmental Protection Agency's Oil and Hazardous Materials Technical Assistance Data System (OHM/TADS) [975] provides useful data on hazardous chemicals, as Figure 12-1 shows. OHM/TADS is also available in printed form as a microfiche product. It consists of: 1.) Material Name Index 2.) CAS Registry Index 3.) Data. OHM/TADS has information on 1400 compounds. Safety precautions, hazard levels, fire protection, explosiveness, handling and storage procedures, disposal information, effects on humans, and effects on the environment are part of the information found in OHM/TADS.

Both CHRIS and OHM/TADS are components of the Chemical Information System. There are several other avenues to the information contained in those databases. They are available as personal computer databases: microCHRIS and microOHMTADS [1770]. There is also a CD-ROM product, CHEMBANK [1766], which combines the two databases with the Registry of Toxic Effects of Chemical Substances (RTECS) [211], a product of the National Institute for Occupational Safety and Health.

RTECS was originally known as the *Toxic Substances List*. It is available as a multi-volume paper set; 1985-86 is the latest edition, with supplements published

CHEMICAL SAFETY OR TOXICOLOGY INFORMATION 253

(1) Accession Number   7216796
(2) CAS Registry Number: 78-93-3
(3) SIC Code: 2892
(4) Material Name: METHYL ETHYL KETONE
(5) Synonyms: 2-BUTANONE; METHYL ACETONE; BUTANONE; ETHYL METHYL KETONE; MEK
(6) Tradename (Company): MEETCO
(7) Chemical Formula: CH3.COCH2.CH3
(8) Species in Mixture: 99% PURE
(9) Common Uses: SOLVENT; SYNTHETIC RESINS; SURFACE COATING INDUSTRY; SMOKELESS POWDER
(10) Transport, Rail (%): 43.6
(11) Transport, Barge(%): 28.8
(12) Transport, Truck (%): 27.3
(14) Containers: TANK CARS AND TRUCKS, METAL DRUMS AND CONTAINERS, GLASS BOTTLES (ALL ICC SPEC.).
(15) General Storage Procedure: DRUM AND OTHER CONTAINER STORAGE IN SEPARATE ROOMS, TRAPPED FLOOR DRAINS PROVIDED. ALL STORAGE AREAS, AUTOMATIC SPRINKLERS OR OTHER ADEQUATE SYSTEM. MECHANICAL EXHAUST ROOM VENTILATION (6 CHANGES/HOUR) STORE AWAY FROM OTHER COMBUSTIBLES, ACIDS, OXIDIZING MATERIALS. AVOID STORAGE IN GLASS CONTAINERS. BULK STORAGE - PROBLEMS OF VENTING, DIKING, SEPARATION ETC.
(16) General Handling Procedure: SPARK RESISTANT TOOLS SHOULD BE USED.
(17) Production Sites: CELANESE CORP. OF AMERICA, PAMPA, TX; HUMBLE OIL AND REFINING CO., LINDEN, NJ; SHELL CHEMICAL CO., DOMINGUEZ, CA; HOUSTON, TX; NORCO, LA; SINCLAIR PETROCHEMICALS, INC., CHANNELVIEW, TX; UNION CARBIDE CORP., CHEMICAL DIV., BROWNSVILLE, TX.
(21) Corrosiveness: SOFTENS AND DISSOLVES PLASTICS.
(24) Detection Limit (Field; Techniques,Ref) (ppm): 200, KETONES, (BNW 420160)
(25) Detection Limit (Lab; Techniques,Ref) (ppm): .0005, C7C, (BNW 420335)
(26) Standard Codes: NFPA - 1,3,0; ICC - FLAMMABLE LIQUID, RED LABEL, 10 GALLON IN AN OUTSIDE CONTAINER; USCG - GRADE C FLAMMABLE LIQUID; IATA - FLAMMABLE LIQUID, RED LABEL, 1 LITER PASSENGER, 40 LITER CARGO.
(27) Flammability: QUITE FLAMMABLE, COMBUSTION PROBABLE.
(28) Flammability Limit(%), Lower: 1.8
(29) Flammability Limit(%), Upper: 10
(30) Toxic Combustion Prod.: SLIGHTLY HAZARDOUS, USE CANISTER TYPE MASK.
(31) Extinguishing Method: WATER MAY BE INEFFECTIVE. "ALCOHOL FOAM", CARBON DIOXIDE, DRY CHEMICALS, AND FOAM. ORDINARY FOAM IN GREATER THAN NORMAL QUANTITIES. WATER SPRAY MAY BE EFFECTIVE TO CONTROL FIRES IN OPEN CONTAINERS AND WATER STREAMS SHOULD BE USED FOR COOLING TANKS AND DRUMS. RESPIRATORY PROTECTION.
(32) Flash Point (C.): -6.11
(33) Auto Ignition Point(C.): 515.56
(34) Explosiveness: STABLE - VAPORS. EXPLOSIVE MIXTURES WITH AIR OVER FAIRLY WIDE RANGE OF CONCENTRATIONS.
(35) Explosive Limit(%), Lower:
(36) Explosive Limit(%), Upper: 12
(37) Melting Point (C.): -86
(39) Boiling Point (C.): 79.6
(41) Solubility (ppm @ 25C): 100000.
(43) Specific Gravity: .805

Fiche #29 G8

Figure 12-1
OHM/TADS Entry for Methyl Ethyl Ketone

thereafter. There is also a microfiche version *RTECS Quarterly Microfiche*, which reproduces the entire file each October. Basic toxicity data are listed for each of the approximately 90,000 compounds in RTECS. (The original journal sources in *CASSI* format are now included with each record.) Data for skin and eye irritation, mutation, reproductive effects, and tumorigenic and toxicity data are part of the records. The vast majority of the compounds have CAS registry numbers in RTECS. Appendices

```
21754. 2-BUTANONE
RTECS: EL6475000   UPDT: 8703   CAS: 78-93-3   MW: 72.12   MF: C4H8-O
SYNONYMS: ACETONE, METHYL- * AETHYLMETHYLKETON (German) * BUTANONE * BUTANONE 2
  (French) * ETHYL METHYL CETONE (French) * ETHYLMETHYLKETON (Dutch) * ETHYL METHYL
  KETONE * ETHYL METHYL KETONE (DOT) * KETONE, ETHYL METHYL * MEETCO * MEK *
  METHYL ACETONE * METHYL ACETONE (DOT) * METHYL ETHYL KETONE * METHYL ETHYL
  KETONE (ACGIH,DOT) * METILETILCHETONE (Italian) * METYLOETYLOKETON (Polish) * RCRA
  WASTE NUMBER U159 * UN 1193 (DOT) * UN 1232 (DOT)
SKIN AND EYE IRRITATION DATA AND REFERENCES:
    eye; human; 350 ppm; JIHTAB 25,282,43 [J Ind Hyg Toxicol]
    skin; rabbit; 500 mg/24H ; MODERATE; JIHTAB 25,282,43 [J Ind Hyg Toxicol]
    skin; rabbit; 402 mg/24H ; MILD; TXAPA9 19,276,71 [Toxicol Appl Pharmacol]
    skin; rabbit; 13780 ug/24H open ; MILD; AIHAAP 23,95,62 [Am Ind Hyg Assoc J]
    eye; rabbit; 80 mg; TXAPA9 19,276,71 [Toxicol Appl Pharmacol]
MUTATION DATA AND REFERENCES:
    sex chromosome loss and nondisjunction;
    S. cerevisiae; 33800 ppm; MUREAV 149,339,85 [Mutat Res]
REPRODUCTIVE EFFECTS DATA AND REFERENCES:
    inhalation;
      rat; TCLo:3000 ppm/7H (6-15D preg); SPECIFIC DEVELOPMENTAL
        ABNORMALITIES(Craniofacial; Urogenital system; Homeostasis); TXAPA9 28,452,74
        [Toxicol Appl Pharmacol]
      rat; TCLo:1000 ppm/7H (6-15D preg); EFFECTS ON EMBRYO OR FETUS(Fetotoxicity);
        SPECIFIC DEVELOPMENTAL ABNORMALITIES(Musculoskeletal system); TXAPA9
        28,452,74 [Toxicol Appl Pharmacol]
TOXICITY DATA AND REFERENCES:
    inhalation;
      human; TCLo:100 ppm/5M; SENSE ORGANS AND SPECIAL SENSES(Other olfaction
        effects; Conjunctiva irritation); LUNGS, THORAX OR RESPIRATION(Other changes);
        JIHTAB 25,282,43 [J Ind Hyg Toxicol]
      mouse; LC50:40 gm/m3/2H; NO TOXIC EFFECT NOTED; 85GMAT -,83,82 [Toxic
        Param Ind Tox Chem Under Single Exposure 1982]
      rat; LCLo:2000 ppm/4H; TOXIC EFFECTS NOT YET REVIEWED; JIHTAB 31,343,49 [J
        Ind Hyg Toxicol]
    intraperitoneal;
      guinea pig; LDLo:2000 mg/kg; LIVER(Other changes); IMMUNOLOGICAL INCLUDING
        ALLERGIC(Other immediate); BIOCHEMICAL EFFECTS(Lipids including transport);
        FCTXAV 15,627,77 [Food Cosmet Toxicol]
      mouse; LD50:616 mg/kg; TOXIC EFFECTS NOT YET REVIEWED; SCCUR* -,6,61 [Shell
        Chem Co Unpubl Rep]
    oral;
      mouse; LD50:4050 mg/kg; NO TOXIC EFFECT NOTED; TOLED5 30,13,86 [Toxicol Lett]
      rat; LD50:2737 mg/kg; NO TOXIC EFFECT NOTED; TXAPA9 19,699,71 [Toxicol Appl
        Pharmacol]
    skin;
      rabbit; LD50:13 gm/kg; TOXIC EFFECTS NOT YET REVIEWED; UCDS** 5/7/70 [Union
        Carbide Data Sheet]
LITERATURE REVIEW:
    ACGIH THRESHOLD LIMIT VALUE REVIEW;
      ★TWA 200 ppm; STEL 300 ppm; 85INA8 5,395,86 [Doc Threshold Limit Values]
    TOXICOLOGY REVIEW;
      ECIET* (3),-,83 [Eur Chem Ind Ecol Toxicol Cent]
STANDARDS AND REGULATIONS:
      ★DOT-HAZARD:FLAMMABLE LIQUID; LABEL:FLAMMABLE LIQUID; CFRGBR
        49,172.101,86 [Code Fed Regul]
      ★MSHA STANDARD-air:TWA 200 ppm (590 mg/m3); DTLVS* 3,29,71 [Doc Threshold
        Limit Values]
      ★OSHA STANDARD-air:TWA 200 ppm; FEREAC 39,23540,74 [Fed Regist]
CRITERIA DOCUMENT:
      ★NIOSH REL TO KETONES-air:TWA 590 mg/m3; MMWR** 34(1S),20S,85 [Morbid
        Mortal Weekly Rep]
STATUS:
      ★EPA GENETOX PROGRAM 1986, Inconclusive: B subtilis rec assay
      EPA TSCA CHEMICAL INVENTORY, 1986
      EPA TSCA 8(a) PRELIMINARY ASSESSMENT INFORMATION, FINAL RULE; FEREAC
        47,26992,82 [Fed Regist]
      EPA TSCA TEST SUBMISSION (TSCATS) DATA BASE, DECEMBER 1986
      NIOSH ANALYTICAL METHODS: see 2-BUTANONE, 2500; in blood, see 2-butanone,
        ethanol, 8002
      NTP CARCINOGENESIS STUDIES;SELECTED, JANUARY 1987
      MEETS CRITERIA FOR PROPOSED OSHA MEDICAL RECORDS RULE; FEREAC
        47,30420,82 [Fed Regist]
```

Figure 12-2
Methyl Ethyl Ketone Entry from the *Registry of Toxic Effects of Chemical Substances* (RTECS)

include "NIOSH Recommendations for Occupational Safety and Health Standards" and "NIOSH Sampling and Analytical Methods," the latter being an index to the methods. The RTECS database can be searched with controlled vocabulary for routes, species, and cell types or for toxic effects and the specific effect or organ system affected.

The *Kirk-Othmer Encyclopedia of Chemical Technology* [551, 916] should never be overlooked as a source of basic information on a chemical or class of compounds. Toxicological references are often found there.

ChemPro [1845] from Fisher Scientific provides information on over 4100 potentially hazardous substances. The microcomputer database allows primary entry by FIRE, SPILL, or POISON and the name of the substance. The record for each indicates firefighting, decontamination, or first aid procedures. Data are condensed from the National Library of Medicine's Hazardous Substances Data Bank (HSDB). The software features a telecommunications link to the larger HSDB if more information is desired.

The CHEMTOX database [1769] covers more than 4600 regulated substances, including all substances regulated by OSHA or other regulatory acts, as well as substances found to be carcinogenic. Included in the records are physical and chemical properties, toxicological data, regulatory information, and recommendations for responding to an emergency. The database is available in a basic or expandable version. The expandable version allows the user to create other files and link them to the CHEMTOX records. Symptoms and properties could be input to find, for example, all substances in the file which cause dizziness and are yellow liquids.

Health Designs, Inc. has created a menu-driven software package, TOPKAT [1953], which predicts toxicity endpoints from the structures of chemicals. Existing models are supplied as a database and new structure-activity-relationship models are available. Mutagenesis, carcinogenesis, teratogenesis, skin irritation and LD_{50} predictions are made by the software. Occupational Health Services' Material Safety Data Sheet Reference File [1955] is a CD-ROM database which includes all substances covered by federal and state occupational health and safety regulations.

Clinical Toxicology of Commercial Products [940] (1984) is a classic work now available both as a database on CIS and as a book. In addition to first aid and general emergency treatment, the work has an "Ingredients Index" for substances commonly found in products used in and around homes and farms. The index provides access both by name and CAS registry number to over 1600 substances. There is also a separate "Trade Name Index" of over 15,000 non-food products which might be ingested. The component of the substance which probably would cause a harmful effect is starred with an asterisk. For example, Easy-Off Oven Cleaner has sodium hydroxide indicated as the probable harmful ingredient if ingested. There is even a section of "General Formulations" where typical ingredients of hundreds of products (for example, oven cleaners) are listed. *Clinical Toxicology of Commercial Products* is especially intended to help the physician with a diagnosis and treatment for acute chemical poisoning.

The Pesticide Databank, the online version of *The Pesticide Manual* [961] (1987), lists pesticides in current use. Included are both chemicals and microbial agents which control pests in crops and animals. Furthermore, plant growth regulators, pest repellents, and other substances are covered in the work. Each entry includes a concise summary of the toxicology of the substance. The substances are indexed by Wiswesser Line-Formula Notation, CAS registry numbers, molecular formulas, as well as chemical, common, and trivial names, and trade marks.

There are two databases on the Chemical Information System which contain material safety data sheets (MSDSs). BAKER [1771] has more than 1600 MSDSs formatted as required under the Occupational Safety and Health Act. A microcomputer version of the database, Saf-T-Manager [1771], is available from the producer, the J. T. Baker Chemical Company. BAKER contains primarily industrial chemicals. MALLIN [1942], produced by Mallinckrodt, Inc., has more than 1400 OSHA-formatted MSDSs for laboratory/electronic chemicals.

Much larger in the scope of its coverage is the Sigma-Aldrich Material Safety Data Sheets database [1956]. Available on CD-ROM, it contains over 30,000 MSDSs. The product is updated quarterly and is searchable by CAS registry number, name, or molecular formula. The text can be exported to any standard word-processing package, and graphic chemical structures can be printed if the WIMP software [99] is available. Another CD-ROM product is CCINFOdisc [1969], a very comprehensive source from the Canadian Centre for Occupational Health and Safety.

12.3.2 Sources Available Only in Printed Format

The *Sigma-Aldrich Library of Chemical Safety Data* [977] (1988) incorporates selected data from the *Registry of Toxic Effects of Chemical Substances*: the LD_{50} data by oral and dermal routes and LC_{50} data for inhalation, threshold limit values, OSHA standards, and results of carcinogenesis studies. The RTECS accession number is included in the entries to facilitate finding more information in RTECS. Mostly acute health effects are listed in the *Sigma-Aldrich Library*. Further information includes decomposition products and ''conditions which may cause a deleterious change in a chemical, sometimes resulting in a hazardous condition'' (incompatibility). Most of the products supplied by Sigma-Aldrich are used in laboratory quantities, but there are numerous entries for chemicals used in industry. Handling and storage tips, information on waste disposal, spills or leaks, and fire-extinguishing media are also included. Over 14,500 chemicals are covered in the second edition.

Patty's Industrial Hygiene and Toxicology [948] is divided into three volumes:

v. 1. General Principles
v. 2. Toxicology
v. 3. Theory and Rationale of Industrial Hygiene Practice.

The second ''volume'' is actually three physical volumes. Chapters are devoted to classes of compounds, such as metals, esters, ethers, and the like. The third part of volume 2 contains a Cumulative Index covering all three parts of that volume. Entries typically have a brief summary of physical-chemical properties followed by extensive reviews of biological effects. Hygienic standards are also listed. For example, under

MEK, it is stated, "The American Conferences of Governmental Industrial Hygienists has established a threshold limit of 200 ppm for methyl ethyl ketone."

A standard work in the field is Sax and Lewis's *Dangerous Properties of Industrial Materials* [976], now in its 7th edition (1989). Over 20,000 substances are covered. Entries include physical and chemical properties and health-related data. Sax and Lewis is now published in three volumes, the first being an introduction (which also contains the "Chemical Name Index," a CAS registry number index, and bibliographic references for the more than 2200 items cited in the complete work). Volumes 2 and 3 contain the actual entries. The authors use a numerical scale to indicate the level of toxicity:

3 = LD_{50} below 400 mg/kg
2 = LD_{50} of 400 - 4000 mg/kg
1 = LD_{50} of 4000 - 40,000 mg/kg.

Aquatic toxicity data are included if available. For those who do not wish to purchase the three-volume seventh edition of Sax and Lewis, their *Hazardous Chemicals Desk Reference* [1798] (1987) or *Rapid Guide to Hazardous Chemicals in the Workplace* [1674] (1986) might suffice. The former covers nearly 4700 of the most hazardous chemicals. Sax also produces serial publications designed to update the more extensive compilations between editions. *Hazardous Chemicals Information Annual* [970] first appeared in 1986.

Verschueren's *Handbook of Environmental Data on Organic Chemicals* [995] (1983) covers not only individual substances, but also mixtures and preparations. For each entry, the name, synonyms, and formula are given, followed by properties, air pollution factors, water pollution factors, and biological effects. Only the physical and chemical properties of direct environmental concern are listed in the *Handbook*. Verschueren presents a detailed discussion of the properties listed in the introduction to the work, including a glossary. Compounds are arranged alphabetically by name, but there is a molecular formula index to assist in identifying the name. The detailed bibliography contains 2355 references.

12.4 SAFE HANDLING, USE, AND DISPOSAL OF CHEMICALS

Of concern in this section are works which deal with generally accepted practices for insuring that risks due to exposure to chemical substances are kept to a minimum. Hence, topics like the design of laboratories, safe laboratory practices, and safe disposal techniques for chemicals are covered in some of the works discussed below.

12.4.1 Books

Bretherick's *Handbook of Reactive Chemical Hazards* [968] (1985) covers about 4900 compounds. The work now includes CAS registry numbers and quantitative information on the energy of decomposition of compounds or mixtures. The author specifically excludes information on toxic hazards since it is available in many other sources.

Bretherick pays particular attention to substances which are highly flammable or which ignite on exposure to air.

A major publication of the National Research Council is *Prudent Practices for Handling Hazardous Substances in Laboratories* [476] (1981). The first section is devoted to "Procedures for Working With Chemicals in Laboratories". A wide range of practical techniques is presented for dealing with toxic substances or substances which are flammable or have a tendency to explode. Protective apparel, safety equipment, and design characteristics (including ventilation) are also topics discussed. The second section, "Procedures for the Procurement, Storage, Distribution, and Disposal of Chemicals," covers all aspects of the life of a chemical substance in an organization, from purchase through disposal. There are also several useful tables or appendices in the book, such as "Threshold Limit Values for Chemical Substances and Physical Agents in Workroom Air," and "Flash Points, Boiling Points, Ignition Temperatures, and Flammable Limits of Some Common Laboratory Chemicals". A companion work is *Prudent Practices for Disposal of Chemicals from Laboratories* [477] (1983). The book is geared toward disposal of the smaller quantities of materials used in the laboratory.

Safety in the Chemical Laboratory [1427] (4 volumes covering 1967-1980) is a collection of articles which originally appeared in the *Journal of Chemical Education*. Each month the *Journal* includes an article on this topic. The American Chemical Society has issued the 4th edition of *Safety in Academic Chemistry Laboratories* [1383] (1985). Other titles in this area are *Managing Safety in the Chemical Laboratory* [1957] (1988) and *Improving Safety in the Chemical Laboratory: A Practical Guide* [1958] (1987). The latter work includes an appendix containing "Laboratory Safety Library Holdings," an annotated bibliography.

Approaching safety from the construction angle are *Guidelines for Laboratory Design: Health and Safety Considerations* [1872] (1987) and *The Chemical Laboratory: Its Design and Operation* [1425] (1987).

12.4.2 Abstracts, Indexes, Databases

Besides the TOXLINE and TOXLIT databases of the National Library of Medicine, there are several specialized abstracting and indexing services and other databases which cover the safe handling and use of chemicals.

CAMEO II [1959], available from the National Safety Council, is a program for "Computer Aided Management of Emergency Operations". Written for the Macintosh, CAMEO II should help emergency response personnel and planners deal with situations involving any of over 2600 commonly transported chemicals. In addition to response information, the database helps to locate chemical storage depots and to evaluate evacuation options.

The standard A&I service for environmental questions is the set formed by *The Environment Abstracts Annual* and *The Environment Index*. Available online as Enviroline [607], the database contains citations to much literature that would not be covered in a more general abstracting service like *Chemical Abstracts*.

The Royal Society of Chemistry produces three relevant works: *Laboratory Hazards Bulletin* [619] (1981-), *Chemical Hazards in Industry* [1960] (1984-), and the Chemical Safety NewsBase [1961] (1984-), which is an amalgamation of the two previously named works, with all duplicates eliminated. The database contained over 17,000 items by the end of 1988 and grows by over 300 items per month.

Other relevant databases are Safety Science Abstracts [1962], HSELINE [1963], a database covering the literature of health and safety in several areas of interest to chemists, and NIOSHTIC [1964], a database produced by the Technical Information Center of the National Institute for Occupational Safety and Health.

12.5 CHEMICAL LAWS AND REGULATIONS: LEGAL REQUIREMENTS FOR THE CHEMICAL INDUSTRY

Laws and Regulations pertaining to chemicals apply to all types and sizes of organizations which produce or deal directly with chemicals. Public Laws in the United States are annually compiled in chronological order in the *U.S. Statutes at Large* and codified (classified) in the *U.S. Code*. Regulations are codified annually in the *Code of Federal Regulations*, a compilation of the rules and regulations promulgated daily in the *Federal Register*. Such rules and regulations originate in various regulatory bodies of the executive branch of government, but have the same weight of authority as laws passed in the legislative branch. Various laws having to do with the transport of materials come into play in the chemical industry in addition to numerous food, drug, and environmental laws. For detailed information, refer to Chapter 16 "Legislation and Regulations" in Wexler's *Information Resources in Toxicology* [48] (1987) and to Chapter 10 "Laws and Regulations" in Webster's *Toxic and Hazardous Materials: A Sourcebook and Guide to Information Sources* [1590] (1987).

Some of the major laws which are frequently mentioned in the popular press are:
- Comprehensive Environmental Response, Compensation, and Liability Act of 1980 (SUPERFUND) [PL 96-510]
- Occupational Safety and Health Act of 1970 [PL 91-596]
- Toxic Substances Control Act of 1976 [PL 94-469].

The last-named act, TSCA, has generated a number of the compilations mentioned earlier in this chapter. In this section works are discussed which help in tracking the various laws and regulations which apply to a chemical substance.

CHEMLIST [1965] is a database on STN which contains information on substances listed in the TSCA Inventory or subject to regulations under TSCA or similar legislation. The following sources are used to build the database from 1979:
- EPA Toxic Substances Control Act Inventory [645, 1062]
- Federal Register [1606]
- Chemical Regulation Reporter [1966]
- EPA Chemicals-in-Progress Bulletin [1967]
- Toxic Substances Control Act Test Submissions (TSCATS) [1948].
- American Petroleum Institute's TOXLIST file.

Thus, in one source STN provides information on over 70,000 substances, with over 2000 additions annually. A summary of the action on the chemical and the source of the data are provided.

The SUSPECT database and its print counterpart, the *Suspect Chemicals Sourcebook* [1775], contain information on regulations from approximately three dozen regulatory programs, covering 3500 chemicals. By entering the name or CAS registry number, all regulations applicable to a given chemical can be found. Another option is to enter a particular regulatory code to find all chemicals covered by that regulation. SUSPECT is searchable on the Chemical Information System. For many of the entries, relevant data are extracted from the regulations and provided in the database.

The Chemical Regulations and Guidelines System (CRGS) [592] covers current regulatory materials and is, in effect, an index to the *Federal Register* for chemical substances. Access is by name, CAS registry number, or a chemical role tag (an indicator of the context in which the substance appears in the regulatory document). CRGS also links together the statutes, regulations promulgated under the statutes, and support documents written prior to the promulgation of a regulation.

A major information producer in this area is BNA, the Bureau of National Affairs. BNA produces many legal reporting services, among them:

- Chemical Right-to-Know
- Chemical Regulation Reporter
- Chemical Substances Control
- Environment Reporter
- Occupational Safety and Health Reporter
- Toxics Law Reporter.

BNA's *Chemical Regulation Reporter* [1966], which contains the texts of U.S. federal chemical regulations as published in the *U.S. Code of Federal Regulations* or updated in the *Federal Register*, plus relevant standards, is available online.

Newsletters are a prime source of information in this area. The Newsnet service [1970] provides full texts of newsletters in many subject areas. Included are a number of relevant sources for tracking safety and environmental regulations.

Hazardline [613], searchable only direct from Occupational Health Services, Inc., contains information on approximately 10,000 substances, including federal and state regulatory information. A full-text source on CD-ROM is ERM Computer Services' ENFLEX INFO [1971]. The only drawback to depending on a CD-ROM product for this type of information is the relative infrequency of updating (quarterly), compared to the weekly or even daily updating of online databases in this area.

12.6 SUMMARY

Information about the best-known toxic or hazardous chemicals is very detailed, ranging from chemical and physical properties to toxicological data and health and safety information. Data on other substances may be harder to find, but there are many and varied sources to consult, including printed handbooks, online databases, CD-ROM databases, newsletters, legal documents, etc. Chemical safety or toxicology information encompasses a tremendously broad field, but one which is rich in information sources.

12.7 SELECTED READINGS

Freeman, Robert R.; Smith, Mona F. "Environmental Information." *Annu. Rev. Inf. Sci. Technol.* **1986**, *21*, 241-305.

Kissman, Henry M.; Wexler, Philip. "Toxicological Information." *Annu. Rev. Inf. Sci. Technol.* **1983**, *18*, 185-230.

Kissman, Henry M.; Wexler, Philip. "Toxicology Information Systems: A Historical Perspective." *J. Chem. Inf. Comput. Sci.* **1985**, *25*(3), 212-217.

Clansky, Kenneth B. "How to Keep Up With Chemical Regs." *The Scientist* **May 18, 1987**, *1*(13), 18-19.

Clansky, Kenneth B. "Info Services for Chemical Regs." *The Scientist* **June 1, 1987**, *1*(14), 18.

Snow, Bonnie. "Monitoring Bioscience Legal and Regulatory News." *Online* **1988**, *12*(6), 107-117.

Van Camp, Ann J. "The TOXNET Gateway." *Online* **1989**, *13*(4), 70-74.

Van Camp, Ann J. "Material Safety Data Sheets: Online and CD-ROM Sources." *Online* **1990**, *14*(2), 97-99.

Chapter

13

CURRENT AWARENESS, RESEARCH IN PROGRESS, BACKGROUND READING, AND DOCUMENT DELIVERY SERVICES

13.1 INTRODUCTION

For decades, librarians and others have attempted to provide services to keep scientists up to date on what is appearing in the new primary literature. The several different activities, printed products, and computer-based services which have this as their goal are collectively referred to as **current awareness services**. Among the types of services provided are customized tables of contents services, printed table-of-contents journals from commercial publishers, computerized SDI (selective dissemination of information) services, and standard interest profiles. Each of those will be explored in detail in this chapter. In some instances, it is necessary or desirable to discover what kind of research is currently being pursued. Sources to determine **research in progress** long before it is published in the primary literature are also treated in this chapter.

The reverse of current awareness is background reading. A chemist may need to explore a topic in depth in an unfamiliar area of chemistry or perhaps even in another scientific discipline. Fortunately, the practice of organizing scientific knowledge into collected secondary works allows relatively easy retrieval of information on many topics. By consulting reviews, encyclopedias, treatises, monographs, multigraphs, and in some cases, dictionaries and textbooks, **background reading** can be explored to apply the older literature to the solution of a current problem. Chemistry has a long tradition of compiling such secondary works to better organize the most important scientific discoveries for easy retrieval and consultation. Treatises were discussed in previous chapters at appropriate places. Therefore, the emphasis in this chapter is on reviews, encyclopedias, dictionaries, and other smaller reference works for background reading.

Once the chemist becomes aware of the existence of interesting primary literature (or even secondary, collected literature), the problem of document delivery arises. If the library of the chemist's home institution actually holds the title, obtaining it is usually straightforward. However, the scope of materials included in relevant databases and other secondary works is becoming ever wider. Since many library materials budgets are inadequate to purchase even the majority of the relevant books and journals, different avenues of **document delivery** must be utilized. Among the choices available are interlibrary loan, commercial document suppliers, reprints obtained from the authors, and text databases of primary literature.

13.2 CURRENT AWARENESS SOURCES

There are a number of variables which must be weighed in the selection or provision of a current awareness service. These include the degree to which the citations provided to the users match their interests (relevance), the amount of labor needed to provide the service, the cost, and the time lag between the appearance of the primary document and the notice of its existence which the current awareness service gives to the user.

13.2.1 Table-of-Contents Services

Many libraries provide to their users a **table-of-contents service** using photocopies of the contents pages of currently received journals. A chemist supplies a list of key journals which usually contain a significant amount of literature of interest. When the latest issue of each title on the list arrives in the library, a photocopy of the table of contents is made and sent to the user. This alerts the chemist to the fact that the issue is in the library and provides a convenient list to scan for items of possible interest. The main advantage of the service is that there is negligible lag time between the appearance of the primary document and the notification of its receipt in the library. The disadvantage from the user's point of view is that the search for information is a random, serial search with perhaps high recall, but low relevance of the total number of items scanned. Furthermore, the library may not be able to afford subscriptions to some relevant journals.

A table-of-contents service is a very labor-intensive activity. For the near future, a way is needed to create from the printed tables of contents an electronic form in a reasonably cost-effective manner. Once digitized, the tables of contents could easily be moved over an electronic mail network to individual users. One option for this variation on traditional table of contents services involves the use of a Cauzin softstrip. Some publishers are now including the softstrip code in their printed journals. The example in Figure 13-1 shows the table of contents and the softstrip code for such an issue, followed by the code converted to an ASCII file by using the Cauzin Softstrip reader.

Eventually, optical character readers will become cheap enough and able to recognize sufficient character fonts that it will be economically feasible for many libraries to convert the printed tables of contents for electronic delivery through optical character recognition techniques. In the interim, the Softstrip code offers an inexpensive option which is hampered only by the absence of the code in the journals themselves.

13.2.2 Printed Table-of-Contents Journals from Commercial Publishers

A number of publishers of printed and computer-based abstracting and indexing services provide printed table-of-contents journals. Two of the better-known such products are *Chemical Titles* and the *Current Contents* series.

Chemical Titles [1744] is one of the oldest printed current awareness services. Published by Chemical Abstracts Service every other week, it includes a key-word index to those issues of about 700 of the main chemical journals which have recently appeared. *Chemical Titles* was the first commercial product to employ a key-word subject index. An example of that index, the author index, and the corresponding title page of a specific journal covered in that issue is shown in Figure 13-2.

The *Current Contents* indexing journals are taken from ISI's very large database of highly significant scientific journals. The following titles are available:

- Agriculture, Biology & Environmental Sciences [759]
- Clinical Medicine [1756]
- Engineering, Technology & Applied Sciences [1758]
- Life Sciences [1755]
- Physical, Chemical & Earth Sciences [1757].

An example taken from a *Current Contents* issue is shown in Figure 13-3.

Each of the *Current Contents* printed issues appears weekly. There is in each issue an index for both authors and keywords. In addition, the contents of important multigraphs are included. A further attractive feature in each issue is the essay section, "Current Comments". Frequently, new ISI products or developments are discussed in these essays, as are citation studies which give insight into the direction science is taking or the relative importance of scientific journals.

The printed table-of-contents journals have several advantages over customized table-of-contents services offered by libraries. First, they cover more journals than the vast majority of chemistry libraries could possibly have on site. Second, they provide subject and author indexes as well as the actual tables of contents. Thus, the sequential browsing of the tables of contents of favorite journals, as is done with the custom library service, is still an option. Finally, the printed services are usually tied to document delivery services offered by the parent company. These are discussed in section 13.5.

266 CHEMICAL INFORMATION SOURCES

Tables of Contents
for
Pergamon Computer Journals
to be published as Softstrips

The Softstrip® (a registered trademark of Cauzin Systems, Inc.) data strip printed on this page includes this issue's entire Table of Contents in a form which can be scanned directly into commonly available computer data bases without typing. The computer data base generated by these data strips will be updated every issue and it will allow you to construct a personal computer-based data base to easily catalog and later search for specific topics.

In 1986 Pergamon Press began using Softstrip System technology. This new system is being broadly adopted by book, journal, software, and magazine publishers. It allows readers to automatically scan and enter data (or computer software) directly from the printed pages of publications into all popular personal computers -- quickly, automatically, and error-free. Data strips, like the one printed on the edge of this page, will be appearing in a wide variety of publications. They may all be scanned by using a low-cost Softstrip Reader which can be obtained from selected computer stores.

The Table of Contents information is structured in a form which will directly enter into personal computer software programs such as Living Videotext ThinkTank or READY! Other data bases may also be used, such as dBASE II, dBASE III, Lotus 1-2-3, and Lotus Symphony.

For detailed information on the structure of the data files, or for more information on the Softstrip System, please write to:

Director of Publishing
Pergamon Press, Inc.
Fairview Park
Elmsford, NY 10523, USA

(914) 592-7700

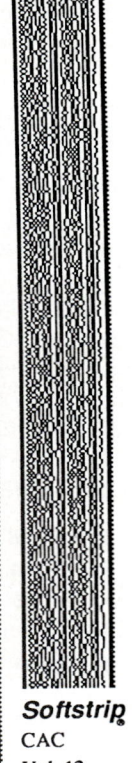

Softstrip
CAC
Vol. 12
No. 4
1988

COMPUTERS & CHEMISTRY

FIGURE 13-1
Table of Contents of a Journal Issue in Cauzin Softstrip Code.

computers & chemistry
An International Journal

contents   vol. 12, no. 4, 1988

	i	Softstrip® data strip containing the table of contents for this issue
J. P. Gastmans, V. E. de Paula et M. Furlan	285	Etude par microordinateur de l'influence des atomes voisins sur les signaux RMN de ^{13}C
Joseph Chrastil	289	Determination of the first order consecutive reaction rate constants from final product
Jaroslaw Kostrowicki, Adam Liwo and Krzysztof Sokolowski	293	A comparative study on some methods for computing equilibrium concentrations
F. T. Chau and K. D. Chan	301	A computer-based temperature measurement system
Concetta De Stefano, Pietro Princi, Carmelo Rigano and Silvio Sammartano	305	The calculation of equilibrium concentrations. ESTIME: a computer program for comparing execution times on different machines
Applications		
H. S. Tan, P. Chiarot and W. E. Jones	317	Microcomputer-assisted temperature programmed desorption studies
S. Lago and C. Vega	343	A generalization for mixtures of a fast algorithm to calculate some intermolecular orientational averages
Note Tom Nicol	357	Performance of the HQRII1 diagonalization routine
Book Review DeLos F. DeTar	359	The Visual Display of Quantitative Information, by Edward R. Tufte

INDEXED IN Current Contents, Computer Contents, BIOSIS Database, Engng Ind. Monthly & Author Index, PASCAL-CNRS Database, Software Reviews on File, INSPEC

Pergamon Press
Oxford   New York   Beijing   Frankfurt   São Paulo   Sydney   Tokyo   Toronto

Printed in Great Britain by A. Wheaton & Co. Ltd, Exeter   ISSN 0097-8485
COCHDK 12(4) 285–360 (1988)

FIGURE 13-1
Table of Contents of a Journal Issue in Cauzin Softstrip Code.

.head 0 + COMPUTERS AND CHEMISTRY; ISSN 0097-8485; V12; N4; 1988
.head 1 +
.head 2 + Articles
.head 3 + pp. 285-287 Etude par microordinateur de l'influence des atomes voisins sur
.head 4 + les signauz RMN de 13C
.head 5 - J. P. Gastmans, V. E. de Paula, et M. Furlan
.head 3 + pp. 289-292 Determination of the first order consecutive reaction rate
.head 4 + constants from final product
.head 5 - Joseph Chrastil
.head 3 + pp. 293-299 A comparative study on some methods for computing equilibrium
.head 4 + concentrations
.head 5 - Jaroslaw Kostrowicki, Adam Liwo and Krzyztof Sokolowski
.head 3 + pp. 301-304 A computer-based temperature measurement system
.head 4 - F. T. Chau and K. D. Chan
.head 3 + pp. 305-315 The calculation of equilibrium concentrations. ESTIME: a
.head 4 + computer program for comparing execution times on different machines
.head 5 - Concetta De Stefano, Pietro Princi, Carmelo Rigano and Silvio Sammartano
.head 2 + Applications
.head 3 + pp. 317-342 Microcomputer-assisted temperature programmed desorption studies
.head 4 - H. S. Tan, P. Chiarot and W. E. Jones
.head 3 + pp. 343-356 A generalization for mixtures of a fast algorithm to calculate
.head 4 + some intermolecular orientational averages
.head 5 - S. Lago and C. Vaga
.head 2 + Note
.head 3 + pp. 357 Performance of the HQRIII diagonalization routine
.head 4 - Tom Nicol
.head 2 + Book Review
.head 3 + pp. 359 The Visual Display of Quantitative Information by Edward R. Tufte
.head 4 - Delos F. DeTar

FIGURE 13-1
Table of Contents of a Journal Issue in Cauzin Softstrip Code.

13.2.3 Databases for Current Awareness

In the fall of 1988, ISI began to offer a new option for current awareness: *Current Contents on Diskette* [1975]. Both Apple Macintosh and IBM PC versions are now available. *Current Contents on Diskette* appeared first in the Life Sciences version, covering the 600 most highly-cited journals in this subject area. (An optional version provides more journal coverage, including about 1200 titles.) Now the Physical, Chemical, and Earth Sciences version is available, and others will appear in 1990.

A great deal of thought went into the software for the database. Capabilities include:

- current profile creation and key-word searching
- automatic scanning
- manual browsing
- printout
- export capabilities for individual articles, for complete tables of contents, or search results
- integration with other ISI products.

Figure 13-2
Sample *Chemical Titles* Entry

(AQ758)

Synthetic Communications
Marcel Dekker Inc.
Articles and Abstracts in English

VOL. 19 NO. 13-14 1989

ION-EL			IRON(I		
CC Pg		J Pg	CC Pg		J Pg
ION-ELECTRON			IRON(III)		
28		2287	120		431
ION-EXCHANGE					477
109		478	125		298
111		2193	IRON(III)-ION		
113		1825	125		297
ION-EXCHANGERS			IRON-CATALYZED		
113		1825	:06		3651
ION-IMPLANTATION			IRON-MANGANESE		
			93		2051
49		1354	IRONE-RELATED		
52		176	122		1400
56		2926	IRRADIANCE		
59		1315	144		93
83		207	IRRADIATED		
"		217	56		3293
"		250	87		346
89		22	89		1
ION-MODERATED			IRRADIATED, NEUTRON		
109		486	46		147
ION-MODIFIED			IRRADIATION		
117		1216	35		730
ION-PAIR			52		181
80		291			243
96		4098	59		1276
122		1171	64		L1521
ION-SOLVENT					L1657
105		920	70		33
IONIC			89		77
35		671	IRRADIATION, NEUTRON		
53		720	64		L1561
73		899	IRRADIATION-INDUCED		
"		R 7			
"		235	70		179
"		337	IRREPS		
"		369	37		3771
"		377	ISATIN		
"		387	137		2255
91		1258	ISING		
93		2106	39		181
105		920	ISING-MODEL		
117		1261	33		1993
120		479	34		200
134		3756	37		L91
IONIC-CONDUCTIVITY			39		
73		241	40		
79		1163			
IONICALLY					
73		417			
IONIZATION					
23		64			
28		25			

Synthesis of N,N'-Diaralkyl 2H-1,2,6-Tetrahydrothiadiazines 2229
J. L. Meisenheimer, J. L. Meisenheimer, Jr.,
D. G. Raya, M. C. Stapleton, and C. J. Goddard

The Synthesis of Amidine Acetals as Intermediates
for the Preparation of Naturally Occurring Cyclic
Amidine Antibiotics 2237
W. K. Anderson and N. Raju

A Novel Synthesis of Mono- and Disubstituted
(1-Cycloalkenyl) Acetic Acid Derivatives via
Ionization/Elimination of β-Lactones 2243
T. H. Black and S. Eisenbeis

Debromination of 4-Bromo-3-arylsydnones
with Sodium Sulfite 2249
K. L. McChord, S. A. Tullis, and K. Turnbull

Synthesis of Anthranilonitriles by Fragmentation
of Isatin Oxime Triflates 2255
J. B. Campbell, Jr. and T. W. Davenport

Facile Palladium-Catalyzed Cross-Coupling of
Monoorganozinc Halides with 3-Iodoanthranilonitriles 2265
J. B. Campbell, Jr., J. Wawerchak Firor, and
T. W. Davenport

A Short and Convenient Synthesis of 5-Alkyl
Substituted 8-Hydroxyquinolines 2273
M. Hojjatie, S. Muralidharan, M. L. Dietz,
and H. Freiser

Synthesis of 2-Aryl-3-cyanophthalimidines
and Related Analogs 2283
J. S. Baum and M. M. Staveski

An Improved Synthesis of (±) Frullanolide.
Application of Selenium Reagents 2293
H. M. C. Ferraz, N. Petragnani, and C. M. R. Ribeiro

Convenient and Optimized Procedure for the Syntheses
of Di- and Trialkoxycarbenium Tetrafluoroborates 2307
U. Pindur and C. Flo

Montmorillonite Supported Borohydride: A New
Reducing Agent for Reductions Under Phase
Transfer Conditions 2313
A. Sarkar, B. R. Rao, and M. M. Konar

CONTINUED

CURRENT CONTENTS®   ©1989 by ISI®   PC&ES, V. 29, #43, Oct. 23, 1989   137

Figure 13-3
Sample *Current Contents* Entry

It is possible to select the fields which will be printed or exported to another computer. With the Mouse Mode set to "Export to File," article citations can be exported simply by pointing to them and clicking the mouse. Keyword searching is also possible with Boolean operators OR or AND. In addition, a profile can be stored for future updates, if desired.

Another ISI option for current awareness searching in a database is Current Contents Search [915]. Offered on both the BRS and DIALOG systems, Current Contents Search includes the five sections of *Current Contents* which cover scientific topics (plus the *Arts & Humanities* and *Social & Behavioral Sciences* editions). Thus, over 4300 scientific journals are accessible in the database. Since the aim of the file is current awareness, only the latest six months to one year of the issues are retained in the database.

Current Contents Search contains two kinds of records:

- table of contents for the entire issue of each journal
- full record for each individual article.

Either the full file or subsets corresponding to the *Current Contents* editions can be searched by subject or other parameters. It is also possible to create a customized table-of-contents service by storing a list of journal titles of interest and running it against the updates to the database.

Other database producers have created special databases for current awareness searching. Chemical Abstracts Service introduced in 1988 the database CApreviews [1746]. The file is truly a preview of what will eventually appear in the CA File. Thus, it provides early access (by at least six weeks) to the bibliographic portion of the CA File records. CApreviews contains about 100,000 records and is updated weekly with 7000 to 9000 new references. However, the file will not grow substantially in size, since the CApreviews record is deleted approximately three months after the complete record (with abstract and full indexing) enters the CA File. The CA database covers more document types than journals, so CApreviews can be used to keep abreast of new patents, dissertations, reports, books, and conference proceedings. Unlike the Current Contents Search database, there is only one record format in CApreviews, corresponding to a complete bibliographic citation. Thus, it is not possible to be sure of getting an entire table of contents for a particular issue of a journal at a given time, because records for that issue may enter the CApreviews database over a period of several weeks.

Another database producer which has introduced a preview file is BIOSIS. Its BioExpress file [1976] contains the latest twelve weeks of references to journal articles in the BIOSIS database. The database is only available through the telecommunications link "The BIOSIS Connection".

13.2.4 Selective Dissemination of Information (SDI)

SDI or **Selective Dissemination of Information** is one of the oldest forms of computer applications to information retrieval. An appropriate database to be searched is first selected. Then the user of an SDI service chooses search terms which reflect areas of interest. A search profile strategy is drawn up and run against the database updates. The search profile may contain subject words, authors' names, journals of interest, etc. When properly refined, the weekly or bi-weekly output from the customized search should contain a high percentage of relevant citations.

There are quite a few ways to obtain an SDI search. All of them have several things in common. They are run at regular intervals to search only the records which were added to the database since the last update. SDI is a customized service. If the output is unsatisfactory, it can be changed to better suit the user. This makes SDI a relatively expensive option for keeping up with the literature, with costs typically in the $300-$500 range each year. Finally, current awareness service through SDI is only as current as the database itself. In general, SDI from an indexing service will be much more current than SDI from an abstracting service. It takes time to read a primary work and prepare an abstract. Hence, there is a built-in delay for primary works to be entered in an abstracting database. It was to minimize this delay that CAS developed the CApreviews database.

Chemical Abstracts Service has another custom SDI option for the *CA* database using software which is totally different from the Messenger search software developed for online searching on the STN system. The service is called the Individual Search Service or ISS [1743]. For 1988, the cost of a single profile for the standard output was $468. An example of the type of output is found in Figure 13-4.

The basic fee for the Individual Search Service covers up to 100 references retrieved from each bi-weekly update. There are additional options such as output of registry numbers ($39), output of CA controlled-vocabulary index terms ($104), and additional copies of the same profile ($26). (Prices are at the 1988 rate.)

Biosciences Information Service (BIOSIS) introduced a new angle to SDI several years ago with its B-I-T-S service [1752]. Instead of printed output, the subscriber receives a diskette each month with a small database of the latest entries which the custom SDI search profile has retrieved from the BIOSIS database. With the proper software (for example, BioSuperfile II [71]), the B-I-T-S entries can be examined, those of lasting interest transferred to the user's own database, and others discarded.

There are other options for providing SDI services. It is possible to set up an SDI profile with an online vendor which will then automatically provide output from the selected database each time the vendor updates the database. Alternatively, a user could simply save a search strategy on a local file and upload it periodically to search only the database updates. Almost all vendors now offer the capability to run a search against the specified updates of a given database.

13.2.5 Standard Interest Profiles

A considerably less expensive current awareness option than SDI is a **standard interest profile** (SIP). Typically, the standard interest profiles cost one-third to one-half the subscription price of an SDI profile. The subscriber must give up the customized search features in exchange for a lower price, however. That is one of the drawbacks to SIPs.

CURRENT AWARENESS, DOCUMENT DELIVERY SERVICES 273

PAGE 5012

PROFILE ID 20101T QUERY NUMBER 001
CA101(9):66175W Journal CASCT CA102001
Analysis of binding surfaces: a methodology appropriate for the investigation of complex receptor mechanisms and multiple neurotransmitter receptors. Rothman, Richard B. (Sect. Brain Biochem. NIMH, Bethesda, MD 20205 USA). Neuropeptides (Edinburgh) [NRPPDD] 1983, 4(1), 41-4 (Eng).

A novel exptl. design is presented for exams. neurotransmitter receptor-ligand interactions: anal. of 3-dimensional binding surfaces. Evidence is presented that a binding surface provides greater information content than does a single binding isotherm. This approach to the design and the anal. of ligand binding studies should facilitate quant. studies of the opiate receptor.

neurotransmitter receptor ligand interaction analysis simulation receptor ligand interaction
COPYRIGHT 1984 BY THE AMERICAN CHEMICAL SOCIETY

PROFILE ID 20101T QUERY NUMBER 001
CA101(9):66185Z Journal CASCT CA102002
Semisynthetic derivatives of glucagon. The contribution of histidine-1 to hormone conformation and activity. Flanders, Kathleen Corey; Horwitz, Edwin M.; Gurd, Ruth S. (Med. Sci. Program, Indiana Univ., Bloomington, IN 47405 USA). J. Biol. Chem. [JBCHA3] 1984, 259(11), 7031-7 (Eng).

Semisynthetic N.epsilon.-acetimidoglucagon [74619-75-3] was prepd. from
1-[de-histidine]-N.epsilon.-acetimidoglucagon [91173-45-4] by coupling N.alpha.-tert-butoxycarbonyl-Nim idazole-DNP-L-histidine N-hydroxysuccinimide ester [91173-46-5] to the peptide in DMF in the presence of 1-hydroxybenzotriazole. The deprotected, purified product was chem. identical to N.epsilon.-acetimidoglucagon and equipotent to N.epsilon.-acetimidoglucagon and native glucagon [9007-92-5] in its ability to activate adenylate cyclase [9012-42-4] and displace [125I]iodoglucagon from rat liver plasma membranes. Semisynthetic [1-phenylalanine]- [91173-43-2], [1-alanine]- [91173-44-3], and [1-de-histidine]-N.epsilon.-acetimidoglucagon prepd. similarly achieved 85, 55, and 35% of the maximal activity and 22, 2, and 6% of the binding potency of N.epsilon.-acetimidoglucagon. The biol. assays indicate that the amino group is involved to a greater extent in transduction than in binding, but the
ABSTRACT CONTINUED
COPYRIGHT 1984 BY THE AMERICAN CHEMICAL SOCIETY

PROFILE ID 20101T QUERY NUMBER 001
CA101(9):66164S Journal CASCT CA102001 CCR 108
Purification of monoiodinated vasoinstestinal peptide (M125I-VIP) by high pressure liquid chromatography (HPLC). Marie, J.C.; Boissard, C.; Rosselin, G. (Unit Rech. Diabetol., Etud. Radio-Immunol. Horm. Proteiques, 75571 Paris, Fr.). Peptides (Fayetteville, N. Y.) [PPTDD5] 1984, 5(2), 179-82 (Eng).

A reverse-phase high-performance liq. chromatog. method is presented which permits the sepn. of 4 forms of M125I-VIP from unlabeled VIP and other iodinated components. The quicker eluting M125I-VIP forms (oxidized and reduced) have a consistently low nonspecific binding with specific target cells of VIP (HT-29, human carcinoma cells) as compared to the late eluting forms of VIP. The retention time is considerably increased when the mol of VIP is fully iodinated.

iodinated VIP purifn HPLC
COPYRIGHT 1984 BY THE AMERICAN CHEMICAL SOCIETY

PROFILE ID 20101T QUERY NUMBER 001
CA101(9):66168W Journal CASCT CA102001
Interlaboratory comparison of 25-hydroxyvitamin D determination. Mayer, Eberhard; Schmidt-Gayk, Heinrich (Abt. Endokrinol., Med. Poliklin. Inn. Med. VI, D-6900 Heidelberg, Fed. Rep. Ger.). Clin. Chem. (Winston-Salem, N. C.) [CLCHAU] 1984, 30(7), 1199-204 (Eng).

This interlab. study on detn. of 25-hydroxyvitamin D (25-OH-D) [64719-49-9] in serum involved 15 labs. in 8 European countries. All could distinguish between normal (50 nmol/L) and grossly increased concns., but for 8 labs. the results for serum samples with low and normal 25-OH-D content overlapped. In general, values were well-reproducible, but interlab. variation in 25-OH-D measurement was large. 24,25(OH)2D3 interfering in most of the assays. The advantages of chromatog. before assay are presented. Liq. chromatog. with UV detection for quantifying 25-OH-D2 [21343-40-8] and 25-OH-D3 [19356-17-3] appears to be an appropriate ref. method, whereas competitive protein binding assay is the method of choice for routine detns. Control sera with subnormal, normal, and above-normal concns. of 25-OH-D are needed for use in standardization of 25-OH-D assays.

hydroxyvitamin D detn serum interlab
protein binding assay hydroxyvitamin D
liq chromatog hydroxyvitamin D
COPYRIGHT 1984 BY THE AMERICAN CHEMICAL SOCIETY

Figure 13-4 Sample CAS Individual Search Service Output

Standard interest profiles are produced for mass consumption. The publisher may hope to attract 100 or more subscribers in order to spread the production costs among them. Consequently, the focus of the SIPs is broader than would be possible with a custom SDI profile. Thus, the SIP subscriber is not likely to find as high a percentage of useful references as might be found with an SDI service from the same database. However, the tradeoff in cost may make it worthwhile to subscribe to a SIP.

The generic name for a whole series of standard interest profiles published by Chemical Abstracts Service is CA Selects [1797]. Priced at $169 per year for the 1988 subscription, there are more than 200 titles to choose from. These include:

High Speed Liquid Chromatography
Chemical Hazards
Photochemistry
Metallo Enzymes & Metallo Coenzymes
Organo-Transition Metal Complexes
Surface Analysis

and many other titles. The CA Selects bi-weekly issues look just like pages from the printed *Chemical Abstracts*. In a sense, the subscriber is getting a re-packaged version of the printed *CA*. However, with CA Selects, as with all true standard interest profiles, there is no indexing in the bi-weekly issues, and no indexes are *ever* produced for the products. They are meant to be scanned cover to cover and thrown away when no longer of use.

There is also a large series of SIPs published by ISI under the title Research Alert [748]. In fact, many abstracting and indexing services produce standard interest profiles. For additional titles, consult *An Index to Standard Interest Profiles in Science and Technology* [977] (1988).

13.3   RESEARCH IN PROGRESS

There are times when even a current awareness service is not current enough. Suppose a chemist wanted to prepare a grant proposal for a funding agency. It would be very desirable to know what related work is currently being pursued by others, but not yet published in the primary literature. Furthermore, a database which lists research in progress and includes the funding agencies might suggest a funding source to which the application should be sent. A tool to help locate databases and printed sources of such information is *Information Services on Research in Progress, A Worldwide Inventory* [382] (1982).

The Fedrip Database [608] is a source for U.S. Federal Research in Progress. Ten federal agencies which fund or support research in the United States contribute to the database. Most of them contribute data monthly, but a few provide only annual updates. Another database of this type is CANCERPROJ [1978], a service of the International Cancer Research Data Bank at the National Cancer Institute. Agriculturally related research sponsored by the U.S. Department of Agriculture and other federal or state agencies is listed in CRIS [1979]. For British research, *Current Research in Britain* [1980] has contributions from over 500 institutions. Separate volumes are published for the *Physical Sciences* and for the *Biological Sciences*.

13.4 SOURCES FOR BACKGROUND READING

The secondary works and databases in the previous sections are all designed to alert the chemist to relevant new primary literature or research in progress. However, the test of time serves as a filter for older literature, allowing the most significant discoveries to pass through and be recognized as worthy of further codification. They take their places among the review articles, treatises, encyclopedias, and single-volume secondary works designed to consolidate and preserve the most important knowledge in science and technology.

13.4.1 Reviews

Recall that there are different types of review articles. Some, directed toward specialists in the field, are written at an advanced level, whereas others, geared for the non-specialist, tend to take a more basic approach. The non-specialist reviews are likely to be rather short in comparison to the specialist reviews, covering only the key literature in the field. Specialist reviews, on the other hand, can be quite long and may include hundreds of references in their bibliographies. These are often published in special review serials, most of which are published annually. They frequently have titles which begin with *Advances in* ...., *Progress in* ...., etc.

In the printed indexes to *Chemical Abstracts*, reviews are coded with a capital R inserted between the volume and abstract number, for example, 74:R48637t. Finding a review article is easy in the CA File or other versions of the online *CA* database. An online search of *CA* which includes the statement => **S L# AND REVIEW** will identify the desired source. However, in such a search, there is no easy way to distinguish between a review written for a specialist and one written for the non-specialist.

The Institute for Scientific Information's *Index to Scientific Reviews* [1616] (1974-) covers mostly the review articles which appear in over 200 review serials. In addition, review articles which are published in the many primary journals covered by *Science Citation Index* are sifted from that source for inclusion in the *Index to Scientific Reviews*. (Recall that *Science Citation Index* also codes its entries by document type, so an online search of the SciSearch database can uncover review articles which were published in primary scientific journals.)

Access to the reviews in *ISR* is provided by the usual ISI Permuterm Subject Index as well as an author index. In recent years, ISI has pioneered the development of **research front specialty** analysis. A **research front** is defined as an area where there is a great amount of research occurring, perhaps converging from separate and distinct disciplines. ISI identifies the research fronts through statistical analyses of citation patterns and assigns a research front specialty number to each of these. The *Index to Scientific Reviews* includes a Research Front Specialty Index to gather into a classified index the references to reviews which have appeared during the period of coverage. With over 30,000 review articles covered each year, the *ISR* is a key source of review articles.

An ambitious project is underway at ISI to produce the *ISI Atlas of Science* [1615] (1987-). The research front specialty concept is being used to help identify the most active areas of research throughout science, medicine, and technology. The series will contain review articles in twelve sections including Biochemistry, Physical Chemistry, and Pharmacology. The fields of analytical chemistry and instrumentation will be a part of the section devoted to Structure, Synthesis, Application.

In a sense, chapters written for multigraphs (composite works) can be thought of as review articles. Finding an appropriate chapter in such a work has been made considerably easier now with the appearance of ISI's *Index to Scientific Book Contents* [1155] (1986-).

13.4.2 Encyclopedias

Not to be overlooked are the major general encyclopedias. Such illustrious sets as the *Encyclopaedia Britannica*, the *Encyclopedia Americana*, the *Academic American Encyclopedia* and others have very good sections devoted to all areas of science, chemistry among them. In fact, the *Academic American Encyclopedia* is available in several electronic formats: as an online database and as a CD-ROM product, the Grolier Electronic Encyclopedia [1889].

Encyclopedias present in a concise and easily accessible manner (invariably in alphabetic order) the whole body of knowledge within the scope of the work. Encyclopedias, like dictionaries, define terms, but encyclopedias go further in attempting to describe and explain the various relationships of the terms and concepts. While they are not meant to cover a topic exhaustively, encyclopedias can lead to more exhaustive treatments through the selected bibliographies which are now commonly a part of encyclopedia articles. Remember that there is a significant time lag for new discoveries to make their way into an encyclopedia. Even with a newly published encyclopedia, it is unlikely that much of the information in the work will be any more recent than 2-5 years prior to the date of publication. (The same can be said of treatises and other collected works of this type.)

One of the best known scientific encyclopedias is the *McGraw-Hill Encyclopedia of Science and Technology* [370] (1987). Published in 20 printed volumes, the set is beautifully illustrated. There is also a CD-ROM edition of the *McGraw-Hill Concise Encyclopedia of Science and Technology* which is packaged with the *McGraw-Hill Dictionary of Scientific and Technical Terms* and marketed as the *McGraw-Hill CD-ROM Science and Technical Reference Set* [1992] (1987). Subsets of the complete printed work are also available, including the *Physical Chemistry Source Book* (1988), the *Spectroscopy Source Book* (1988), etc.

A new work is the *Encyclopedia of Physical Science and Technology* [1981] (1987). Under the guidance of a distinguished advisory board, the authors produced 15 volumes containing over 550 articles, many of them on topics of interest to chemists.

The most famous chemistry encyclopedia is the *Kirk-Othmer Encyclopedia of Chemical Technology* [551]. Published in 25 volumes between 1978 and 1984, the set is undoubtedly the single most valuable resource in a chemistry library's reference collection. It can now be searched online through several vendors. Alternatively, a library may choose to purchase the CD-ROM edition [916] or the *Kirk-Othmer Concise Encyclopedia of Chemical Technology* [1993] (1985).

Another important encyclopedia from Wiley is the *Encyclopedia of Polymer Science and Engineering* [1054] (1984-89). The *Encyclopedia* provides a comprehensive overview of polymer and plastics technology. Physical, mechanical, and biological properties are covered, as are the morphology and compatibility of the substances. Methods of production and uses of the polymeric materials are also considered. This, too, is a tool which can be purchased in CD-ROM format.

In the area of industrial chemistry, one can consult the *Encyclopedia of Chemical Processing and Design* [1644] (1976-) or *Ullmann's Encyclopedia of Industrial Chemistry* [1982] (1985-). The latter work is now in its 5th edition, and for the first time, it is being published entirely in English.

It is important to be aware that a given edition of a scientific encyclopedia may appear over several years before it is complete. Therefore, the publication date of the volume being used should be noted in order to judge the currency of the information in the volume. There are other important encyclopedias included in the Chemistry Reference Sources Database.

13.4.3 Dictionaries

As noted earlier, dictionaries differ from encyclopedias in one major respect: for the most part, they do not attempt to show the interrelationships among the terms which are defined. Given the difficulty of pronouncing many scientific terms, it is surprising to find that most scientific dictionaries (as well as most scientific encyclopedias) do not include the proper pronunciation of the terms. Unabridged general dictionaries of the English language should be consulted for pronunciation, since they contain a multitude of scientific terms.

McGraw-Hill publishes dictionaries in many areas of science, including the *McGraw-Hill Dictionary of Chemistry* [1983] (1985). Other standard chemical dictionaries are *Hawley's Condensed Chemical Dictionary* [199] (1983), *The Facts on File Dictionary of Chemistry* [444] (1981), the Oxford *Concise Dictionary of Chemistry* [440] (1985), *Longman Illustrated Dictionary of Chemistry* [1389] (1982), and *Miall's Dictionary of Chemistry* [448] (1981).

There are many other types of dictionaries, among them, thesauruses, dictionaries of acronyms, and translating dictionaries. For these, the Chemistry Reference Sources Database should be searched using terms like **ACRONYMS** or **ABBREVIATIONS** or **THESAURUS**. For translating dictionaries, include in the search statement the **NAME** of the foreign language **AND DICTIONARY**.

13.4.4   Treatises, Monographs, and Multigraphs

As noted earlier, treatises are generally discussed in appropriate sections of earlier chapters in this book. Multi-volume **treatises** exhaustively cover the whole of a subdiscipline of chemistry, such as organic, inorganic, or physical chemistry, or even a narrower area than that, such as organometallic chemistry. Treatises are arranged in a classified order and provide full documentation to the references in the primary literature from which they draw their facts. It is frequently the case that a treatise has the word "comprehensive" in the title. However, the nomenclature of science is not always as clear as it should be, so we find that the *Encyclopedia of Fluid Mechanics* is really a treatise! Other titles simply imply that the works are treatises, such as *The Enzymes* [889] or *The Peptides* [1984].

Monographs and **multigraphs** are not hard to distinguish from treatises, since they are invariably published in one volume. These are the true "books" of science. They exhaustively treat a narrow topic, employing the same type of classified arrangement of the subject matter and full documentation of references found in treatises.

Books are not usually well indexed in abstracting or indexing journals. An exception is the *Index to Scientific Book Contents* [1155] (1986-), which covers the individual chapters of multi-authored scientific books (multigraphs). All scientific disciplines are included in *ISBC*. Produced by ISI, the printed *Index to Scientific Book Contents* has a Permuterm Subject Index and a Category Index, which classifies the books into about 100 subject categories. There is also an author/editor index. The database was available on a European vendor (DIMDI) in 1989.

13.4.5 Finding Out About Books

In most libraries the card catalog or online public access catalog (OPAC) is generally the first major reference source a user consults to find a book on a topic of interest. There are other databases or printed tools which can serve a similar purpose. The Bowker Company, publisher of *Books in Print*, has produced a CD-ROM version of that product, *BIP+* [1985]. A subset of the Bowker database with a better focus for scientists is *Scientific and Technical Books and Serials in Print* [583]. The annual set has subject, title, and author listings, as well as a directory of publishers. In 1982, Bowker published *Pure and Applied Science Books, 1876-1982* [584], with a similar format and indexes as found in *Books in Print*.

For the area of chemistry, about 750 book titles are found in *Guidelines and Suggested Title List for Undergraduate Chemistry Libraries* [1] (1982). Included are most of the classic works in modern chemistry. The "New Books" section of the "Lab Guide" [1011], a special August issue of the journal *Analytical Chemistry*, lists all of the titles which have been critically reviewed during the previous year in *all* of the American Chemical Society journals. Another effort related to the ACS is the "Book Buyers' Guide" [1986], published in the September issue of the *Journal of Chemical Education*. This is a classified list of in-print books which were exhibited at the latest meeting of the American Chemical Society. Finally, the American Chemical Society publishes the monthly *Chemical Books in Brief: Acquisitions Aid* [1987], designed to assist in the selection and purchase of current books in chemistry and related areas.

A number of scientific news journals, such as *Nature*, *Science*, and *Chemical and Engineering News* have extensive listings of new books received.

13.4.6 Finding Out About Serials

The largest source specifically for bibliographic information on chemistry serials is *CASSI, Chemical Abstracts Service Source Index, 1907-89* [229]. This work is extensively described in Chapter 14, section 14.4.1. Suffice it to say here that *CASSI* includes references to the many serials covered by *Chemical Abstracts* throughout its history and even covers older journals prior to the establishment of *CA*. One of the most important aspects of CASSI is that it provides a **union list of serials**, linking the journal titles to the libraries which hold them, primarily in the U.S. and Canada, but also in certain foreign libraries. Chemical Abstracts Service also produces the *CASSI Keyword-Out-of-Context Index* [1293] in microfiche to serve as a rough subject index to the titles. CASSI is also searchable online through ORBIT.

A special section of *CASSI* is the "List of 1000 Journals Most Frequently Cited in Chemical Abstracts." This list is not an analysis of authors' citations in the sense discussed in the chapters dealing with the *Science Citation Index*. Rather, it is a ranking of the journals contributing the largest numbers of entries to *CA* in a given year.

A list of journals developed through the technique of citation analysis is the *Journal Citation Reports*, published each year as the last volume of *Science Citation Index*. The *Journal Citation Reports* [1989] provide insights to the most important journals in science, technology, and medicine. In the *JCR*, the section known as the "Journal Ranking Package" lists journals by subject area, ranked by the so-called impact factor. In addition to the "Chemistry" sections, entries in the "Biochemistry and Molecular Biology," "Biophysics," "Crystallography," "Electrochemistry," "Physics: Atomic, Molecular and Chemical," and "Spectroscopy" sections could be of interest in selecting appropriate journals. For more recent English-language journals, the "New Journals Review" [1856], usually published in the first October issue of *Nature* each year, has reviews of titles which began publication the previous year.

Older journals are sometimes hard to identify. Some of the older guides to the literature included long lists of such works. Among such guides are the early edition of Crane and Patterson's *A Guide to the Literature of Chemistry* [8] (1927), Dyson's *A Short Guide to Chemical Literature* [1990] (1958), and Mellon's *Chemical Publications; Their Nature and Use* [10] (1965). Neufeldt's *Chronologie Chemie, 1800-1980* [1396] (1987) also has a list of older serial titles. The abbreviations found in older *Beilstein* volumes can be perplexing unless one knows to consult the abbreviations list in the front of the H (Hauptwerk) volumes.

13.5 DOCUMENT DELIVERY OPTIONS

There are a number of ways to obtain material which is not held by a local library. One of those is **interlibrary loan**, the process by which one library either borrows a copy of a book or other work from another library or receives a photocopy of a journal article, book chapter, or selected pages of a collected work. In the past, interlibrary loan has been a free service. Increasingly, there are charges attached to the use of the service. More and more, both libraries and users seem to be turning to **commercial document delivery services** from organizations which have as their main or a subsidiary business the supplying of various library materials which are not locally available.

13.5.1 Commercial Document Delivery Services

Document Retrieval: Sources and Services [1991] (1989) is billed as a complete directory of public and private document retrieval suppliers. In it are descriptions of the services offered by commercial document suppliers in over 100 cities throughout the world. There is considerable variation in the scope of services offered by information brokers who specialize in document delivery. Some claim to be a "one-stop shopping center," capable of providing any type of material requested. Others provide only certain years of selected titles, while still others limit their service to those journal titles available in a particular collection.

Both DIALOG and ORBIT have offered online document ordering for many years. BRS does not provide direct online document ordering, but does serve as a switching station for orders through MCI Mail. A recent list of document suppliers from DIALOG includes over 75 companies, most of which are suppliers of journal articles. For journal articles, the Institute for Scientific Information, Carolina Library Service, FIND/SVP, The Information Store, and Information on Demand are typical suppliers.

The Institute for Scientific Information (ISI) has long made available a journal article supply service. The service was first known as OATS (*O*riginal *A*rticle *T*ear *S*heets). Now, it is called The Genuine Article [1750]. Nearly 7000 journals are included in the service, covering the current year and three previous years. In 1990 ISI expanded its document delivery service and can now fulfill orders for nearly 3000 more science journals, some of which date back to the 1800s. The 1990 cost was $9.50 for the first ten pages, and $2.00 for each additional ten pages, including almost all royalty fees. Rush and call-in services (with telefacsimile as an option) are available, and various payment options are provided, among them, monthly billings, deposit accounts, and volume contracts (for 800 or more articles per year). ISI often provides an original article torn from one of the copies of the journal, multiple subscriptions of which are ordered by the company.

FIND/SVP, The Information Store, and Information on Demand are typical of the private, multi-purpose information providers that have proliferated in the last few decades. All of these provide an optional rush service, with normal charges ranging from $4 to $14 per article. However, publishers' royalty fees and the suppliers' page charges may inflate the minimum costs. Information on Demand, the leader in the document delivery business, is now owned by SCI/TECH Publishing Services, a Pergamon affiliate.

One of the first not-for-profit organizations to provide document delivery service was Carolina Library Services. They will supply items from the affiliated libraries of the University of North Carolina (Chapel Hill), Duke University, as well as the United States Environmental Protection Agency and National Institute of Environmental Health Sciences Libraries (Research Triangle Park) at prices somewhat lower than the four commercial suppliers discussed above.

A more recent supplier from the not-for-profit ranks is Chemical Abstracts Service (CAS). The 1990 cost of the basic CAS Document Delivery Service ranged from $9 to $18 depending on the option used to order the document. Copyright fees are paid, but if the copyright status is unclear or CAS does not have permission to copy an item, the original document is sent. Although CAS's prices are somewhat higher than other sources, most "fugitive" items (those hard to obtain from other sources) can now be procured with certainty from Chemical Abstracts Service. Prior to the institution of the service, some of the items cited in the *Chemical Abstracts* database were virtually inaccessible in the United States.

There are, of course, many other document delivery services available for a fee from for-profit and not-for-profit vendors in this country. To discuss all of them would be impossible in this book, but the examples selected should give a sense of the range of services available.

13.5.2   Author Reprints

Most publishers will not provide direct reprints from their journals. However, the authors of the articles included in the journals may obtain a certain number of reprints which they can send to those who request them. In order to facilitate the process of obtaining copies from the authors, the Institute for Scientific Information (ISI) makes available a stock of cards called "Request-A-Print". The cards are printed with a removable return address label. When used in conjunction with ISI's *Current Contents Address Directory: Science and Technology* [1754] (formerly, *Who Is Publishing In Science*), one can request reprints directly from the authors. However, such direct requests may not be a reliable form of document delivery because chemists move so frequently in today's competitive scientific world.

Another option for document delivery which is offered by the American Chemical Society is linked to their Single Article Announcement Service. With the Single Article Announcement Service, a subscriber receives contents pages of all ACS journals and may order reprints of individual articles using a form provided by the service. Similarly, the American Institute of Physics (AIP) offers *Current Physics Index*, a guide to the current journal contents of AIP journals. Reprints of the articles are sold by the publisher.

13.5.3   Translations

It is lamentable that the foreign-language translation skills of the many scientists who are native speakers of English have declined so dramatically in the last few decades. Nevertheless, a large number of services and reference sources have been developed to help fill the void in this area. The most expensive option for translation on a word-for-word basis is to hire a professional translator to do a custom translation. While there have been great strides in translation by computers in recent years, few commercial applications are yet discernible. Two other possibilities for obtaining translations are shared collections of translations and cover-to-cover translation journals.

Many freelance translators and professional translation firms will provide custom translations for a fee. Translators with the appropriate language skills and technical background may be located through *Translation and Translators: An International Directory and Guide* [1998] (1979) or the *Translator Referral/Translation Services Directory* [1999] (1982-).

Once a document has been custom-translated into English, it could be used by others, provided there were a means of finding out about it and, of course, the customer were willing to share it. Cooperation in this area has led to the establishment of major collections of translation clearinghouses. Often an organization which contributes translations to such a collection establishes a discount credit line which helps "pay" for other translations it obtains from the clearinghouse. For example, an organization might be required to contribute twenty translations in order to receive one for free.

One of the most successful translation cooperatives is the National Translations Center. Originally established by the Special Libraries Association, the Center was for many years located in Chicago. In 1989, the National Translations Center was formally transferred to the Library of Congress in Washington, D.C. Beginning in the mid-1980s, the NTC's major indexing journal, the *Translations Register Index* [2000], was merged with a similar publication, the *World Translations Index* [1164] of the International Translations Centre in the Netherlands. After the Library of Congress took over the NTC, a decision was made to once again publish a separate index. The holdings of the NTC for the period 1969-86 are found in the *Consolidated Index of Translations into English* [1140].

Another important provider of translations on the North American continent is the Canada Institute for Scientific and Technical Information (CISTI). They maintain a database showing the location of completed translations in Canada, the United States, the United Kingdom, and several other countries. All translations prepared in Canada and reported to CISTI are passed on to the National Translations Center for listing in their index. Another organization with substantial activities in the translation area is the British Library. The British Library Document Supply Centre regularly receives copies of translations which are then listed in *British Reports, Translations, and Theses* [2001]. Their collection includes copies of almost everything which is in the National Translations Center in the United States.

The British Library is also a major producer of **cover-to-cover translation journals**. Such journals include translations for the entire series of articles published in a foreign-language journal. In the United States, major publishers of cover-to-cover translations are Consultants Bureau (a division of Plenum Press) and the American Institute of Physics. Cover-to-cover translation journals present several problems, not the least of which is the determination of their existence and when they began to be published. Of help in that regard are *Journals in Translation* [2002] (1988) and such holdings lists as *Journals with Translations Held by the Science Reference Library* [2003] (1985) or *Soviet Serials in Translation Held by the Center for Research Libraries* [1128] (1983). Another possibility is to check the listing of the original foreign-language journal in *CASSI, Chemical Abstracts Service Source Index* [229]. *CASSI* is quite good for identifying the existence of a cover-to-cover translation journal and has the advantage of being online through ORBIT. A second problem with the cover-to-cover translation journals is that the pagination of the translation rarely, if ever, matches that of the original. Unfortunately, creative scientific authors sometimes combine the titles of the original versions with the page numbers of the translation journals, creating frustration among interlibrary loan librarians who try to verify the citations.

In summary, locating and purchasing an existing translation is likely to be expensive, both in terms of time and money. If a custom translation must be done, expect the cost to be even higher. The conventional wisdom is to glean as much information as possible from English-language abstracts for the work. (Remember to search for equivalents of foreign-language patents which may have been granted in English-speaking countries.) For those who want to spend a bit of time learning enough of the fundamentals of a language to be able to decipher more than the numeric data, graphs, and chemical structures, the following should prove useful:

Reading the Russian Language: A Guide for Librarians and Other Professionals [1303] (1974)
Russian for the Scientist and Mathematician [1305] (1984)
Chemistry Through the Language Barrier: How to Scan Chemical Articles in Foreign Languages, with Emphasis on Russian and Japanese [1394] (1970)
Scientific German [929] (1979).

13.6 SUMMARY

A chemist who is seeking either the very latest published primary information or ancillary background information has a number of valuable services and collections to choose from. The choice of current awareness services for the most recent literature is limited only by the available funds and the time to utilize the services. Likewise, a search of the older literature can be done in reviews, treatises, monographs, encyclopedias, and other types of collected works, depending on the needs of the researcher. Searches in both directions are likely to produce interesting literature references for documents which are not held by the local library. In such instances, document delivery services draw upon the larger universe of available resources.

13.7 SELECTED READINGS

Lavendel, Giuliana A. "SDI in Scientific and Technical Libraries: An Overview of the Options." *Sci. Techn. Libr.* **1981**, *2*(1), 1-2.

Campbell, Robert M.; Stern, Barrie T. "ADONIS — A New Approach to Document Delivery." *Microcomput. Inf. Management* **1987**, *4*(2), 87-107.

Pinzelik, John. "Chemistry." In *Selection of Library Materials in the Humanities, Social Sciences, and Sciences*; McClung, Patricia A., Ed.; American Library Association: Chicago, 1985; pp 261-268.

Garfield, Eugene. "Reviewing Review Literature. Part 1. Definitions and Uses of Reviews." *Curr. Contents* **1987**, (18), 36.

Garfield, Eugene. "Reviewing Review Literature. Part 2. The Place of Reviews in the Scientific Literature." *Curr. Contents* **1987**, (19), 3-8.

Chapter 14

THE PERSONAL LIBRARY AND SCIENCE WRITING AIDS

14.1 INTRODUCTION

A large part of a scientist's time is devoted to writing of one sort or another. Long before work begins in the laboratory, much time has been spent taking notes on information in the primary and secondary literature which may prove useful in the research project. Good habits in this initial stage will pay off in time saved when the final research reports and journal articles are being written. As research progresses, detailed laboratory notebooks are maintained by the researcher. These may prove to be very important in the determination of the patentability of a new product. Laboratory notebooks may also become part of a specialized collection in a corporate library, available for the internal use of other researchers. Before a lecture is presented at a conference, it is common for the scientist to prepare a written text, a **preprint**, which may be distributed to others on request long before the appearance of the research results in the formal primary literature. Thus, even before submitting a manuscript of a journal article, the scientist has done a considerable amount of writing about the research project.

There are definite standards for scientific writing which have evolved over the years. One of these standards has to do with the way **bibliographic citations** or **bibliographic references** are written. A bibliographic citation has the information necessary to compile a bibliography of references in a standard format. It contains enough information to allow the original source document to be located. Unlike authors in many other fields, writers in most scientific disciplines use standard abbreviations for the references in footnotes and bibliographies rather than the fully spelled-out titles of journals and other works. Likewise, there are standard scientific symbols, standards for terminology and for nomenclature of compounds, and other types of standards with which today's scientific writer must contend. Furthermore, there are stylistic expectations on the part of editors, reviewers, and even the readers. To assist the scientist in writing and in maintaining personal notes, a number of reference works and other types of aids have been developed in recent years. The most important of them will be discussed in this chapter.

14.2 SOFTWARE FOR PERSONAL DATABASES AND WORDPROCESSING

Many scientists have constructed elaborate card files of useful references and other information. As such files grow in size they become harder and harder to maintain. Retrieving information from manual index files is sometimes quite difficult, even though scientists often develop their own classification schemes. Some might also code notches around the outer edges of the cards to allow manual sorting of the file. Today, however, the personal computer gives the scientist a powerful new option to maintain a file of personal references. Before the computer is used for such purposes, considerable thought should be given to how a personal documentation or **personal information management system (PIMS)** should be constructed, in particular, to the makeup of the records and fields and how the database records can be re-used with wordprocessing software.

Stibic's *Personal Documentation for Professionals; Means and Methods* [128] (1980) not only describes ways to organize the *records* of a personal collection, but also offers various options for filing the *material* itself (books, reports, journals, clippings, photocopies, slides, microfiches, etc.). Stibic describes and evaluates both computer-based systems and card files. File construction techniques such as classification and indexing are also discussed.

At this point in the history of microcomputers, a considerable amount of software exists for the creation and maintenance of small personal bibliographic databases. A very desirable feature of a *chemical* information storage and retrieval system is the ability to handle both bibliographic citations and chemical structural drawings. Factors which should be taken into account in selecting such software include price, operating system and equipment requirements, ease of use, flexibility, maximum size of the database, and speed of operation. There are some personal information management software packages which are designed especially for chemistry. Others are more general in their approach, but have significant applications of interest to chemists. The special features of some of those systems are described below.

Molecular Design Ltd.'s ChemBase [73] is one component of a set of integrated software dubbed the Chemist's Personal Software Series. ChemBase is the component used to create personal databases for storing and retrieving molecules, reactions, and associated data on a personal computer. This software is also compatible with data from the *Index Chemicus* database and comes with a choice of two files of compounds.

(See Table 11.2). The package includes a wordprocessing program, ChemText [81], which can import images of molecules either from ChemBase or from other databases. This allows the insertion of chemical structures into the text of files produced by the wordprocessor. ChemText comes with several commonly-used partial structures, or structures can be drawn freehand with a mouse. Stereochemical bonds can be indicated, and the user has several choices with regard to the display of hydrogen atoms. ChemText permits both text and graphics to be controlled and manipulated in one software package. ChemTalk/ChemHost [80] is a communications package which links the IBM PC to a host computer running Molecular Design Ltd.'s software for reaction and structure searching.

With Compress's ChemFile [76] a database of structures, data, and text can also be created. Another Compress software package, ChemLit [76], produces files which contain abstracts associated with structures. Thus, searching by either substructure of the molecule or by abstract terms is possible. Searches in ChemFile can also be done by correlating data with structures. For example, all compounds which contain a -$CONH_2$ group and exhibit a particular biological activity within certain limits can be found.

Molecular Presentation Graphics (MPG) [90] is a chemical structure drawing program for the IBM PC or Macintosh. MPG works with many popular wordprocessing software packages (Microsoft Word, WordPerfect, WordStar, etc.). Hawk Scientific Systems, Inc., the producer of MPG, also offers DATALYST, a text-oriented database manager. The software allows the creation of an authority list of search terms. DATALYST also includes a molecular formula calculator to generate a compound's molecular weight, a feature of many such chemical database managers. SoftShell International has developed software to integrate chemical structures into documents and presentations. For the IBM PC, the product is named ChemWindow [2005], with a counterpart, ChemIntosh [75], available for the Macintosh. Two other Macintosh-based products, ChemDraw [74] and Chem3D [2006], are from Cambridge Scientific Computing. By using the Macintosh Clipboard, any drawing created with ChemDraw can be included in other applications or transferred to Chem3D. Chem3D can also read and write text files for use with other computer programs, for example a structure created by MM2 on a mainframe computer. WIMP, Wisconsin Interactive Molecule Processor [99], available from the Aldrich Chemical Company, draws both structures and reaction schemes which can be printed on a variety of devices, including laser printers and plotters.

Alchemy II [1901] promises to become a household word among chemists because of the addition of CONCORD 3D coordinates to many of the compounds in the Registry File on STN International. By using STN Express, the coordinates can be downloaded and drawn in three dimensions on the PC with Alchemy II. The molecular images can then be output to a file and integrated into WordPerfect or other wordprocessing programs that accept HPGL graphics files. Alchemy is much more than just a software system for drawing molecules. Crystallographic input, fitting by least squares techniques, and energy minimization and MM2 options are standard features of the program.

PSIDOM [93] from Hampden Data Services is a PC-based system for the retrieval of molecular structures, associated data, and the printing of structures. Hampden Data Services is the producer of STN Express, with which PSIDOM is fully compatible. The company has been very active in the development of a standard **chemical structure drawing interface** which would enable the creation and use of chemical structure and reaction records for local and remote databases. Up to 20,000 chemical structures can be created in a single database with PSIDOM. Structures can be printed on a variety of devices. There are components of PSIDOM which make it capable of extremely fast substructure searching on a PC and allow it to convert files for loading directly into larger mainframe databases. With the PsiORAC module, details of a chemical reaction can be laid out in a format suitable for uploading into ORAC, a mainframe chemical reaction database.

On a much smaller scale, ChemSmart [79], which is available from ISI, can store and retrieve chemical structures, physical constants and other data. The system permits searching by chemical structure or substructure. Data for up to 300 compounds can be placed on a 5 1/4'' IBM PC diskette. Furthermore, the product includes a basic library of 250 organic molecules taken from the *Index Chemicus* database (see Table 11.2). The user can add other compounds. Structures are displayed in graphic form and can be rotated or enlarged.

There are literally dozens of software packages which can be obtained to create personal bibliographic databases. These range in cost from under $50 to well over $1000. Some of these database managers, such as The Scientific Reference System [2010], Reference Manager [2011], REF-11 [2012], and Pro-Cite [92], have some very interesting features and options. The capability to recognize field tags in records which have been downloaded from BRS, DIALOG, STN International, etc. and to import those records into the database manager is one such feature. Another is the ability to take selected records as they exist in the database on the PC and transform them into various bibliographic styles required by journal publishers.

A good example of such software is Pro-Cite, which can store up to 32,000 records on a hard disk of an IBM PC. (A Macintosh version is also available.) The producer of the software, Personal Bibliographic Systems, has a front-end product, Pro-Search [1887], which simplifies searching on DIALOG, BRS, and other vendors' systems. With the optional Biblio-Links software, records from those and other database vendors can be loaded into Pro-Cite. There are style sheets for many organizations built into the Pro-Cite software, including the requirements of the American Chemical Society. Pro-Cite also has facilities for entering the records into the correct position in a manuscript where footnotes and bibliographies appear.

Most commercially-available wordprocessing packages handle only the normal alphanumeric characters and punctuation. Obviously, scientific writers need a much wider range of symbols to produce a scientific manuscript. To answer that need, scientific wordprocessing software which can encode and print symbols such as Greek characters, mathematical symbols, equations, and even complex diagrams have been developed. Among the packages available for microcomputers today are MASS-11 [88], T^3 [98], The Egg [83], and Spellbinder Scientific [97].

14.3 AIDS TO WRITING

Collecting data from the literature, reading background studies, and storing and retrieving that information occupy only part of a scientist's time. It is the laboratory work and ultimately the publication of the results of the experiments which give the real reward to the scientist. The recognition by one's peers of truly significant research is valued very highly in all fields. There is a better chance of such recognition if the scientist has developed skills in writing. In this section are presented works which aim to improve and sharpen the scientist's writing skills.

14.3.1 Laboratory Notebooks

Howard M. Kanare's *Writing the Laboratory Notebook* [120] (1985) attempts to teach the principles of proper scientific notekeeping. In addition to practical tips on organizing and writing the notebook, Kanare provides examples of notebook entries. He has included a chapter on the legal and ethical aspects of writing as well as a chapter on patents and the protection of invention. Concluding with a sketch of the electronic notebook, he compares the advantages and disadvantages of the electronic notebook with those of the conventional handwritten notebook. A few companies which sell microcomputer software suitable for electronic laboratory handbooks and software for **laboratory information management systems** (**LIMS**) are also noted. LIMS are computerized laboratory information systems with management components (for sample backlog, scheduling, workload, personnel evaluation) and analytical components (for data reduction, graphical display of data, and report writing).

14.3.2 Style Guides

The ACS Style Guide; A Manual for Authors and Editors [110] (1986) is intended to be a true **style guide**, providing practical information and guidance on scientific writing in general. On the other hand, the ACS *Handbook for Authors* (1978 and 1967) which it supersedes, concentrated on instructions to authors who were contributing to various American Chemical Society (ACS) publications. Although such information is also included in the revised work, *The ACS Style Guide* adds material of a more general nature, for example, preparation of manuscripts in machine-readable form and a brief guide to the chemical literature. A basic chapter on the scientific paper itself reviews such areas as the components of the paper and the types of presentation. Chapter 2, "Grammar, Style, and Usage," comprises half of the text, excluding appendices. On pp. 106-114 of Chapter 2 is a section which lays out the standard format of the references to be used in bibliographies. Every chemist should be familiar with these pages. The variations in the requirements of different ACS publications are noted.

The ACS Style Guide states that the minimum data required in references for journals are the author, *abbreviated* journal title, year of publication, volume number, and initial page of the cited article. (The complete page span is preferred.) Note that the title of the article and the issue number and month of publication are not included.

Therefore, the chemist who attempts to find an article in a bound journal is more likely to know the page numbers on which it appears than the issue number or month of publication of the issue in which the article is located. Since most science journals today number their pages continuously from the first to the last issue of a volume, the page numbers included in a bound journal should be printed on the spine. This should be done even at the expense of omitting inclusive issue numbers and months, if necessary.

Omission of the title of the article from journal citations in the bibliographies of scientific works is something long bemoaned by interlibrary loan librarians. However, the most striking difference between library records and the approved minimum form of a journal citation in chemistry is the use of the *abbreviated* journal title. The *ACS Style Guide* lists abbreviations for a number of the most commonly cited journals, but a more nearly comprehensive listing can be found in *CASSI*, discussed in section 14.4.1.

It should be noted that journals not published by the American Chemical Society and particularly journals in biochemistry might have different style requirements for their authors. Although most biochemical journals, regardless of their publisher, now use the *Chemical Abstracts* system of journal abbreviation, some followed a different standard in the past and may continue to do so.

Books can be cited in ACS publications with the minimal data of author or editor, book title, publisher, city of publication, and year of publication. One considerable difference between the current *ACS Guide* and the 1978 *Handbook* is that book titles are now italicized instead of being enclosed in quotation marks. The *Guide* notes that other types of publications should be cited with enough information to identify and locate the source. Any numbers found on the title page or cover of a publication other than a book or a journal are usually very important and should be listed. This is especially true for patents, reports, and government publications, as the examples on pp. 112-113 of the *Guide* (and in Figure 1-2 of this book) show.

Other chapters of *The ACS Style Guide* cover illustrations and tables, copyright and permissions, and making effective oral presentations. The appendices contain much useful information on topics such as element symbols, atomic numbers and atomic weights, symbols for commonly-used physical quantities, proof-readers' marks, and a brief review of all ACS publications and databases which were in existence in 1986. Search **INDX = "STYLE GUIDE"** in the Chemistry Reference Sources Database for additional materials of this type.

14.3.3 "How-to-Write" Books

Fieser and Fieser's *Style Guide for Chemists* [112] (1960), available as a 1972 Krieger reprint, is limited to problems involved in the writing of chemical papers. One of the unique features of the book is Chapter 12 "Pronunciation," which covers both chemical names and the names of famous chemists. Even the pronunciation of some Latin and Greek prefixes is included.

An amusing book which aims to instill confidence in scientific writers is Schoenfeld's *The Chemist's English* [123] (1986). The little book is full of examples illustrating both correct and incorrect chemical writing. Nestled in chapters like "An Investigative Examination of Driveliferous Jargonogenesis" and "Is You Is or Is You Ain't My Data?" are some very practical tips for the chemist who is writing in English.

Day's *How to Write and Publish a Scientific Paper* [109] (1988) is intended to help scientists and students in all disciplines. The author stresses certain basic principles which are common to most disciplines and provides the specific and practical information needed to write effective scientific papers. In short, this is a cookbook for the scientific paper, though not simply a book of recipes. Like Schoenfeld, Day gives examples of the misuse of English in scientific writing, while describing the essential components of a scientific paper in logical order. These are the *introduction* (statement of the problem, background information, rationale for the study), the *materials and methods section* (full details of how the problem was studied), the *results section* (the data), and the *discussion section* (what the findings mean).

Day advises authors to write out references to the literature in full as they are collected. Then all of the data will be available to be fitted into the many different permutations required in the bibliographic stylistic rules of different journals. Appendix I of Day's book is a list of "Selected Journal Title Word Abbreviations" which follow the American National Standards Institute (ANSI) Z39.5-1969 standard for abbreviating periodical titles. Chapters on the design of tables and the preparation of illustrations are supplemented with practical advice on submitting the manuscript and dealing with editors and printers. Formats other than the traditional primary journal article are also dealt with. These are the review paper, conference report, book review, and thesis.

There are other important style guides and "How-to-Write" books in related disciplines. Search in the Chemistry Reference Sources Database for **INDX = "STYLE GUIDE" AND NAME** of the discipline or **INDX = "WRITING AID" AND NAME** of the discipline in order to find them.

14.4 BIBLIOGRAPHIC STANDARDS FOR SERIALS AND LIBRARY RECORDS

It has already been noted that library records and the records in the bibliographies of published scientific works often disagree. In this section we will see how to translate from one system of citing to another.

14.4.1 *CASSI* and Serial Abbreviations

The most important reference work for the standard abbreviation of scientific journals in chemistry is *Chemical Abstracts Service Source Index* [229] or, as it is commonly called, *CASSI*. The 1907-89 cumulative version is the latest edition. Quarterly supplements will appear for the period after 1989, and each year the final quarterly supplement will be cumulated into an annual supplement. Between publication of the supplements, new journal abbreviations are listed every other week in the even-numbered issues of *Chemical Abstracts* in the last section, "CA Abstracted Publications: Additions and Changes". *CASSI* is also available as a database through the ORBIT Search Service.

CASSI provides detailed and accurate bibliographic data on serials and other types of publications using national and international standards. *CASSI* covers not only chemistry publications, but also many publications in the biological, engineering, and physical sciences. It also lists publications which were abstracted by the first abstracting journal, *Chemisches Zentralblatt*, which began in 1830. Publications which are cited in the *Beilstein Handbook of Organic Chemistry*, even those prior to 1907, are also included. One should be aware, however, that the old journal abbreviations from the printed *Beilstein* may not be in the same form as those used in *CASSI*, and the compilers of *CASSI* have not supplied "SEE" references in such cases. Take for example the use of *B.* in *Beilstein* as the abbreviation for the old German-language journal *Berichte der Deutschen Chemischen Gesellschaft*, the forerunner of *Chemische Berichte*. This journal can be found in *CASSI*, but not by looking under *B*. It is listed as **Ber-DtschChemGes**! The full journal title is apparent in the actual bibliographic entry in *CASSI*, as depicted in Figure 14-1. Although it makes the printed *CASSI* somewhat difficult to use, the characters in lighter type must be ignored and the bold-face characters, which are the abbreviations to be used in compiling bibliographies of cited works, must be run together in order to find items in the correct alphabetical sequence. This may at times make it appear that some items are out of alphabetical sequence. For example, Journal of the American Chemical Society is found in *CASSI* before Journal of Structural Chemistry since *J.Am.Chem.Soc.* files before *J.Struc.Chem.*. Over 60,000 entries are thus placed in alphabetical order in *CASSI*.

What kind of information does *CASSI* include? First, it gives the full name of the publication as it appears on the printed primary source. The standard abbreviation to be used in citing the work in bibliographies for ACS publications (and other publications which follow this standard) is indicated in dark type. The **CODEN**, a six-character standard abbreviation for the journal, and the **International Standard Serials Number**, **ISSN**, are important for online searching. Furthermore, a publication history of the journal is included. Since journals frequently split into sections, merge, absorb other journals, and often change their titles, this information is very valuable.

14.4.2 Library Forms of Serial Entries: *CASSI* vs. *AACR*

Another important bit of information in *CASSI* is the title printed in capital letters. That is the form of the entry specified by the *Anglo-American Cataloging Rules* [184],

Berichte der **Deut**schen **Chem**ischen **Ges**ellschaft. BDCGAS.
    ISSN 0365-9496. In Ger. v1 1868 - v61 1928.
    DEUTSCHE CHEMISCHE GESELLSCHAFT. BERICHTE. HEIDELBERG.
Divided into **Ber.Dtsch.Chem.Ges.A**, which see and
Ber.Dtsch.Chem.Ges.B, which see.

FIGURE 14-1
Bibliographic Portion of an Entry from *Chemical Abstracts Service Source Index (CASSI)*.

(AACR). Those rules are followed by most libraries in the United States. AACR II, the second edition of the *Anglo-American Cataloging Rules*, made a major change in the way libraries now catalog serials. Under the previous rules, if the name of an organization was included in the title of a journal, the organization name certainly would have become the **main entry**, the first line on the card catalog card. This determines which letter to search under. Hence, *Journal of the American Chemical Society* would have been cataloged:

AMERICAN CHEMICAL SOCIETY. JOURNAL.

Under the new rules, the main entry would be:

JOURNAL OF THE AMERICAN CHEMICAL SOCIETY.

Since this journal was published for decades before the appearance of the AACR II rules, most libraries would have cataloged it under "A". Therefore, *CASSI* provides a link between abbreviated journal titles and the records as used in many library catalogs. Be aware, however, that with the production of computerized serials holdings lists in the last decade or so, many libraries have updated their older serial entries. Thus, even the old German journal *B*. may be listed as:

BERICHTE DER DEUTSCHEN CHEMISCHEN GESELLSCHAFT

in a library's catalog instead of the *CASSI* AACR indication in capital letters in Figure 14-1. One of the better things about this change in AACR II is that it allows the journals to be filed strictly by title in a form close to that which scientists are accustomed to using. In the many science libraries which ignore call numbers for journals and elect to file them alphabetically by title as it appears on the publication (and on the spine of the bound journal), this is a very positive change. About half of the science libraries in the United States prefer to file journals by title rather than by call number. However, such an arrangement can have its disadvantages, as any chemist can attest who has tried to track down *Chemical Communications* in its different successive-entry title variations where journals are filed by title.

In summary, *CASSI* is a very important tool for questions related to the format of bibliographic entries for serials and for translating abbreviated journal titles to their library equivalents. Even publications from meetings (conference proceedings) and edited collections (what we have called "multigraphs") are included in *CASSI* now. The use of *CASSI* related to document delivery is discussed in chapter 13, section 13.4.6.

14.5 BOOKS AND OTHER WORKS ON THE NOMENCLATURE OF CHEMICAL COMPOUNDS

The International Union of Pure and Applied Chemistry (IUPAC) periodically issues official nomenclature books. Listed in Table 14.1 are the most recent editions. Between separate editions, recommendations are published in *Pure and Applied Chemistry*, IUPAC's official journal. In fact, the previous edition of the last work listed was published in its entirety in *Pure and Applied Chemistry* **1979**, *51*, 1-41.

TABLE 14.1
IUPAC Books on Chemical Nomenclature

Compendium of Chemical Terminology [161] (1987)

Compendium of Analytical Nomenclature [157] (1988)

Guide to Trivial Names, Trade Names, and Synonyms for Substances Used in Analytical Chemistry [162] (1978)

Nomenclature of Inorganic Chemistry [165] (1989)

Nomenclature of Organic Chemistry [171] (1979)

Quantities, Units, and Symbols in Physical Chemistry [168] (1988)

In the field of biochemistry, the International Union of Biochemistry (IUB), has published the following authoritative books:

Biochemical Nomenclature and Related Documents [164] (1978)
Enzyme Nomenclature [158] (1984).

The IUB publishes recommendations as they are made in the *European Journal of Biochemistry*, the official journal of the Federation of European Biochemical Societies (FEBS).

IUPAC and IUB nomenclature rules allow a choice of names in many cases. On the other hand, a major abstracting organization such as Chemical Abstracts Service (CAS) strives for a single unique and unambiguous name for a compound which will be used in its printed and online products. In recent years, Chemical Abstracts Service has moved much more toward the use of names which depict the structure. Both the IUPAC and CAS systems of nomenclature are the results of decades of trying to improve and standardize the naming of chemical compounds through the use of **systematic nomenclature rules**. These rules aim to exactly describe the composition of a compound and as much as possible to depict its structure. Trivial names are often assigned before the structure of a compound is known, thus covering all variants of a compound due to tautomerism, etc. Trivial names may be derived from the source (for example, lactic acid), the shape (cubane), or simply for other, sometimes quizzical reasons, (for example, nil).

Most of the authoritative nomenclature books for chemical compounds have been issued through the International Union of Pure and Applied Chemistry or the International Union of Biochemistry. However, there are some books which give a broader introduction to the subject of chemical compound nomenclature.

A classic book on the subject is Cahn and Dermer's *Introduction to Chemical Nomenclature* [181] (1979), now in its fifth edition. Both inorganic and organic compounds are dealt with by the authors. An appendix lists "Important Recent Changes from IUPAC Nomenclature by Chemical Abstracts Indexes".

Lees and Smith have gathered contributions from several experts to compile *Chemical Nomenclature Usage* [180] (1983). Essentially a survey of the problems and requirements of various nomenclature systems, the book attempts to bridge the gap between experts on chemical nomenclature and those who must use chemical names on a day-to-day basis. Although there is some emphasis on British practices, there are general chapters covering such topics as biochemical nomenclature and the use of trivial, generic, and common names. The appendix contains a useful classified bibliography covering the history of nomenclature, agro-chemical nomenclature, and pharmaceutical nomenclature, as well as listing many IUPAC recommendations.

The *CRC Handbook of Chemistry and Physics* [179] (annual) includes sections on the nomenclature of inorganic chemistry and organic chemistry at the beginning of the respective sections in the handbook. The organic section includes a few pages of basic organic ring compounds. "Symbols, Units, and Nomenclature in Physics" is taken from a definitive document published by the International Union of Pure and Applied Physics. The recommended symbols for physical quantities can be found there. *Lange's Handbook of Chemistry* [178] (1985) also contains discussions of nomenclature for inorganic and organic chemistry plus IUPAC physical and chemical symbols and terminology. In addition, *Lange's* has a brief section on steroid nomenclature. Biochemical abbreviations, alchemical symbols used in biology and botany, latin terms, and other biochemical terminology are found in some of the miscellaneous tables at the end of the *Merck Index* [177] (1989).

The latest "Index Guide" [176] to *Chemical Abstracts* includes an appendix on "Chemical Substance Index Names". The appendix is also available separately as CAS Reprint 298 with the title "Naming and Indexing of Chemical Substances for *Chemical Abstracts*" [175]. One can find in the publication a "Selected Bibliography of Nomenclature of Chemical Substances". The journals *Pure and Applied Chemistry* and *European Journal of Biochemistry* should be consulted for more recent nomenclature recommendations.

14.6 OTHER STANDARDS FOR NUMERICAL DATA, SYMBOLS, AND TERMINOLOGY

In addition to the physical data sections in handbooks mentioned in the previous section, there are other places where standards for numerical data, symbols, and terminology can be located. Terminology recommended by IUPAC for physical, inorganic, organic, macromolecular, and analytical chemistry can be found in the *Compendium of Chemical Terminology: IUPAC Recommendations* [161] (1987). In addition, an organization known as CODATA (the Committee on Data for Science and Technology of the International Council of Scientific Unions) is heavily involved in establishing standards for numeric data. CODATA has published a number of helpful guides in the *CODATA Bulletin*, as shown in Table 14.2. *CODATA Bulletin* no. 9 (1973) also includes references to available guidelines for reporting data in various other areas, such as crystallography, infrared spectra, optical spectra, thermal conductivity, thermochemistry and thermodynamics, gas chromatography, and thermal analysis. Many of these guidelines were published as primary journal articles.

In recent years recommendations of this type have often been published in *Pure and Applied Chemistry*, the official journal of the International Union of Pure and Applied Chemistry. (See Table 14.3.)

TABLE 14.2
Selected Standards for Numerical Data, Symbols, and Terminology Published in the CODATA *Bulletin*

No.	Year	Title
44	1981	Calorimetric Measurement on Cellular Systems; Recommendations for Measurements and Presentation of Results
32	1979	Guide for the Presentation in the Primary Literature of Numerical Data Derived from Observations in the Geosciences
30	1978	Guide for the Presentation in the Primary Literature of Physical Property Correlations and Estimation Procedures
25	1977	Biologists' Guide for the Presentation of Numerical Data in the Primary Literature
20	1976	Recommendations for Measurement and Presentation of Biochemical Equilibrium Data
13	1974	The Presentation of Chemical Kinetics Data in the Primary Literature
9	1973	Guide for the Presentation in the Primary Literature of Numerical Data Derived from Experiments

TABLE 14.3
Selected Standards for Numerical Data, Symbols, and Terminology Published in *Pure and Applied Chemistry*

Vol.(no.)	Year	Title
60(7)	1988	Glossary of Terms Used in Photochemistry
58(10)	1986	Recommendations for the Presentation of Thermodynamic and Related Data in Biology
57(1)	1985	Names, Symbols, Definitions and Units of Quantities in Optical Spectroscopy
57(3)	1985	Definition of pH Scales, Standard Reference Values, Measurement of pH and Related Terminology
57(10)	1985	Recommended Terms, Symbols, and Definitions for Electroanalytical Chemistry
57(11)	1985	Nomenclature for Thermal Analysis—IV
55(8)	1983	Glossary of Terms Used in Physical Organic Chemistry

14.7 SUMMARY

It is important to follow accepted standards in scientific writing, whether that be in the naming of chemical compounds, the use of symbols for physical constants, or abbreviations for journals. In this chapter we have discussed a number of reference works and other sources which can help the scientific writer adhere to such standards. Books which offer practical tips on the process of writing and computer software for maintaining personal information files were also surveyed. The important reference tool *CASSI* which serves as a bridge between library records and records as they appear in bibliographies of scientific journals or in abstracting or indexing services was also presented. The chemist who consults the works discussed in this chapter and who adopts the practical advice found in many of them is sure to become a better scientific writer.

14.8 SELECTED READINGS

"Technical Wordprocessors for the IBM PC and Compatibles." (A Report by the PC Technical Group of the IBM PC Users Group of the Boston Computer Society). *Am. Math. Soc. Notices* **1986**, *33*(1), 8-37.

Tenopir, Carol. "Software Options for In-House Bibliographic Databases." *Lib. J.* **1987**, *112*(9), 54-55.

Tenopir, Carol; Lundeen, Gerald W. "Software Choices for In-House Databases." *Database* **1988**, *11*(3), 34-42.

Wachtel, Ruth E. "Personal Bibliographic Databases." *Science* **1987**, *235*(4792), 1093-1096.

James, Geoffrey. *Document Databases*. Van Nostrand Reinhold: New York, 1985. [350]

Norris, A. C.; Oakley, A. L. "Electronic Publishing and Chemical Text Processing— I. A Survey of Software for Two-Dimensional Chemical Structure Editing." *Comput. Chem.* **1988**, *12*(3), 245-251.

Norris, A. C.; Oakley, A. L. "Electronic Publishing and Chemical Text Processing— II. Character and Vector Representations of Chemical Structures." *Comput. Chem.* **1988**, *12*(3), 253-255.

"Scientific Word Processors: Formulas for Success." *PC Magazine* **1988**, *7*(13), 251-256, 260.

Lundeen, Gerald. "Software for Managing Personal Files." *Database* **1989**, *12*(3), 36-48.

Mills, Ian M. "The Choice of Names and Symbols for Quantities in Chemistry." *J. Chem. Educ.* **1989**, *66*(11), 887-889.

Chapter 15

MISCELLANEOUS INFORMATION SOURCES

15.1 INTRODUCTION

Although numerous specialized information sources are found in the previous chapters, questions often arise which fall outside the scope of those sources. Whether it be an address for a chemical supplier, historical or biographical information, chemical business information, or some other need, many sources of assistance are available. The works in this chapter supplement the chemist's information bank with the best choices for some common information needs.

15.2 HISTORY OF CHEMISTRY

The modern science of chemistry is well into its third century, and a rich collection of literature chronicles the historical development of chemistry. Important historical figures are of interest in this field of study, but so too is the history of the techniques and instruments which are milestones on the path to modern chemistry. Some guides to the literature of the history of science which may prove useful are Sarton's *A Guide to the History of Science* [375] (1952), Corsi and Weindling's *Information Sources in the History of Science and Medicine* [2014] (1983), and Jayawardene's *Reference Books for the Historian of Science* [2015] (1982). Furthermore, Sturchio's *The History of Chemistry: A Critical Bibliography* [1813] (1985), Multhauf's annotated bibliography *The History of Chemical Technology* [2016] (1984), and other bibliographies in the Garland Publishing Company series "Bibliographies of the History of Science and Technology" may be of use.

Another important work is *A Guide to the Culture of Science, Technology, and Medicine* [1631] (1984). *Recent Developments in the History of Chemistry* [1627] (1985), a product of the Royal Society of Chemistry, *Chemistry in America, 1876-1976* [1626] (1985), and Cole's *Chemical Literature, 1700-1860: A Bibliography with Annotations, Detailed Descriptions, Comparisons, and Locations* [2017] (1988) are also significant, recent works.

The Beckman Center for the History of Chemistry has published *Corporate History and the Chemical Industries: A Resource Guide* [1782] (1985) and *Guide to Archives and Manuscript Collections in the History of Chemistry* [1817] (1986). The Center for the History of Chemistry in Philadelphia was formed in 1982 to help preserve the heritage of chemistry. A joint venture of the American Chemical Society, the American Institute of Chemical Engineers, and the University of Pennsylvania, the Center's mission is to discover and disseminate information about historical resources and to encourage research, scholarship, and popular writing in the history of chemistry, chemical engineering, and the chemical process industries.

A truly great accomplishment is Sieghard Neufeldt's *Chronologie Chemie, 1800-1980* [1396] (1987). Presenting a timetable of developments in chemistry, the book summarizes by year the most important writings and developments, with references to the original articles. Important histories of chemistry from 1842 to the present are noted in an appendix, as are Nobel Prize winners in physics, chemistry, and medicine. Other appendices deal with topics like the development of chemical nomenclature, chemical societies, and older scientific periodicals (up to 1900).

Ihde's *The Development of Modern Chemistry* [1395] (1984), Asimov's *A Short History of Chemistry* [1634] (1965), and Jaffe's *Crucibles: The Story of Chemistry From Ancient Alchemy to Nuclear Fission* [1855] (1948) cover the whole of chemistry. Subdisciplines of chemistry are the topics of works such as:

Essays on the History of Organic Chemistry in the United States [1457] (1986)
Essays on the History of Organic Chemistry [1698] (1987)
A History of Biochemistry [717] (1972-83)
History of Chemical Engineering [1352] (1980)
A History of Analytical Chemistry [1432] (1977)
History of Analytical Chemistry [1433] (1966)
History of Clinical Chemistry [1525] (1983).

For collections of original material, see *Landmarks of Science* [1630] (1967-76), a microprint collection, and *A Source Book in Chemistry* [1629], published in two volumes (1952-68) and covering the periods 1400-1900 and 1900-1950.

15.3 BIOGRAPHICAL SOURCES

Biographical compilations typically include details of birth and death, education, and the main publications or research interests of important chemists or scientists in other disciplines. Scientists are covered in more general biographical sources, but there are also many specialized sources which concentrate on scientists. The most important biographical reference works will be found among the nearly 700 sources listed in *Biographical Sources: A Guide to Dictionaries and Reference Works* [1791] (1986).

Probably the best-known source of biographical information on American and Canadian scientists is *American Men and Women of Science* [376] (1989), the first edition of which was published in 1906. The tradition followed in all editions of *American Men and Women of Science* is that only living scientists are included in the current edition. To assist in finding entries in previous editions, the *American Men and Women of Science Editions 1-14 Cumulative Index* [2018] (1983) should be consulted. Since the 14th edition was published in 1979, the index is an avenue to nearly three quarters of a century of biographical entries in the various editions of *AMWS*. The current 17th edition covers over 125,000 scientists working in ten broad scientific disciplines and 164 sub-disciplines. Information in the entries (which are also searchable as an online database) includes basic biographical data, positions held, honors and awards, research specialties, and, in most cases, mailing addresses.

For scientists who died prior to the publication of the first edition of *AMWS*, the *Biographical Dictionary of American Science: The 17th Through the 19th Centuries* [380] (1979) is available. Billed as a retrospective companion to *AMWS*, the work has 300-400 word entries for about 600 scientists who were born in the period 1606-1867.

Poggendorff's biografisch-literarisches Handwörterbuch der exakten Naturwissenschaften [1143] (1863-) is the classic example of a **bio-bibliographical compilation**, containing brief biographical entries and a detailed bibliography for the most important scientists. The work has been appearing in parts since 1863, with each part covering certain periods of time. (See Table 15.1.) Volume 7A covers only Germany, Switzerland, and Austria and includes a supplement which updates the entries for the most important German scientists of all time periods. Volume 7B covers all other nationalities. In addition to the usual biographical information and lists of principal publications, obituary notices are cited.

By far the most important English-language compilation is the *Dictionary of Scientific Biography* [1622] (1970-1980). Living scientists are excluded from the work. The full edition of the set contains 16 volumes, but a one-volume abridgment is available in the *Concise Dictionary of Scientific Biography* [379] (1981). About 5000 individual biographies are included in the full set for physical and biological scientists and mathematicians. There is an index volume designed to trace the evolution of problems, concepts, and subjects through the alphabetically-arranged work.

TABLE 15.1
Chronological Coverage of Scientists in *Poggendorff's Bio-Bibliographical Handbook of the Exact Sciences*

Volume	Coverage
1-2	up to 1858
3	1858-1883
4	1884-1903
5	1904-1922
6	1923-1931
7A-B	1932-1953

Information on twentieth century scientists can be found in *McGraw-Hill Modern Scientists and Engineers* [1790] (1980). The work covers about 1100 winners of prestigious international awards since the 1920s. Included are 200 Nobel Prize winners through 1978. There is an extensive topical and analytical index and cross references to related articles. Selected bibliographies are included.

Biographical information appears in many other formats, including journal articles, standard encyclopedia articles, books, and festschrifts. A **festschrift** is a collection of papers assembled to honor a scientist. It invariably includes an introductory biographical essay, often with a comprehensive bibliography of the scientist being honored. Another important category is **biographical memoirs**. An example of such publications is *Biographical Memoirs of the National Academy of Sciences*.

Often the best source leading to biographical information on an important scientist is the card or online catalog of a major research library, where one should search for the personal name as a subject. For additional sources, search the Chemistry Reference Sources Database for **BIOGRAPHY**.

15.4 DIRECTORIES

There are literally thousands of directories in existence today, many of them specifically designed for use in science and technology. A **directory** provides a link between an individual and another person or organization or company. (There are other types of publications with the word ''directory'' in the title which we have considered ''guides'' in this book, such as the *Directory of Computerized Data Files* and the *CODATA Directory of Data Sources for Science and Technology*.) A work which serves as a guide for over 1200 scientific directories is the *Directory of Technical and Scientific Directories* [1144] (1988). A more general guide is *The Directory of Directories* [1788] (1988). Since directory information is of little or no use if it is outdated, the works discussed below are frequently updated and many of them have computer-readable versions which are even more up-to-date than the printed product.

15.4.1 Directories of People

There are many reasons for needing to find a current address or phone number for a scientist, including the desire to write for a reprint of a publication, to find a seminar participant, to request a reference for someone who is being considered for employment, etc.

Current Contents Address Directory [1754] (annual) is a compilation of addresses of authors who publish each year in the thousands of publications included in the coverage of *Current Contents*. Formerly titled *Current Bibliographic Directory of the Arts and Sciences*, the *Current Contents Address Directory* is now published in separate editions for *Science and Technology* and *Social Sciences/Arts and Humanities*. Not only are the authors' addresses given, but also a brief description of the publications written by each author that year is reproduced. The latter information includes journal title, volume and issue number, page span, and year. Also, an organization and geographic approach is possible with the *Current Contents Address Directory*.

INDIANA UNIVERSITY
Bloomington, Indiana 47405
Tel. (812)855-9043
FAX (812)855-8300

Department of Chemistry

Paul A. Grieco, Chairman

Degrees Offered: B.S., M.S., Ph.D., M.A.T.

Fields of Specialization: Analytical, Biological, Inorganic, Organic, Physical, and Nuclear Chemistry.

Interdisciplinary Program: Chemical Physics (Department of Physics), Biochemistry (Department of Biology), Combined Degree in Medical Sciences (I.U. School of Medicine), Biophysics (Department of Physics), Molecular, Cellular and Developmental Biology (Department of Biology), Environmental Chemistry and Geochemistry (School of Public and Environmental Affairs; Department of Geology), and Master of Library Science–Information Specialist in Chemistry (School of Library and Information Science).

ALLERHAND, ADAM (b. 1937), Professor. B.S., 1958, State Tech. Univ., Chile; Ph.D., 1962, Princeton Univ. Res. Assoc., 1962, Univ. of Illinois. *Physical Chemistry.* Nuclear magnetic resonance and relaxation; biochemical applications of Fourier transform NMR. Tel. (812)855-5513.

Adam Allerhand and Steven R. Maple, Requirements for ultrahigh resolution NMR of large molecules on high-field instruments, J. Magn. Reson., 76, 375-9(1988).

Steven R. Maple and Adam Allerhand, Ultrahigh-resolution NMR on a 500 MHz instrument, J. Magn. Reson., 80, 394-9(1988).

Steven R. Maple and Adam Allerhand, Analysis of minor components by ultrahigh resolution NMR. 2. Detection of 0.01% diacetone alcohol in pure acetone and direct measurement of the rate of the aldol condensation of acetone, J. Am. Chem. Soc., 109, 6609-14(1987).

Steven R. Maple and Adam Allerhand, Analysis of minor components by ultrahigh resolution NMR. I. Evidence for the detectability of weak resonances near peaks which are 10,000 times larger, without suppression of the large peaks, J. Magn. Reson., 72, 203-10(1987).

Steven R. Maple and Adam Allerhand, Analysis of minor components by ultrahigh resolution NMR. III. Effect of field-frequency lock on signal-to-noise ratio, J. Magn. Reson., 75, 147-52(1987).

Steven R. Maple and Adam Allerhand, Ultrahigh resolution NMR. 6. Observation of resolved carbon-13-nitrogen-15 scalar splittings in carbon-13 NMR spectra of samples of natural isotopic composition, J. Am. Chem. Soc., 109, 56-61(1987).

Adam Allerhand and Steven R. Maple, Ultra-high resolution NMR, Anal. Chem., 59, 441A-444A, 446A, 448A, 450A, 452A(1987).

Steven R. Maple and Adam Allerhand, Detailed tautomeric equilibrium of aqueous D-glucose. Observation of six tautomers by ultrahigh resolution carbon-13 NMR, J. Am. Chem. Soc., 109, 3168-9(1987). •••

NOVOTNY, MILOS VLASTISLAV (b. 1942), Rudy Professor of Chemistry. B.S. equiv., 1962; Ph.D., 1965, Univ. of Brno, Brno, Czechoslovakia. 1965-67, Czechoslovak Academy of Sciences; 1968-69, Royal Karolinska Inst., Sweden; 1969-71, Univ. of Houston. *Analytical Chemistry.* Chromatography; chromatography/mass spectrometry; capillary electrophoresis; analysis of biological and environmentally important compounds; mammalian pheromones; supercritical fluids. Tel. (812)855-4532.

F. Andreolini, S. C. Beale and M. Novotny, Determination of serum metabolic profiles of bile acids by microcolumn liquid chromatography/laser–induced fluorescence, HRC CC, J. High Resolut. Chromatogr. Chromatogr. Commun., 11, 20-4(1988).

Milos Novotny, Recent advances in microcolumn liquid chromatography, Anal. Chem., 60, 500A-502A, 504A-510A(1988).

*Paul David and Milos Novotny, Separation of alpha-ketoacids by supercritical fluid chromatography as their quinoxalinol derivatives, J. Chromatogr., 452, 623-9(1988).

Debra R. Luffer, Leonard J. Galante, Paul A. David, Milos Novotny and Gary M. Hieftje, Evaluation of a supercritical fluid chromatograph coupled to a surface-wave–sustained microwave–induced–plasma detector, Anal. Chem., 60, 1365-9(1988).

Cheryl L. Flurer, Claudio Borra, Franca Andreolini and Milos Novotny, Separation of proteins by microcolumn liquid chromatography based on the reversed–phase and size–exclusion principles, J. Chromatogr., 448, 73-86(1988).

Cheryl Flurer, Claudio Borra, Stephen Beale and Milos V. Novotny, Fused silica packed microcolumns as micropreparative tools in protein analytical studies, Anal. Chem., 60, 1826-9(1988).

Stephen C. Beale, Joseph C. Savage, Donald Wiesler, Shawn M. Wietstock and Milos Novotny, Fluorescence reagents for high-sensitivity chromatographic measurements of primary amines, Anal. Chem., 60, 1765-9(1988).

Karl Erik Karlsson and Milos Novotny, Separation efficiency of slurry-packed liquid chromatography microcolumns with very small inner diameters, Anal. Chem., 60, 1662-5(1988).

Leonard J. Galante, Mark Selby, Debra R. Luffer, Gary M. Hieftje and Milos Novotny, Characterization of microwave–induced plasma as a detector for supercritical fluid chromatography, Anal. Chem., 60, 1370-6(1988).

Michal Roth, Josef Novak, Paul David and Milos Novotny, Thermodynamic studies into a sorption mechanism within the cross–linked polysiloxane stationary phases, Anal. Chem., 59, 1490-4(1987).

Stephen C. Beale, Donald Wiesler and Milos Novotny, Characterization of the phenolic fraction obtained from fossil-fuel materials by microcolumn liquid chromatography and its ancillary techniques, J. Chromatogr., 393, 391-406(1987).

Mark E. Philips, Mark R. Deakin, Milos V. Novotny and R. Mark Wightman, Effect of added water on voltammetry in near–critical carbon dioxide, J. Phys. Chem., 91, 3934-6(1987).

Franca Andreolini, Claudio Borra and Milos Novotny, Preparation and evaluation of slurry–packed capillary columns for normal–phase liquid chromatography, Anal. Chem., 59, 2428-32(1987).

F. Andreolini, B. Jemiolo and M. Novotny, Dynamics of excretion of urinary chemosignals in the house mouse (Mus musculus) during the natural estrous cycle, Experientia, 43, 998-1002(1987).

Franca Andreolini, Claudio Borra, Donald Wiesler and Milos Novotny, High-efficiency separation and characterization of high-molecular–weight heterocyclic sulfur compounds in fossil fuels by microcolumn liquid chromatography and related techniques, J. Chromatogr., 406, 375-88(1987).

Michal Roth, Joette L. Steger and Milos V. Novotny, Diffusion of isomeric polycyclic aromatic hydrocarbons in compressed propane, J. Phys. Chem., 91, 1645-8(1987).

Bozena Jemiolo, Franca Andreolini, Donald Wiesler and Milos Novotny, Variations in mouse (Mus musculus) urinary volatiles during different periods of pregnancy and lactation, J. Chem. Ecol., 13, 1941-56(1987).

Susan V. Olesik, Joette L. Steger, Nobutoshi Kiba, Michal Roth and Milos V. Novotny, High-precision apparatus for physicochemical measurements by capillary supercritical fluid chromatogrphy, J. Chromatogr., 392, 165-74(1987).

Claudio Borra, Soon M. Han and Milos Novotny, Quantitative analytical aspects of reversed–phase liquid chromatography with slurry–packed capillary columns, J. Chromatogr., 385, 75-85(1987).

M. Novotny, Capillary chromatography in the condensed mobile phases: from the concepts to analytically useful separations and measurements, HRC CC, J. High Resolut. Chromatogr. Chromatogr. Commun., 10, 248-56(1987).

Claudio Borra, Donald Wiesler and Milos Novotny, High-efficiency microcolumn liquid chromatography separation and spectral characterization of nitrogen-containing polycyclics from fossil fuels, Anal. Chem., 59, 339-43(1987).

M. F. Yancey, Metabolic abnormalities in the diabetic disease process as quantitated by high-performance chromatographic techniques. (D).

S. D. Harvey, Chemical investigations of the pheromone systems of the house mouse. (D).

Figure 15-1
Sample Entry from the *ACS Directory of Graduate Research*

The American Chemical Society compiles a number of directories which are quite useful. Published biannually is the *ACS Directory of Graduate Research* [457]. The geographic scope of the publication is limited to the U.S. and Canada. It includes listings for departments of chemistry, biochemistry, chemical engineering, pharmaceutical/medicinal chemistry, clinical chemistry, and polymer science. Departments are arranged alphabetically by institution, and each faculty member is listed with an indication of area of specialization. The example in Figure 15-1 illustrates the additional kinds of information found in the *ACS Directory of Graduate Research*, which is also available online.

The American Chemical Society also compiles *College Chemistry Faculties* [472] (1990). Although not published as frequently as the *ACS Directory of Graduate Research*, the directory nevertheless provides reasonably current directory information for those teaching in smaller academic institutions. *Chemical Research Faculties* [469] (1988), another ACS publication, is international in scope. 1922 academic departments in 107 countries are listed.

Other professional societies sometimes publish useful directories, such as the American Institute of Chemical Engineers' *Chemical Engineering Faculties* [1032] (annual). Another example is the *Directory of Atomic, Molecular, and Optical Scientists* [398] (1986), which lists scientists both alphabetically by name (with full addresses and phone numbers) and by research specialty. For other directories, search the Chemistry Reference Sources Database for **DIRECTORY AND CHEMISTS**.

15.4.2 Directories of Organizations

One of the most useful directories of organizations is the *Directory of American Research and Technology; Organizations Active in Product Development for Business* [979] (annual). Formerly titled *Industrial Research Laboratories of the U.S.*, the work is arranged by company and has a geographic and personnel index. *Scientific and Technical Organizations and Agencies Directory* [1690] (1985) is billed as a descriptive guide to new and established organizations, including associations, societies, government agencies, research centers, publishers, educational institutions, libraries, foundations, and trade or commercial bodies. With such a wide range of organizations covered, it is not surprising that in excess of 12,000 entries are in the work. The *Directory of Technical and Scientific Directories* [1144] (1988), mentioned previously, should be consulted for directories of organizations not discussed in this section.

A service which is probably under-utilized outside the immediate Washington, D.C. area is the Library of Congress's National Referral Center [2019]. The NRC originally concentrated on scientific and technical organizations, providing a link between them and potential clients who were in need of information. Thus, a number of independent research organizations are included in the National Referral Center's directory file, as well as such organizations as information analysis centers. The NRC publishes a *Directory of Federally-Supported Information Analysis Centers* [390] (1980).

A similar online service is DIRLINE [2020], a National Library of Medicine directory database. For the most part, this is a subset of the National Referral Center's files in the areas of medicine and toxicology. However, the NRC entries are supplemented with information from a number of other federal agencies. A paper compilation available from NLM is *Health Hotlines* [2021] (1988), which lists all organizations in DIRLINE having toll-free telephone numbers. Close to 200 organizations, including poison control centers, are found in *Health Hotlines*. A subject index groups together related organizations. NLM's DBIR database [1974] concentrates on biotechnology and molecular biology.

15.4.3 Directories of Suppliers of Chemicals and Chemical Laboratory or Plant Equipment

There are many printed and computer-based sources which lead to suppliers of chemicals. In some cases, the supplier is also the producer of a chemical, but it is often the case that a supplier will market the chemicals of several different producers.

ChemQuest [360] is a database of over 50 chemical catalogs containing over 60,000 chemicals. Suppliers in the USA, Europe, and Japan are included. The database can be searched by inputting molecular formula, CAS registry number, the name of a compound, or by drawing its chemical structure. Once retrieved, the substances can be ordered online. Such information as the smallest amounts which will be sold by each supplier is among the facts associated with the records in ChemQuest. The *Fine Chemicals Directory* [252] is another option for identifying suppliers of small quantities of chemicals. *Chem Sources-USA* and *Chem Sources-International* [1016] (annual) are produced in both printed form and as a combined database. The printed versions are published annually and collectively cover more than 130,000 chemicals available from over 1600 suppliers or producers throughout the world. Chemicals are classified by type of usage, for example, reducing agents. Furthermore, chemical names, trade names, and the names of the companies can be searched. Addresses and telephone numbers of the companies are included. The major chemical supplier Aldrich Chemical Company offers a CD-ROM product Aldrichem Data Search [2026] which includes a database of 50,000 chemicals and 6000 laboratory equipment products. Various search possibilities (names, structure fragments, molecular formulas, physical data, chemical class, etc.) can be utilized on either an IBM PC or a Macintosh.

An American Chemical Society contribution to the chemical supplier literature is *Chemcyclopedia* [1039] (annual). Chemicals are categorized by the type of use or function which they serve, for example, surfactants, catalysts and process chemicals, fine chemicals, etc. Entries include common or industrial chemical names, trade names, form of availability, and potential applications. A company and supplier index is provided.

SRI International publishes its *Directory of Chemical Producers* [2022] (annual) in three versions: *USA*, *Canada*, and *Western Europe*. However, the *OPD Chemical Buyer's Directory* [1040] is a much cheaper and more widely available directory. The *OPD* directory is commonly referred to as the "Green Book". It includes a brochure/catalog section, a section on "Chemicals & Related Materials," which lists some 15,000 substances, a "Chemicals Supplier Index," and a "Chemicals Shipping & Storage" section with about 100 services listed. A more frequently published work is the weekly *Chemical Marketing Reporter* [2023], the source from which the *OPD Chemical Buyer's Directory* is taken.

Various news magazines provide annual supplements containing directory information for chemical suppliers. Among them are:

Chemicalweek Buyers' Guide Issue [1010]
Nature Directory of Biologicals [959]
Analytical Chemistry Lab Guide Issue [1011].

The *Analytical Chemistry* Lab Guide Issue is a particularly good guide to all sorts of U.S. products, from books to laboratory chemicals to instrument suppliers. In Great Britain, the *Chemical Industry Directory and Who's Who* [1009] is published annually. For equipment, the *Guide to Scientific Instruments* [385] comes with the journal *Science*. Another important source of information on scientific instruments is the journal *Review of Scientific Instruments* [2025].

Finally, a general directory, the *Thomas Register of American Manufacturers* [965] (annual) should be mentioned. The printed version of the *Thomas Register* is divided into three sections: "Products and Services," "Company Names and Addresses," and "Catalog of Companies," the last also known as *Thomcat*. Approximately 150,000 North American manufacturing companies and their products are covered in the sixteen-volume set. The online version is augmented with descriptive text and indexing from other Thomas publications. There are more than 50,000 product classes and over 110,000 brand names in the directory. Over two-thirds of the records now include FAX numbers for rapid transmission of orders or inquiries.

15.5 TRADE LITERATURE

Trade literature refers to a broad category of materials published by industrial companies. To one degree or another, trade literature is meant to help market their products or services. Thus, it includes suppliers' catalogs, detailed technical brochures or pamphlets, and some trade journals. In the last category, the journal publications largely consist of advertising and are often supplied free of charge. Typically there is a section devoted to new materials or products in the industry. There is often a section called "literature," which describes free brochures, handbooks, catalogs, bibliographies, etc. available from the companies in that industry. A good guide, which, though written from a British point of view, contains much valuable general information on trade literature, is *Finding and Using Product Information: From Trade Catalogues to Computer Systems* [42] (1986).

Trade literature is poorly covered by abstracting and indexing services. It is best to find the addresses of companies in which you have an interest and to write them directly to solicit their trade literature. Such companies might also provide information on relevant trade journals where their products are routinely advertised. A prime example of one type of trade literature is the catalog issued by the Aldrich Chemical Company, *Aldrich Catalog Handbook of Fine Chemicals* [1037] (annual).

15.6 INFORMATION SOURCES FOR THE CHEMICAL INDUSTRY

Many of the information needs in the chemical industry are in the realm of business or marketing. There are many specialized sources to consult for such information. Some useful guides are Strauss's *Handbook of Business Information* [2027] (1988), Daniells' *Business Information Sources* [2028] (1985), and *Marketing Information: A Professional Reference Guide* [2029] (1984). It is beyond the scope of this book to discuss secondary sources which are devoted to comprehensive coverage of business or marketing in all areas. Instead, sources which concentrate on the chemical industry are presented here. Dorman's *Chemical Industries: An Information Sourcebook* [347] (1987), Peck's *Chemical Industries Information Sources* [11] (1979), and Sittenfield's *Handbook for Obtaining Chemical Use and Related Economic Information* [23] (1980) are appropriate guides to the literature. A dated, though still quite useful source is Bourton's *Chemical and Process Engineering Unit Operations; A Bibliographic Guide* [6] (1967). All of these delve considerably into the technological aspects of the chemical industry.

The Royal Society of Chemistry publishes two products aimed at rapid dissemination of chemical business information. *Chemical Business Update* [2030] (1987-) is a monthly abstract journal which emphasizes the European markets for the chemical industry. The Royal Society's *Chemical Business Bulletins* [2031] is a series of weekly current awareness publications for twelve specific areas like agrochemicals, cosmetics and toiletries, paints and coatings, specialty chemicals, etc. Some commercial information is also covered in the RSC's *Chemical Engineering Abstracts* [530] (1970-), but their main source devoted to chemical business news is the Chemical Business Newsbase [525] (1984), an abstract database derived from the *Chemical Business Bulletins* and other sources. Topics such as mergers, production, new products, sales, etc. are covered, with about 70 percent of the content devoted to European business news. The rest of the coverage is from U.S. and Japanese sources. Common names are used for indexing chemical substances, and CAS registry numbers may also be used in the search strategy. There are controlled-vocabulary index terms for trends, market sectors, and legal aspects. It is possible to search by broad geographic area or by specific country in the Chemical Business Newsbase. The file covers the literature from October 1984 and is updated with about 1000 new entries per week.

Chemical Industry Notes [591] (1968-) is an American Chemical Society abstracting service. Although coverage is somewhat slanted to American chemical business information, the coverage of *Chemical Industry Notes* is worldwide. Facilities, people, market data, resources and their use, products and processes, and unit cost and price information are among the topics covered in the abstracting service. It is available online from 1974 with over three-quarters of a million records.

Another British source is CHEM-INTELL [2032], a database with worldwide information on manufacturing plants and trade and production figures for over 100 organic and inorganic chemicals. Production information for the last ten years, as well as export and import figures are provided. Trade is broken down by major trading partner, including a detailed synopsis of the latest two years. About 15,000 plant records and 5000 statistical records are found in CHEM-INTELL.

The *Chemical Economics Handbook* [529] (1955/65), a product of SRI International, is now available as a full-text database. The work provides data on over 1300 chemicals and the chemical industry, as well as economic indicators. The printed handbook is updated monthly by the *Manual of Current Indicators*, which serves as the updating vehicle for the database also. The *CEH* is an excellent source for both current and historical information. While it concentrates on the U.S., there is some international information in the book.

Of course, there are more specific sources of information on particular chemical industries, such as petrochemicals, fertilizers, cosmetics, etc. The guides to the literature mentioned in the introductory paragraph to this section will lead to those more specialized sources.

15.7 STANDARDS AND SPECIFICATIONS

In the United States, there are many organizations involved in setting industrial standards. On the governmental side is the National Institute of Standards and Technology (formerly the National Bureau of Standards). Industries in the U.S. support a number of other standard-setting bodies, such as ASTM (the American Society for Testing and Materials), ANSI (the American National Standards Institute), Underwriters Laboratories, and others. This is in contrast to the situation in most other industrialized countries where a single official standardization organization is the norm, as with the British Standards Institution. On the international level, standards are established by the ISO, the International Organization for Standardization, and many other bodies.

The terms "standards" and "specifications" are sometimes used interchangeably in vernacular speech. However, in commerce they have very distinct meanings. In general, **standards** relate to the *minimal* requirements which have been established in an industry for the quality or size or level of performance for a product. **Specifications** are used in a much narrower scope of activity in commerce and may be related to a particular order for a product. Thus, it may be the case that a specification sets more stringent requirements for the delivery of a product than the standard itself has established for that product. In chemistry, standards are found for chemical names, analytical or test methods, properties of substances, packaging, labeling, transport, and disposal of, as well as exposure to, chemicals, etc.

Often standards are referred to by a code consisting of an abbreviation for the standards organization and an alphanumeric designation. Thus, an ASTM standard as found in the *Annual Book of ASTM Standards* [1742] might be cited as ASTM (D 129-64) 05.01.

In the United States there are two organizations which attempt to provide easy access to the standards of many organizations. These are the National Standards Association and Information Handling Services. Both are commercial firms and both of them index and provide copies of standards. The National Standards Association publishes a microfiche and print collection of standards from many organizations in the U.S. Their database, Standards & Specifications [639], covers nearly 75,000 military, federal, and industry (voluntary) standards. Information Handling Services has a microform service, VSMF (Visual Search Microfilm), available in either microfilm or microfiche formats. The VSMF Industry Codes and Standards set [2033] actually consists of 17 subsets of standards from major associations and societies. The database, Industry and International Standards [2034], also produced by Information Handling Services, covers over 65,000 standards.

For organizations which prefer to consult a smaller printed index, the *Index and Directory of U.S. Industry Standards* [978] (annual) or the *World Industrial Standards Speedy Finder* [2035] (1983) may be of use. The *Index and Directory* includes over 55,000 standards from over 400 societies worldwide. The *Speedy Finder* covers over 53,000 standards from five major industrial countries and the International Organization for Standardization. Nearly 12,500 chemical standards are among those in the latter work.

15.8 TEACHING OF CHEMISTRY

There is a large body of pedagogical literature, both for the teaching of science in general and for chemistry in particular. In recent years, the application of computers to the teaching of science has greatly increased the approaches available to the teacher. In this section, a few of the most important pedagogical tools as well as sources of computer-based materials are presented.

Some important journals are the *Journal of College Science Teaching*, published by the National Science Teachers Association, *The Crucible*, the journal of the Science Teachers' Association (Canada), and the *Journal of Chemical Education*, the official journal of the Division of Chemical Education of the American Chemical Society. For a British slant, see the Royal Society of Chemistry's *Education in Chemistry*. *Biochemical Education* is a quarterly publication of the International Union of Biochemistry. A fuller listing of relevant journals is given in the *Bibliography of Chemical Education Journals* [1073] (1981). A few newsletters should also be mentioned, such as *Computers in Chemical Education Newsletter*, *Project SERAPHIM News*, the *International Newsletter on Chemical Education* (an IUPAC publication) and the *CHED Newsletter*, the newsletter of the ACS Division of Chemical Education.

Teaching Science: A Teaching Skills Workbook [1312] (1985) is designed to teach how to teach through a series of activities which focus both on practical skills and issues in science teaching. The *Unesco Handbook for Science Teachers* [2036] (1980) is another practical guide. Unesco has also produced *Teaching School Chemistry* [1413] (1984). Newbury's *The Teaching of Chemistry* [1411] (1965), while dated, is still a valuable source of practical teaching tips.

The largest database for pedagogical literature is ERIC, the Education Resources Information Clearinghouse database [2037]. ERIC abstracts and indexes both reports and journal articles in all fields of education. The entry in Figure 15-2 is typical. The ERIC database draws on two printed products: *Current Index to Journals in Education* (CIJE) and *Resources in Education* (RIE). ERIC is perhaps the cheapest database to search online and is available for more recent years in CD-ROM format from several vendors.

Audiovisual materials of relevance to chemistry are reviewed in the *Journal of Chemical Education*'s special feature columns "AV Review" or "Overhead Projector Demonstrations." A major audiovisual compilation is the Royal Society of Chemistry's *Index of Chemistry Films* [471] (1984). The A-V Online database [2038] is available for online searching on DIALOG. *Science Books and Films* [2039], a review journal, is another important source for this type of information. Reviews from the journal have been consolidated in *Films in the Sciences: Reviews and Recommendations* [2040] (1980) and *The Best Science Films, Filmstrips, and Videocassettes for Children* [2041] (1982).

An instructor who desires to incorporate computer-based training into a course now has an incredible variety of products from which to choose. The ACS Division of Chemical Education's Committee on Computers in Chemical Education will lend a collection of over 100 commercial software packages to any school, college, university, or agency for previewing. There is a charge for this service. Free or nearly free programs are available from several sources. Project SERAPHIM [1779] has been in operation since 1982 as a National Science Foundation-sponsored clearinghouse for information on instruction in chemistry by microcomputers. Project SERAPHIM collects and distributes computer-based materials and even trains teachers in their use. Over 600 programs are available. A spinoff from Project Seraphim is the *Journal of Chemical Education: Software*, which reviews and publishes programs and printed documentation. The phone number for Project SERAPHIM (now located at the University of Wisconsin, Madison) is 608-263-2837.

Other organizations which distribute relatively inexpensive educational software are WiscWare and ISAAC.

Wisc-Ware [2042] is a consortium of colleges and universities that distributes research and instructional software for IBM microcomputers. Three types of licenses

AN —	EJ270158
CHAN—	SE532079
TI —	Recent Developments in Teaching Chemistry in the United Kingdom, II
AU —	Hudson, M.J.
SO —	Journal of Chemical Education; v59 n10 p841-42 Oct 1982 (Oct 1982)
IS —	CIJFEB83
DT —	080; 141
LA —	English
IT —	*Chemistry; *College Science; Computer Oriented Programs; Educational Trends; Foreign Countries; Higher Education; *Instructional Innovation; *Science Curriculum; Science Education; *Science Instruction; Secondary Education; *Secondary School Science; Science and Society; *United Kingdom
AB —	Reviews developments in chemistry curriculum/instruction, providing lists of chemistry cassettes, monographs, educational literature from the Royal Society of Chemistry, new teaching packages, and course materials for science in society. Extends information presented in a previous article (EJ 235 051) appearing in this journal. (JN)

FIGURE 15-2
Typical Entry from the ERIC (Education Resources Information Clearinghouse) Database.

are available: class license, site license, and individual license. The class and site licenses are available only to staff and faculty members at campuses that are members of the WiscWare consortium. The individual license fee is lower for individuals at member campuses. The tollfree number for WiscWare is 800-543-3201 (BITNET: WISCWARE@WISCMACC). A representative program is CRYSTALLAB, which provides instruction in the basic concepts of crystal structure and diffraction patterns.

ISAAC (Information System for Advanced Academic Computing) [2043] is a clearinghouse for information about the use of IBM-compatible hardware and software in higher education. Access to the system is free; users can connect to the computer which houses ISAAC through a toll-free number. Alternatively, BITNET or Internet/ARPAnet can be used. The system includes a bulletin board and numerous databases. The collective databases contain nearly 6000 entries with descriptions of projects which have used IBM computers, as well as listings of over 2800 software packages suitable for instruction or research. There is also general information about the use of computers and a guide to additional sources of information, including journals, associations, directories, and databases. The "academic software packages" category on the database covers a number of distributors, among which are Wisc-Ware, CONDUIT, the National Collegiate Software Clearinghouse, and COMPress.

CONDUIT [1822] is a nonprofit publisher of educational software at the University of Iowa. Since 1971 CONDUIT has reviewed, tested, packaged, and distributed many instructional computer programs and related printed materials. CONDUIT also attempts to assist authors in developing computer-based software. The software is intended primarily for higher education. COMPress [2044], a division of Queue, Inc., is a commercial publisher with a number of chemistry titles in their catalog.

Finding suitable experiments for demonstration purposes or for use in a teaching laboratory can sometimes be a frustrating experience. The frustration can be considerably reduced by consulting some of the sources described below.

CHEMLAB [1765] is a database of articles describing laboratory experiments in the *Journal of Chemical Education* from 1957 onwards. Experiments can be sorted by major fields (analytical, inorganic, physical, radiochemistry, biochemistry, polymer science, etc.). Project SERAPHIM distributes the microcomputer version of the database. A printed version has now appeared in its 2nd edition as *Annotated List of Laboratory Experiments in Chemistry from the Journal of Chemical Education, 1957-1984* [2045] (1986).

The American Chemical Society has published two volumes of *Chemical Demonstrations: A Sourcebook for Teachers* [1417] (1985-87). Included are 108 teaching demonstrations for introductory programs in chemistry in volume 1 and an additional 112 demonstrations covering all areas of general chemistry in volume 2. The Modular Laboratory Program in Chemistry [1781] is from Chemical Education Resources, a company which specializes in the publication of undergraduate chemistry experiments. Shakhashiri's *Chemical Demonstrations: A Handbook for Teachers of Chemistry* [1416] (1985-86) is a compilation written by the master teacher who founded ICE, the Institute for Chemical Education, at the University of Wisconsin. Finally, Alyea's *Microchemistry Projected: The 200 Best of the 1000 TOPS Experiments Which Were*

Published in the *Journal of Chemical Education* [1871] (1981) presents a selection of the TOPS, Tested Overhead Projection Series.

For those who seek help in formulating examination questions in chemistry, VIS-AID Devices [2046] has handbooks of test questions in all fields of chemistry, including polymer chemistry. All questions are selected from original tests and examinations in North American colleges and universities, that is, from the original instructors' tests. About 70 percent are multiple-choice questions. Either printed or IBM-compatible diskette versions of the handbooks are available. The venerable Schaum's Outline series has several titles in chemistry areas, which could also serve as sources of examination questions: general chemistry [2054], analytical chemistry [1449], biochemistry [2055], organic chemistry [1459], physical chemistry [1696], and thermodynamics [2056].

Additional sources for this section can be found in the Chemistry Reference Sources Database under such terms as **EDUCATION, TEACHING, EXPERIMENTS**, etc.

15.9 THE STUDY OF CHEMISTRY/CAREERS IN CHEMISTRY

The choice of a profession is one of the most important decisions one can make. Works in this section assist in the selection of institutions of higher education at which to pursue chemical studies and in the investigation of various facets of chemistry as a profession.

Colleges and universities which offer bachelor's degrees certified by the American Chemical Society must follow the guidelines established in *Undergraduate Professional Education in Chemistry: Guidelines and Evaluation Procedures* [3] (1983). Formulated by the ACS Committee on Professional Training (CPT), the 1983 guidelines were supplemented in May 1984 by "Course Suggestions as Appendices to the 1983 CPT Guidelines" and reprinted with extensions in 1988. The "Course Suggestions" were prepared with the assistance of various divisions of the American Chemical Society in order to help plan courses in the main areas of the chemical curricula. Over 580 schools have met the test for the ACS-approved curricula. The "ACS List of Approved Schools" is available on request from the ACS Office of Professional Training, 1155 16th Street, N.W., Washington, D. C. 20036.

The ACS publications *College Chemistry Faculties* [472], the *ACS Directory of Graduate Research* [457], and *Chemical Research Faculties* [469] are quite useful in selecting a college or university to attend. The first two are restricted geographically to the U.S. and Canada, but *Chemical Research Faculties* is worldwide in its coverage. *Research in Chemistry at Undergraduate Institutions* [474] notes in the introduction that predominantly undergraduate departments in public institutions are the fastest growing source of professional chemists. Research in progress at 151 such chemistry departments is reported in the 1985 edition of that work. *Graduate Programs in Chemistry* [473] (1983), another ACS publication, is designed to provide the undergraduate student with preliminary information about graduate programs in the U.S. The *Guide to Chemical Education in the U.S. for Foreign Students* [1870] (1981) will help those unfamiliar with the American system of education. It covers both undergraduate and graduate chemistry education in U.S. colleges and universities. The *Guide* also describes the kind of educational preparation students should have in order to be chemistry majors in the U.S. A brief discussion of the U.S. educational system, typical course contents, suggested texts, and typical examination questions are found in the *Guide*. Of broader scope than chemistry is *The American University: A World Guide* [1298] (1984). A traditional source for information on American colleges and universities in the Peterson's Guides series [2047] (annual). These are available in both printed and online versions.

There are general sources of information which can assist in the choice of a career. The *Occupational Outlook Handbook* [2048] has brief descriptions of several hundred occupations, including salary ranges, working conditions, and future job outlook. A companion periodical is the *Occupational Outlook Quarterly*. The *Dictionary of Occupational Titles* [2049] contains current detailed information on the characteristics of many jobs.

In chemistry, *Chemical & Engineering News* has several regular features each year which can be of great help in selecting a career. Toward the end of each October the "Employment Outlook" survey [2050] is published in *C&EN*. Information on the demand for chemical professionals in the job market, salaries of beginning B.S., M.S., Ph.D. and experienced chemists, and career planning information is included. In the last category, a number of brochures and directories published by the American Chemical Society and others are listed. Everything from manpower and salary surveys to addresses of professional organizations is found there. Furthermore, a number of free brochures available from the ACS are listed, such as "Careers in Chemistry: Questions and Answers," "I Know You're a Chemist, But What Do You Do?" "Futures Through Chemistry: Charting a Course," "Chemical Careers in the Life Sciences," and "Careers in Chemical Education". The American Chemical Society periodically publishes studies on the salaries of chemists in separate, more detailed publications like *Starting Salaries* [2051], a survey of recent graduates in chemistry and chemical engineering, and *Salaries* [467], a comprehensive survey which considers factors like age, race, education, and gender, and how they affect salaries in chemistry. Recently, the ACS reports have been presented in separate publications:

Salaries of Non-Academic Chemists [912]
Salaries of Academic Chemists [911]
Salaries of Non-Academic Chemical Engineers [910]
Employment Status and Demographic Characteristics of ACS Members [913].

Some other career guides which can be consulted are *Opportunities in Chemistry Careers* [1682] (1987) and *Opportunities in Chemistry* [1409] (1985).

Once the formal educational process in chemistry is nearing completion, practical tips on finding a job can be found in *Entering Industry: A Guide for Young Professionals* [1537] (1975) and *Guide to the Chemical Industry: Technology, R & D, Marketing, and Employment* [1557] (1983). Another source which may be useful is *The Best Resumes for Scientists and Engineers* [2052] (1988).

15.10 SUMMARY

There are many facets to chemistry. A field this rich in historical background and so diverse in its choice of occupations and fields of specialization needs many reference tools in order to fully investigate an area of interest. The materials presented in this final chapter round out the study of the reference materials with which every chemistry student should be familiar. With this beginning, you have come far toward mastering the fascinating world of chemical information science.

15.11 SELECTED READINGS

"Computers in Undergraduate Chemistry Education." *Chem. Eng. News* **1988**, *66*(43), 30-36.

Bohning, James J. "Integration of Chemical History into the Chemical Literature Course." *J. Chem. Inf. Comput. Sci.* **1984**, *24*(2), 101-107.

Pimentel, George C.; Coonrod, Janice A. *Opportunities in Chemistry: Today and Tomorrow*; National Academy Press: Washington, DC, 1987. [2053]

APPENDICES

I. Index to Some Library of Congress Classification Numbers Relevant to Chemistry

II. ACS Recommended Journals

III. Instructions for Using the CRSD Database with Pro-Cite

IV. Subject Term Authority List for the Chemistry Reference Sources Database (CRSD)

APPENDIX I

INDEX TO SOME LIBRARY OF CONGRESS CLASSIFICATION NUMBERS RELEVANT TO CHEMISTRY

Alchemy	QD23.3-27
Aliphatic Compounds	QD301-315
Alkaloids	QD421
Amino Acids	QD431-436
Analysis of Organic Compounds	QD271-272
Analytical Chemistry	QD71-145
Analytical Chemistry Textbooks and Treatises	QD75, QD81, QD101
Analytical Spectroscopy	QD95-96, QC450-467, QD272.S6
Antibiotics	QD375-377
Aromatic Compounds	QD330-341
Atomic Physics	QC170-197
Atomic Theory	QD461
Benzene Rings (Condensed)	QD390-395

317

Biochemistry	QD415-431.7, QH345, QP501-801
Biochemistry Textbooks and Treatises	QD415
Biography of Chemists	QD21-22
Bioorganic Chemistry	QD415-441
Biophysics	QH505
Carbohydrates	QD320-327
Chemical Compounds (Physical Aspects)	QD471-481
Chemical Elements	QD466
Chemical Engineering	TP155-59
Chemical Information Science	QD8.5
Chemical Nomenclature	QD7, QD291, QD149, QD902.5, QD451.5
Chemical Reactions	QD501-505
Chemistry Dictionaries	QD5
Chemistry Directories	QD23
Chromatography	QD117.C5, QD79.C4
Colored Compounds	QD441
Computers	QA76-78
Constitution and Properties of Matter	QC170-220
Crystallography	QD901-999
Crystals	QD931-951
Electrochemical Analysis	QD115
Electrochemistry	QD551-571, QD261, TP250-261
Electrochemistry of Organic Compounds	QD273
Enzymes	QP601-619
General Chemistry	QD1-69
General Chemistry Textbooks	QD31-35
Genetic Engineering	QH442
Genetics	QH426-531
Handbooks, Tables, Etc. of Physiochemical Property Values	QD65
Heterocyclic Compounds	QD399-405
History of Chemistry	QD11-18
Inorganic Chemistry	QD146-197
Inorganic Chemistry Textbooks and Treatises	QD151
Lasers	QC680-688, QD701-731, TA1675-1680
Macromolecules	QD380-388
Metals	QD171-181
Molecular Biology	QH506
Molecular Theory	QD461
Natural Product Chemistry	QD416-436
Neurophysiology	QP351-495
Non-metals	QD161-169
Nuclear Chemistry	QD601-655
Nuclear Magnetic Resonance	QC762
Nucleic Acids	QD431-436, QP620-625
Organic Chemistry	QD241-449
Organic Chemistry Textbooks and Treatises	QD251-253
Organometallic Compounds	QD410-412
Peptides	QD431-436

Periodic Table	QD467
Photochemistry	QD701-731
Physical Chemistry	QD450-731
Physical Chemistry Textbooks and Treatises	QD453
Polymers (Inorganic)	QD196
Polymers (Organic)	QD380-388
Proteins	QD431-436, QP551-552
Qualitative Analysis in Chemistry	QD81-98
Quantitative Analysis in Chemistry	QD101-121
Quantum Chemistry	QD462
Quantum Electrodynamics	QC679-699
Radiochemistry	QD601-655
Reaction Mechanisms (Organic)	QD281
Salts (Inorganic)	QD189-193
Solid State Chemistry	QD478
Solutions and Solubility	QD540-549
Spectrum Analysis SEE: Analytical Spectroscopy	
Steroids	QD426
Surface Chemistry	QD506-508
Synthesis (Organic Chemistry)	QD262
Synthetic Methods (Inorganic)	QD156
Synthetic Methods (Organic)	QD262, QD281
Teaching of Chemistry	QD40-49
Terpenes	QD416
Theoretical Chemistry	QD450-731
Thermochemistry	QD510-536
Transition Metals	QD172.T6
X-Ray Crystallography	QD945

APPENDIX II

ACS Recommended Journals

The Committee on Professional Training (CPT) of the American Chemical Society has published a list of recommended journal titles. This will be used as part of the evaluation of departments seeking ACS accreditation of their undergraduate degree programs in chemistry. The CPT recommends that libraries in those institutions hold at least twenty titles in the second two categories. See the CPT *Newsletter* no. 4 (Spring 1987): 23. (A revised list was produced in September 1987, and the titles below reflect those revisions.)

	Title	Publisher
I.	**Publications Readily Accessible**	
	Chemical Abstracts	Chemical Abstracts Service
	Chemical and Engineering News	American Chemical Society (ACS)
II.	**Top Priority Journals**	
	Accounts of Chemical Research	ACS
	Analytical Chemistry	ACS
	Angewandte Chemie (International edition in English)	VCH Publishers
	Biochemistry	ACS
	Inorganic Chemistry	ACS
	Journal of Biological Chemistry	American Society for Biological Chemistry
	Journal of Chemical Education	Division of Chemical Education, ACS
	Journal of Chemical Information and Computer Sciences	ACS
	Journal of Chemical Physics	American Institute of Physics
	Journal of Organic Chemistry	ACS
	Journal of Physical Chemistry	ACS

Journal of the American Chemical Society	ACS
Journal of the Chemical Society (London):	Royal Society of Chemistry
Chemical Communications	
Dalton Transactions	
Faraday Transactions	
Perkin Transactions	
Macromolecules	ACS
Organometallics	ACS
Tetrahedron	Pergamon
Tetrahedron Letters	Pergamon

III. **Highly Recommended Journals**

Bioorganic Chemistry	Academic
Canadian Journal of Chemistry	National Research Council of Canada
Chemical Physics Letters	Elsevier
Chemical Reviews	ACS
Chemische Berichte	VCH Publishers
Chemistry Letters (Japan)	Chemical Society of Japan
E.S.&T.; Environmental Science and Technology	ACS
Helvetica Chimica Acta	Sweizerische Chemische Gesellschaft
Industrial and Engineering Chemistry Research	ACS
Journal of Catalysis	Academic Press
Journal of Chromatography	Elsevier
Journal of Coordination Chemistry	Gordon & Breach
Journal of Electroanalytical Chemistry	Elsevier
Journal of Organometallic Chemistry	Elsevier
Journal of Polymer Science	Wiley
Langmuir	ACS
Magnetic Resonance in Chemistry	Wiley
New Journal of Chemistry; Nouveau Journal de Chemie	Societe Chimique de France
PNAS (Proceedings of the National Academy of Sciences)	National Academy of Sciences
Quarterly Reviews (sic) (Chemical Society Reviews)	Royal Society of Chemistry
Spectrochemica Acta	Pergamon

APPENDIX III

INSTRUCTIONS FOR USING THE CRSD DATABASE WITH PRO-CITE

1. INTRODUCTION

The Chemistry Reference Sources Database (CRSD) includes over 2150 records, many more items than are discussed in the text of *Chemical Information Sources*. Each record contains full information about the printed work, database, software or other item.

In the case of books, serials, and audiovisual materials, the full bibliographic citation is listed. Some of the records have an additional field, the call number. This indicates that the items are held by the Indiana University Chemistry Library and also that they are available to instructors of chemical information courses in the U.S. and Canada for a two-week loan period. A fee equivalent to one hour at the hourly rate of the federal minimum wage plus postage will be assessed for any item borrowed direct. (The material is also available through the regular interlibrary loan channels, and those who do not wish to order direct are encouraged to use that borrowing mechanism.) Direct requests must include the call number for the item and the name of the course for which the borrower is the instructor. The borrower must also agree to pay for a replacement copy in case the item is lost, stolen, or damaged. Send direct requests to: Gary Wiggins, Chemistry Library, Indiana University, Bloomington, In 47405. (FAX 812-855-6611)

The CRSD database is normally supplied in Pro-Cite format with the Search-Only version of Pro-Cite. Contact the author if an ASCII version of CRSD is desired. If the CRSD database is obtained as an ASCII file, it could be searched with any wordprocessing software which has a FIND capability. Likewise, it could easily be edited to work with other database management software. Each record in the ASCII file of the database is separated from others by four asterisks, ****.

2. GETTING STARTED WITH PRO-CITE

The following is a brief description of how to use the Search-Only version of Pro-Cite with the CRSD database. Because this is a Search-Only version, certain features of Pro-Cite will not be accessible. Some options will be dimmed on the screen. These are options which would be available with the full Pro-Cite program.

1. Create a Directory

Create a directory, copy all of the files from the disks into the directory, and follow the procedures below.

2. Run the SETUP Program

Run the SETUP program by typing **SETUP**, pressing the [ENTER] key, and answering the questions about your computer, printer, and monitor. To set the values for the CONFIG.SYS file, type **y** when asked about the file and press [ENTER] in response to the first prompt. Then name the disk drive of your boot disk. If the CONFIG.SYS file is properly established, the program will simply ask you to press the space bar to continue. If additional prompts appear, including one which asks if you want to save the changes made, type **y** and press [ENTER].

    Important Note: If you have created or made changes to an existing CONFIG.SYS file using the steps above, you must now reboot your computer by pressing [CTL][ALT][DEL] simultaneously. This will activate the new settings in the file.

3. Help is Available

Help with Pro-Cite is available at any time when running the program. Press the [HOME] key of the numeric keypad, read the information, and press [ESC] to dismiss the help screen.

4. Moving Around in Pro-Cite

Several special keys speed your interaction with Pro-Cite. To select an item on a menu, use the arrow keys to move the highlighted bar to the item and press [ENTER], or type the capitalized letter in the highlighted word. To return to the previous screen or menu, use [ESC]. To scroll lists or pages, use the arrow keys or [Pg Up]/[Pg Dn] keys.

5. Explore Pro-Cite

To run the Pro-Cite Search-Only version, type **procite** and press [ENTER]. Open the CRSD database by typing **o** for Open a Pro-Cite Database. (CRSD is the only available database, so the pointer already points to it.) Press [ENTER].

When the Main Menu appears, familiarize yourself with it by pressing the [HOME] key for further explanation. Press [ESC] to dismiss the Help screen.

Notice the line that reads "2156 Records 2156 Selected" at the top right corner of the screen. The first number indicates how many records are in the database. Because you cannot save any changes to the Search-Only CRSD database, this number will never change. The "2156 Selected" indicates how many records in the database are selected, that is, in the answer set when a search is run or there as a result of other special activities. Because you will be searching or selecting records, this number will change.

6. View Sample Records

To view the CRSD records, move into the Edit mode by typing **e** for Edit. The record in the sample screen will appear.

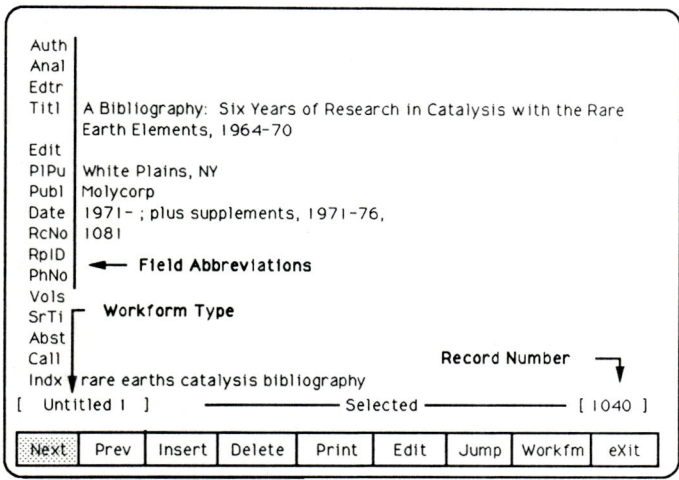

All of the records in the CRSD database have been entered into a user-defined workform named "Untitled 1". The Pro-Cite record number is located at the lower right corner, and the column of field abbreviations is along the left side. The CRSD record number is in the RcNo field. It is those numbers which are listed in this book, not the Pro-Cite record numbers.

The following explains the field codes which are used in the CRSD database:

Field Abbrev.	PRO-CITE Field #	Field Title—Explanation
Auth	1	Author—entered as Smith, James T.; Jones, A. B. No corporate authors are used in this work.
Anal	4	Title (Analytic)—title of specific volumes in a collected work
Edtr	7	Editor—editors' names, entered in the same order as authors' names
Titl	9	Title (Main)—title of a book or collected work or name of a database[1] or software[2]
Edit	15	Edition—the numbered edition of a printed work
PlPu	18	Place of Publication—city of publication or the place where the database or software producer is located
Publ	19	Publisher Name—publisher's name or the name of the producer of a database or software. The street address is included for software.
Date	20	Date of Publication—year(s) of publication or beginning year of a database or serial publication
RcNo	22	CRSD Record Number—the record number corresponding to the number entered in brackets in the book, for example, [925]. All records have a number, even if the items are not mentioned in the book.[3]
RpID	23	Report I.D.—a report number
PhNo	25	Phone Number—used for the telephone number of certain companies
Vols	26	Volumes—total numbered volumes in a complete multi-volume set
SrTi	32	Series Title—the name and number of a volume in a series
Call	44	Call Numbers—if present, this is the call number of a work available for borrowing, either director through interlibrary loan.
Indx	45	Index Terms—subject terms or phrases, assigned according to the subject term authority list (file INDEX.LST on the diskette or Appendix IV in the book)

---

[1] Records for databases have the word "database" (or, in the case of printed works which are also available as databases, the phrase "also database") appended to the Title field. In addition, the vendors which make the databases available for searching are included as index terms in the records. Thus, to find all STN International databases in the CRSD, search **INDX=STN AND DATABASE**

[2] Records for software generally have the full address, including city mailing code, and phone number. Each has the word "software" appended to the name in the Title field. This makes it possible to search for various kinds of software by combining subject words in an AND statement, for example, **GRAPHICS AND SOFTWARE AND MACINTOSH**

[3] To find the record using the CRSD Record Number, enter the search statement: **#22=925**

3. SEARCHING THE CRSD DATABASE

From the Main Menu, type **s** to Select Records and **s** again to Search Database. You will see the Pro-Cite search window and function keys at the bottom of the screen.

With Pro-Cite, search terms or phrases may be qualified by a field label, for example, **auth=Mellon**. However, some of the customized field abbreviations (for example, Anal, RcNo, PhNo, and Edtr) are not recognized by Pro-Cite. In order to limit searches to those fields, the corresponding field numbers should be used in expressions such as #4= , #22= , #25= , #7= , etc. For example, to find CRSD Record Number [1081], search **#22=1081**.

An easy way to find the field numbers without having to refer to this guide is to press the INS (Insert) key while in the search expression window. The CRSD Workform pops up. Try it. By moving the cursor to the appropriate field abbreviation and pressing the [ENTER] key, the field number is automatically inserted into the search expression. (At least one blank space must be at the end of the search expression at the point where the INS key is pressed, or this must be done in the first space of the search expression.) Use the arrow keys to move the pointer to Publ in the CRSD workform. Press [ENTER], and you will see that field #19 is entered in the search window. Type **McGraw-Hill** and press [ENTER]. A total of 53 records is found.

An unqualified search will attempt to match words in all fields. The strategy

GUIDE AND CA

retrieves guides to the use of *Chemical Abstracts*, but also picks up those records which are guides published in California. They may have nothing to do with Chemical Abstracts, but the abbreviation CA for California is used in the Place field, thus resulting in false drops. A better strategy to find guides to the use of Chemical Abstracts would be: **INDX=GUIDE AND INDX=CA**, using the field label for the controlled vocabulary in the subject index field.

For the most precise subject searching, locate the appropriate terms or phrases in the authority list (file INDEX.LST or Appendix IV of this book). Label the terms or phrases from the index with the field code INDX= or #45= . Within the Search mode of Pro-Cite, holding down the [ALT] key and pressing "i" (ALT-I) will pop up the index list. You can move down the list of controlled-vocabulary subject terms by entering the first character of the term you want to search. Once found, the term or phrase is automatically entered into the search strategy by pressing the [ENTER] key. Note that phrases must be enclosed in double quotation marks when searched in Pro-Cite. (This occurs automatically when the ALT-I option is used.) A search term or phrase in quotation marks can be up to 80 characters long.

Return to the Select menu and press **a** to Select All Records. Enter the strategy

INDX=GUIDE AND INDX="GENETIC ENGINEERING"

to locate the one guide in the CRSD database to information sources in biotechnology.

Try the other examples of searches in Pro-Cite which are listed below. Remember that it is necessary to select ALL records before beginning a new search. If that is not done, the second search is performed on the records selected in the first search, thus providing a way to refine the original search.

> PHYSICS AND GUIDE AND NOT STYLE
> (TEACH* OR EDUCATION*) AND CHEMISTRY
> #25=800*

The truncation symbol in Pro-Cite is: * In the last example, all records with 800 phone numbers are retrieved. The asterisk can be placed at the beginning or end of a term or phrase. Nesting related terms in parentheses is supported.

A Quick search is restricted to searches of the first author, first word of the title, and date. Consequently, for subject searches, it is not recommended. The [F5] key on an IBM PC can be used to rapidly initiate a Quick author search. This inserts ''Auth='' into the search expression window. The [F6] key functions similarly for a Quick title search.

Restore all of the records by typing **a** in the Select menu and pressing [ENTER]. Type **q** to bring up the Quick Search window and press [F5]. ''Auth='' appears in the search window. Type **Dean** and press [Enter]. The search finds 3 records. Return to the Main Menu and type **p** to Print Selected Records to Screen. The next section tells more about how to manipulate the records in the answer set.

4. VIEWING OR PRINTING AN ANSWER SET

Return to the Select menu and press **a** to Select All Records. Perform a subject search on the word ''absorption''. Thirteen records should be selected. Answer sets can be formatted with Pro-Cite in a variety of ways. Automatically formatting a bibliography in any standard punctuation style is one of the special features of Pro-Cite. Return to the Main Menu of Pro-Cite by pressing the [SPACE] bar to continue and typing **x**. To format the bibliography of 13 answers in the ANSI (American National Standards Institute) style, type **p** to select Print Selected Records to Screen from the Main Menu. Use the arrow keys or Pg Up/Pg Dn keys to move through the bibliography. When finished, press [ESC] to dismiss the printed set.

To change the layout of the bibliography, return to the Main Menu and type **o** for Options. The screen shown in the example will appear.

To make the following changes, use the arrow key to move the pointer to Display with Record Numbers? and press [ENTER] to change No to Yes.

Use the arrow key or type **p** to move the pointer to Punctuation. Press [ENTER] to change ANSI to Other.

Press [F10] to finish and return to the Main Menu. Type **p** to Print Selected Records to Screen. You will see a pop-up window with a list of punctuation files.

Move the pointer to Science and press [ENTER]. Another formatted bibliography displays on the screen, this time in *Science* journal format. Scroll through the bibliography and notice the differences between the Science and ANSI formats. Press [ESC] to dismiss it.

Type **p** again and change the bibliography layout as many times as you wish.

```
Options                                    Use: ↑ ↓ → →   PgUp PgDn; F10 to finish
▶Display with call numbers?     [   No]    Top margin                    [    4]
 Display with index terms?      [   No]    Bottom margin                 [    5]
 Display with abstract field?   [  Yes]    Left margin                   [    5]
 Display with record numbers?   [   No]    Maximum line width            [   65]
 Use short form?                [   No]    Lines per page                [   66]
 Modify fields in short form ...            Line spacing                  [    1]
                                            Lines between citations       [    1]
 Alphabetic order?              [  Yes]    Indent citations?             [  Yes]
 Punctuation                    [ ANSI]    Indent how far                [    5]
 Number citations?              [  Yes]
 First citation number          [    1]    Output:                     [ Screen]
 Pages numbered?                [  Yes]    Style ...
 First page number              [    1]    Format page?                  [  Yes]
 Heading on first page?         [  Yes]    Modify authority lists ...
 Heading: [Bibliography]

 Ready printer ...                          Wait between pages?           [   No]
 Underline character            [ _ ]       Form feeds after citations?   [   No]

 Renumber record numbers ...                Scratch disk                  [   @]
 Alphabetize record numbers ...             Record number interval        [   10]
 Modify user defined workforms ...          Printing Order            [Ascend.]
```

Notice that the full record in the customized CRSD database will not display with some of the available formats, notably, ACS and other journal-oriented formats. However, the formats which display the full bibliographic information from the records provide ample selection for most purposes.

If you wish to send the bibliography to the printer or to a disk as an ASCII text file, return to the Options menu from the Main Menu. Move the pointer to Output in the Options menu and press [ENTER] until the desired destination appears.

When finished, return the settings of the Options menu to their original settings, and press [F10] to return to the Main Menu.

Type **s** to Select Records. The selected records can also be viewed in the BROWSE mode for an abbreviated record, then expanded to the full record with the VIEW option. Type **b**, followed by **v** when the windows appear on the screen. Press the [Esc] key to get back to the Browse mode after the record is VIEWed. To move through the list of selected records in the answer set, choose the FORSEL (Forward to the next selected record) option within Browse. Records which are not desired in the final answer set can be Deselected while in Browse mode.

5. SORTING THE BIBLIOGRAPHIES

It is possible to do some sorting on the bibliographies selected with Pro-Cite (or even to sort the entire database if desired). However, keep in mind that the workform used to create the CRSD database is a custom workform, so some of the Pro-Cite sort routines may result in inaccurate sorting. With that in mind, let us sort the thirteen records found in the search for "absorption".

To sort, at the Main Menu, type **r** for soRt and **o** for Select Fields for Sorting. The workform for CRSD appears. The database as supplied is sorted first by author or editor, then by title, and finally by date. (Titles without an author or editor are alphabetized by the first word encountered in the title.)

```
┌─────────────────────────────────────────────────────────────────────┐
│ Fields to Sort                 Use: ↑ ↓ →   PgUp  PgDn  Del         │
│ Workform: Untitled1                        Unused: 140 of 200       │
│                                                                     │
│        ▶ 1  Auth ◀          Auth              SrRo                  │
│             AuRo            AuRo              SrTi                  │
│             Affl            PlPu              SrVo                  │
│          2  Anal            Publ              Srls                  │
│             Medm          3 Date              DcTy                  │
│             CoPh            Copy              CoPh                  │
│          1  Edtr            RcNo              Aval                  │
│             AuRo            RpID              StLc                  │
│          2  Titl            IsID              CODN                  │
│             Ltto            PhNo              ISSN                  │
│             Titl            Vols              ISBN                  │
│             DtMt            PaMe              Note                  │
│             PlMt            Size              Abst                  │
│             Medm            RpRt              Call                  │
│             Edit            SrEd              Indx                  │
│                                                                     │
│                       Author                                        │
│                                                                     │
│         F8 Sort Sizes     F9 Workfm      F10 Done                   │
└─────────────────────────────────────────────────────────────────────┘
```

To change the sorted order, point to the Auth field and type the number 2 in its place. Repeat this process with the Edtr field.

Point to the Title field which has a 2 next to it, press [SPACE] to delete that number. Do not type another number. Repeat this process for the Anal field.

Point to the Date field which has a 3 next to it and type the number 1.

You have just defined the new sort. Press [F10] to finish. To sort the records type **s** for Sort Selected Records. Pro-Cite sorts the records by date and then author or editor. Press [SPACE] and type **x** to Return to the Main Menu. Verify the sorted order by typing **p** for Print Records to Screen. Scroll through the bibliography and notice that the dates appear (roughly) in ascending order. Finally, press [ESC] to dismiss the bibliography, type **r** to Sort Selected Records, then type **c** to Clear Sort.

Type **o** and return the Fields for Sorting to their original settings (Auth and Edtr, 1; Anal and Titl, 2; Date, 3). Press [F10], then type **x** to Return to Main Menu.

6. CONCLUSION

This ends the tutorial for Pro-Cite. You have successfully learned how to search the database, format a bibliography, and sort the records. As soon as you quit Pro-Cite, any changes you made (except changes to the Options screen) are canceled and the Search-Only version is ready to be used again.

There are many Pro-Cite features available that have not been explained in this tutorial. You have the ability to explore most of them with the Search-Only disk. Simply return to the Main Menu and press the [HOME] key for information on different commands.

When you are finished exploring Pro-Cite, press the appropriate key(s) to return to the Main Menu. Type **q** to Quit. This will close the CRSD database and return to DOS.

Copying the Pro-Cite Search Only disk for resale is prohibited. However, you are welcome to make duplicate copies as long as you acknowledge PBS as the originator and conform to copyright restrictions.

If you have any problems running Pro-Cite, please re-read these instructions and make sure you have run the SETUP program. If you cannot find a solution, call the PBS Technical Support Staff at (313) 996-1580.

For more information on the complete line of IBM and MacIntosh information management products available from PBS, call a sales representative at (313) 996-1580.

Pro-Cite is a registered trademark of Personal Bibliographic Software, Inc. IBM is a registered trademark of International Business Machines Corporation.

APPENDIX IV

SUBJECT TERM AUTHORITY LIST FOR THE CHEMISTRY REFERENCE SOURCES DATABASE (CRSD)

abbreviations
absorption
abstracts
acids
acronyms
ACS
adhesives
adsorbents
agriculture
AIP
air
alchemy
alcohols
alkaloids
alkyls
alloys
aluminum
AMA
American Chemical Society, use: ACS

American Institute of Physics, use: AIP
American Mathematical Society, use: AMS
American Medical Association, use: AMA
American Society for Testing and Materials, use: ASTM
amines
amino acids
ammonia
AMS
analysis
analytical
Analytical Abstracts
animals
AOAC
aqueous, use: water
argon
Asian
Association of Official Analytical Chemists, use: AOAC

333

ASTM
astronomy
atomic
audio-visual, use: AV
Auger electron spectroscopy
authors, see also: patentees, patent assignees
AV (used for works containing any audiovisual media, including slides and audiocassettes)
awards
azeotropes
bases
Beilstein
bibliographic instruction
Bibliographic Retrieval Service, use: BRS
bibliography
biochemistry
biochemists
biography
biology
biosciences
BIOSIS
biotechnology, use: genetic engineering
books
brain
Britain
BRS
buffers
business
CA
CA File
CAI
calculus
calorimetry
Canada
cancer
carbohydrates
carcinogens
careers
CAS ONLINE
catalog
cataloging
catalysis
CBE
CD-ROM, use: CDROM
CDROM
cell chemistry, use: cytochemistry
cells
Center for Research Libraries, use: CRL
ceramics
certification
Chemical Abstracts, use: CA
chemical engineering
chemical engineers
chemical equations
Chemical Hazards Response Information System, use: CHRIS
chemical industry
Chemical Information System, use: CIS
Chemical Journals Online, use: CJO
chemical nomenclature
Chemical Substances Information Network, use: CSIN
chemicals
chemistry
chemists
CHEMLINE
children
Chinese
CHRIS
chromatography
CIS
citation index
CJACS
CJO
classification
clinical chemistry
coatings
CODEN
collection development
collisions
colloids
color
colorimetry
combustion
Committee on Professional Training (ACS), use: CPT
composition
compounds, use: chemicals (in most instances); see also: organometallic compounds, coordination compounds
comprehensive index (used for numeric data compilations which exist in more than one set or are large multi-volume titles)
computational chemistry
computer-assisted instruction, use: CAI
computer science
computers
conductivity
conferences
constitutional chemistry
consultants
coordination compounds
corrosion
Council of Biology Editors, use: CBE
coupling constants
CPT

critically evaluated data
CRL
crystal structures
crystallographers
crystals
Current Abstracts of Chemistry and Index
        Chemicus, use: Index Chemicus
current awareness
cyclic
cytochemistry
DARC
database (used mostly in the Title field as a
        qualifier)
database management system, use: DBMS
databases
DBMS
DEC
demonstrations, use: experiments
design
DIALOG
dictionary
dielectrics
diffraction
diffusion
directory
disposal
dissertations (used for Ph.D. level only; see
        also: theses)
dissociation constants, use: ionization constants
distillation
DNA
document delivery
design
drugs
dyes
education
electrochemistry
electrode potential
electrolytes
electron diffraction
electronic
electronics
electrophoresis
elements
employment
encyclopedia
energy
energy levels
engineering
English
entomology
environment
enzymes
eponyms

equilibrium
equilibrium constants
equipment
essential oils
estimation
Europe
examinations (used for academic tests or
        quizzes)
exercises
experiments
explosives
extraction
fats
filmstrips, use: AV
flame
flash points
fluorescence
food
formatter
formula index
formulary
Fourier transform infrared, use: FTIR
French
front-end software
FTIR
functional groups
fusion
gas
gateway software
genetic engineering
geochemistry
geology
German
Germany
GL (indicates a work cited in Guidelines and
        Suggested Title List for Undergraduate
        Chemistry Libraries)
glass
Gmelin
government
government publications
graduate education
grants
graphics
Great Britain, use: Britain
guide
halides
halogens
handbook (used mostly for one-volume data
        compilations)
handbuch (used for large multi-volume German
        data compilations)
hardware
hazards

heat
helium
herbicides
heterocyclic compounds
high performance liquid chromatography, use: HPLC
high pressure liquid chromatography, use: HPLC
high temperature
history
Houben Weyl
HPLC
hydrocarbons
hydrogen
IBM
ICP
identification
immunology
index
Index Chemicus
indexes
inductively coupled plasma, use: ICP
industry
information analysis centers
information science
infrared, use: IR; see also: FTIR
inorganic
instruments
International Standard Serial Number, use: ISSN
International Union of Biochemistry, use: IUB
International Union of Pure and Applied Chemistry, use: IUPAC
ionization constants
ionization potentials
ions
IR
isotopes
ISSN
Italian
IUB
IUPAC
Jahn-Teller effect
Japanese
journals
ketones
kinetics
labelling
laboratories
laboratory information management systems, use: LIMS
Landolt Bornstein
lasers
LC
LEED
librarianship

libraries
Library of Congress, use: LC
LIMS
lipids
liquids
Lockheed, use: DIALOG
low energy electron diffraction, use: LEED
low temperature
Macintosh
macromolecular
markets
Markush
mass
mass spectrometry
material safety data sheets, use: MSDS
materials
mathematics
measurements
medical libraries
medicine
MEDLINE
meetings, use: conferences
melting points
MeSH
metabolic pathways
metabolism
metals
methanol
methods
microcomputer, use: PC
microwave
minerals
mixtures
modeling, use: molecular modeling
molecular formula
molecular modeling
molecular orbitals
molecular sequence databank
molecular spectra
molecular weights
molten salts
molybdenum
Mossbauer effect
MSDS
mutagens
name reactions
named effects
National Bureau of Standards, use: NBS; see also: NIST
National Institute of Occupational Safety and Health, use: NIOSH
National Institute of Standards and Technology, use: NIST
National Library of Medicine, use: NLM

SUBJECT TERM AUTHORITY LIST FOR CRSD 337

natural products
NBS
neurochemistry
NIH-EPA Chemical Information System, use: CIS
NIOSH
NIST
NLM
NMR
nomenclature
nomographs
nuclear magnetic resonance, use: NMR
nucleic acids
numeric data
Numerica
nursing
nutrition
OCLC
OHMTADS
Oil and Hazardous Materials Technical Assistance Data System, use: OHMTADS
oils
online chemical dictionary
online databases
online searching
optical resolution
ORBIT
organic
organometallic compounds
oxidation
oxidation reduction potentials
oxides
oxygen
ozone
PAH
paints
paper
patent assignees
patent family data
patentees
patents
PC
peptides
periodic table
personal computer, use: PC
pesticides
petroleum
pH
pharmaceuticals
pharmacy
phase diagrams
phase transitions
phenols

photochemistry
physical
physical properties
physics
plants
plastics
platinum
polarography
pollution
polycyclic aromatic hydrocarbons, use: PAH
polymers
preparation, use: synthesis; see also: reactions
private libraries
proceedings, use: conferences
processes
profession
proteins
purification
QSAR
quantitative structure-activity relationship, use: QSAR
quantum chemistry
quantum mechanics
Questel
radiation
radiochemistry
Raman
ranking, use: rating
rare earths
rating
reaction diagrams
reaction rates
reactions
reagents
RECON
redox potentials, use: oxidation reduction potentials
refractive index
Registry File
registry numbers
Registry of Toxic Effects of Chemical Substances, use: RTECS
regulations
reports
research
research in progress
resonance
resumes
retardants
reviews
ring, use: cyclic
ring index
rotational
Royal Society of Chemistry, use: RSC

RSC
RTECS
rubbers
Russian
safety
salaries
salts
SCI
science
Science Citation Index, use: SCI
scientific units
scientific writing
scientists
SCISEARCH
screens
SDC, use: ORBIT
SDI
sealants
selective dissemination of information, use: SDI
semiconductors
separation
sequences
serials
shipping
SI
SIP
software (used mostly in the Title field as a
        qualifier. In the Index field, the term
        is usually part of a phrase, as in:
        "front-end software" or
        "telecommunications
        software.")
solids
solubility
solvents
songs
Soviet Union, use: USSR
Spanish
specifications
spectra
spectroscopy
stability constants
standard interest profile, use: SIP
standard reduction potentials, use: oxidation
        reduction potentials
standard reference data
standard reference materials
standards
statistics
steam
stereochemistry
steroids
STN
structures

style guide
subject index
subject term authority list, use: thesaurus
substructure searching
sugars
superconductivity
SuperSearch
suppliers
surface
surfactants
symbols
synchrotron radiation
synthesis
System Development Corporation, use: ORBIT
tables (used mostly for multi-volume data
        compilations)
teaching
technical reports, use: reports
technology
techniques, use: methods
telecommunications software
terpenes
terpenoids
textbook
textiles
Theilheimer
thermal
thermodynamics (used also for thermal
        properties)
thesaurus
theses (use for master's level only; see also:
        dissertations)
thin-layer chromatography, use: TLC
titles
titration
TLC
toxic
Toxic Substances Control Act, use: TSCA
toxicology
trade literature
trademarks
tradenames
transition metals
transition probabilities
translations
translators
transport
treatise
TSCA
ultraviolet, use: UV
undergraduate education
union list
units
US

USSR
UV
vacuum
vapor pressure
VAX
vibrational
visible
vitamins

water
Wilsonline
Wiswesser Line Notation, use: WLN
WLN
wordprocessing
writing aid
X-rays
zoology

INDEX

abbreviations (CA)
    SEE: CA File database—abbreviations
abbreviations, journals
    SEE: journals—abbreviations
abstracting journals
    SEE: journals—abstracting
abstracts 5, 99, 145
    example 5, 48, 49
Academic American Encyclopedia 95
accession numbers 4, 5
accession numbers, patents
    SEE: patents—application numbers
acronyms
ACS Directory of Graduate Research 304, 312
    example 303
ACS Style Guide 289-290
ADJ logical operator 97
Alchemy software 136, 287
Aldrich Catalog/Handbook of Fine Chemicals 191, 306
Aldrich Microfiche Library of Chemical Indices 204-205
American Men and Women of Science 301
analyte 177
Analytical Abstracts 177-178
analytical chemistry
    SEE: constitutional chemistry
Analytical Chemistry 175, 278, 306
AND logical operator 35
Anglo-American Cataloging Rules 292-293
Annual Book of ASTM Standards 308
ASCII code 39, 40-41
assignee
    SEE: patent assignee
ASTM 168, 308
attribute commands (STN) 151
audiovisual materials 310
author searching 53, 59, 67, 69-72

authors' names, hyphenated 63, 69-70
authors' names, initials 69, 70-71
Automated Patent System 107

background reading 263-264, 275-279
basic index 52
    SEE ALSO: CA File—Basic Index; Registry File—Basic Index; CJACS File—Basic Index; SciSearch database—Basic Index
basic patent 108
baud rate 40, 41, 42
Beckman Center for the History of Chemistry 300
Beilstein 157, 185-191
    abbreviations 188, 279, 292
    access 187-191
    arrangement 186-187, 213
    chronological coverage 186
    contents 157, 185-186, 191-195, 213
    example 192-195
    guide 187, 210
    indexes 185-186
    page references 189, 191
    subject coverage 185, 191
    system 186
Beilstein database 157-158, 191-192
    DISPLAY options (STN) 195-196
    example 197, 216-217
    physical properties 191-197
    reaction chemistry 213-217
bibliographic citation 61, 285, 289-290, 291
bibliographic databases
    SEE: databases—types
bibliographic reference
    SEE: bibliographic citation
bibliographic searching
    SEE: databases—searching—bibliographic

bibliographies 12, 94-95
bibliography manager software
   SEE: PIMS software
bio-bibliographical compilation 301
biochemistry 173-175, 237
   reviews 237, 238
   treatises 223
biographical memoirs 302
biographical sources 300-301
BIOSIS Connection 271
BIOSIS database 127, 271, 272
BioSuperfile software 272
bits 39
B-I-T-S SDI service 272
BNA 260
BONd command (STN) 150, 152, 154
books 277-278
   citation format 290
   guides 278
Books in Print 278
Boolean search operators 35, 54, 96
   order of precedence 37, 54
   SEE ALSO: ADJ, AND, NOT, OR, WITH logical operators
BRS 18, 22, 37, 43, 122, 128
Bulletin of Chemical Thermodynamics 206
Bureau of National Affairs
   SEE: BNA
business 307-308
bytes 40

CA File database 46, 50, 89, 227
   abbreviations 84-85, 88, 117, 225
   abstract search 99
   author search 53, 69-70
   Basic Index 48, 49, 52, 85-86, 128, 136
   controlled-vocabulary 52, 83, 85, 136
   corporate source search 71, 72
   DISPLAY command 48-49, 50-51
   document types 91-92
   journal coverage 60
   records 48, 49
   reviews 275
   section code searches 86
      example 91
CA Selects 274
CAOLD File database 46, 50, 89, 127, 145
CApreviews database 271
carcinogenic 245
careers
   SEE: chemistry—careers

CASREACT database 225-228, 243
   example 27
   link to CJO database 228
CASSI 279, 282, 291-293
CASSIS 107-108
Cauzin softstrip 265, 266-268
CD-ROM
   SEE: databases—CD-ROM
ChemBase software 286
ChemConnection software 157
Chemical Abstracts 86-89
   abbreviations 84-85
   Author Index 69-70
   Chemical Substance Index 83, 116-117, 134, 224, 225
   classification 86-87
   document type codes 91-92, 112
      example 5, 6, 9
   Formula Index 120-122
      example 121
   General Subject Index 83, 88, 116, 224
   history 28
   Index Guide 82, 83-85, 116-117, 224, 295
      example 82, 117
   Index Name 117, 121, 136
   Index of Ring Systems 123
   Patent Index 111
      example 111
   patents in 101, 110-112
   reviews 275
   Subject Index 89, 116
   volume indexes 89
   weekly issues 6, 86-87
      Author Index 87
      classification 87
      Keyword Index 87-88
      Patent Index 87
Chemical Abstracts Service Source Index
   SEE: *CASSI*
Chemical Business Newsbase database 307
chemical dictionaries
   SEE: online chemical dictionaries
Chemical Economics Handbook 307
chemical engineering 307
Chemical Hazards Response Information System
   SEE: CHRIS
Chemical Industries: An Information Sourcebook 307
chemical industry 27, 28, 277, 307-308, 313
Chemical Industry Notes 307
chemical information science 1, 26
Chemical Information System 130-131, 142, 170-171, 179, 249-251

Chemical Information Systems, Inc. 22, 142, 171, 250
Chemical Marketing Reporter 306
Chemical Name field (REG) 132
   example 133
chemical names 89, 118-119, 294-295
   IUPAC 293-294
   searching 116, 119, 132, 136-137, 223-225
chemical names *(CA)* 85, 117, 118, 223-225, 294, 295
   alphabetizing rules 117
   Keyword Index 77, 88
chemical names (CA) 89, 136-137
chemical names (REG) 89, 132, 133, 135
Chemical Patents Index 108-109
Chemical Publications: Their Nature and Use 26, 279
Chemical Reactions Documentation Service database
   SEE: CRDS database
chemical structure drawing interface 288
Chemical Titles 77, 82, 265
   example 269
chemicals, disposal 258
chemicals, flammable 258
chemicals, hazardous 245-261
chemicals, indexing *(CA)* 223-225
chemicals, nomenclature
   SEE: chemical names
chemicals, preparation 211-243
   indexing *(CA)* 213-218, 223-225
   searchable fields (Beilstein) 213-214
   searchable fields (CASREACT) 226
   searching (Beilstein) 213-217
   searching *(CA)* 224-225
   searching (CA) 225
   searching (CASREACT) 225-228
      example 227
   searching with registry numbers 128, 225
chemicals, purification 168-169
chemicals, regulations 259-260, 261
chemicals, safety 257-259
   abstracts 258-259
   guide 246-247
chemicals, shipping 306
chemicals, suppliers 305-306
chemicals, toxic 246, 252-257
   guide 246-247
chemistry,
   careers 313
   history 26, 27, 299-300, 314
   study 312-313
   sub-disciplines 2-3
   teaching 309-312

Chemistry Reference Sources Database
   SEE: CRSD
CHEMLAB database 311
CHEMLINE online chemical dictionary database 130, 247
ChemQuest database 158, 305
ChemSources 305
CHRIS 250, 252
Chronologie Chemie, 1800-1980 279, 300
citation searching 97-98
   patents 112
   SEE ALSO: co-citation searching
cited reference 61, 72
   anonymous 63
   example 65
citing reference 61
   anonymous 63
CJACS File database 97-99, 127
   Basic Index 97-98
CJO File database 97-99, 228
   DISPLAY options 98
claims 102
CLAIMS database 110, 111
classification 4, 186, 198, 212, 213, 235
   SEE ALSO: patents—classification
classification *(CA)* 5, 86-88
   example 6, 86, 87
classification (LC) 4
   example 6
classification (REG) 132
classification (SCI)
   SEE: research front
Clinical Toxicology of Commercial Products 250, 255
co-authors 61, 64
co-citation searching 66
CODATA 182, 295
CODATA Bulletin 295, 296
CODATA Directory of Data Sources for Science and Technology 182
CODEN 71, 72, 292
collection development 28, 278-279, 283
College Chemistry Faculties 304, 312
command-driven software 44
communications software
   SEE: telecommunications software
Compendium of Chemical Terminology: IUPAC Recommendations 295
compound class identifiers (REG) 132
compounds
   SEE: chemicals
comprehensive index to numeric data
   SEE: numeric data compilations— comprehensive indexes

Comprehensive Inorganic Chemistry 219, 220
Comprehensive Organic Chemistry 220
Comprehensive Treatise of Electrochemistry 166
CompuServe 41
CONCORD 136, 287
conference proceedings 8-9, 93
connection codes 125-126
constitutional chemistry 3, 161, 178
    guides 162, 173
    handbooks 168-169
    nomenclature 169
    reviews 175-176
controlled-vocabulary subject index 7, 31, 82
convention patents
    SEE: patents—equivalent
corporate source searching 66-67, 71-72
CRC Composite Index for CRC Handbooks 206
CRC Handbook of Chemistry and Physics 191, 200-202, 295
CRC Handbook of Data on Organic Compounds 204
    SEE ALSO: HODOC database
critically evaluated data 182-183, 210
cross-database searching
    SEE: databases—searching—cross database
cross reference 6, 82, 83, 117
CRDS database 234
CRSD 16, 25, 31, 34, 40, 72, 95, 106, 162, 174, 182, 187, 277, 290, 291, 302, 304, 312
crystallography 72-173, 179
current awareness 80-82, 263, 264-274
    SEE ALSO: journals—current awareness; standard interest profiles; SDI
Current Chemical Reactions 230-232
    databases 230, 232
    example 231
Current Contents 60, 64, 67, 77, 82, 265
    example 269
Current Contents on Diskette 268, 271
CV (controlled vocabulary) field (CA)
    examples 48, 49

Dangerous Properties of Industrial Materials 257
DARC 139, 144, 159
DARC CHEMLINK front-end software 157
data compilations
    SEE: numeric data compilations
databank 26
database producers 23, 33
database vendors 18, 22-23, 33
databases
    CD-ROM 18, 39, 95

    coverage, subjects 34
    coverage, time periods 34
    guides 23-25, 32, 100
    SEE ALSO: databases—locater fields
    indexes
    SEE: inverted dictionary files
    locater fields 130, 134, 174
    non-bibliographic 15
    personal 286-288, 297
    searching
        advantages 15-16
        benefits 34-35
        bibliographic 33-38
        commands 37
        costs 35-36, 135, 143-144
        cross-database 34
        equipment 39-40
        novice
        SEE: default; front-end software; gateway software
        numeric 196-197
        options 18
        problems 36-37, 207
        software 39, 43-46
    types 14
        bibliographic 15
        dictionary 17
        directory 17
        numeric 16
        text 17, 95-96
default 43-44
    STN 147
defensive publication
    SEE: statutory invention registration
deferred examination 104
demonstrations
    SEE: teaching—demonstrations
Derwent 38, 107, 108-109, 158, 232
descriptor 46
DIALINDEX 24-25
DIALNET 41
DIALOG 22, 25, 33, 37, 43, 60, 63, 65 67, 72, 112, 128, 129-130, 158, 200, 209
DIALOGLINK software 112, 157, 208
dictionaries 11, 169, 187, 277
dictionary databases
    SEE: databases—types
Dictionary of Organic Compounds 127, 202-203
    example 203
    SEE ALSO: HEILBRON database
Dictionary of Scientific Biography 301
DIPPR database 208-209
directories 17, 27, 302-306

directory databases
  SEE: databases—types
Directory of American Research and Technology 304
Directory of Technical and Scientific Directories 302, 304
DISPLAY ALL command (STN) 91, 98
DISPLAY BROWSE command (STN) 97, 98
DISPLAY command (STN) 36, 43, 50-51, 134-136, 148, 195
  SEE ALSO: Beilstein database—DISPLAY
    CA File database—DISPLAY
    CJO File database—DISPLAY
    Registry File database—DISPLAY
DISPLAY HISTORY command (STN) 90
DISPLAY IND command (STN) 85
DISPLAY OCC command (STN) 98, 99, 195
DISPLAY SAMPLE command (STN) 50
disposal
  SEE: chemicals, disposal
Dissertation Abstracts International 34, 92, 95
dissertations 92
  example 9
document delivery 16, 38, 98, 264, 279 283
Document Delivery Service (CAS) 281
document delivery services (commercial) 279-281
document type 9, 91-92
  codes 9, 77-78
    books 6, 92
    patents 69, 92, 112
  databases devoted to 92
documents 16
dot-disconnected formulas
  SEE: formulas, dot-disconnected
downloading 38-39, 46
duplex 41-42

EasyNet 41, 46, 47
ECDIN 251
Education Resources Information Clearinghouse
  SEE: ERIC
Eight Peak Index of Mass Spectra 170
electronic publishing 297
element count searches 129, 130, 134
element symbols as search terms 196
element term searches (STN) 137
elements (*Gmelin*) 199
elements, spectroscopy 169
employment 313
Encyclopedia of Physical Science and Technology 276

Encyclopedia of Polymer Science and Engineering 277
Encyclopedia of Electrochemistry of the Elements 166
Encyclopedia of Industrial Chemical Analysis 164
Encyclopedia of Polymer Science and Engineering 95
encyclopedias 11, 276-277
end-user 38
engineering
  SEE: chemical engineering
environment 246-247, 248, 251, 258, 261
Environmental Chemicals Data and Information Network
  SEE: ECDIN
Enzyme Commission Number 133
equipment 306
equivalent patents
  SEE: patents—equivalent
ERIC database 309
  example 310
European Patent Convention 104
examinations 312
exercises 26, 27, 28
EXPAND command (STN) 75-76, 83
  examples 53, 70
expanding 53, 74

false drops 35, 38
festschrift 302
field-specific operator
  SEE: LINK logical operator
fields 46
Fieser and Fieser's Reagents for Organic Synthesis 239
films
  SEE: audiovisual materials
foreign language 283
format
  SEE: document type
formula index 28, 200, 202, 218, 222, 228
formula index (*CA*)
  SEE: *Chemical Abstracts*—Formula Index
formula weight 133
formulary 239
formulas 119-122, 137
  SEE ALSO: Hill System (formulas)
formulas (*CA*)
  SEE: *Chemical Abstracts*—Formula Index
formulas (REG) 134
formulas (SciSearch) 80
formulas, dot-disconnected 122

fragment codes 109, 124
front-end software 45-46, 57, 156-157

gateway software 45-46, 57
Genuine Article 280
German 187, 188-189, 214-215
GFI database 200, 201
Gmelin 197-200, 217-218
    access 199-200
    arrangement 198-199
    chronological coverage 198
    indexes 199-200
    subject coverage 197-198
    system numbers 199
Gmelin Formula Index database
    SEE: GFI database
Government Reports Announcement and Index 94
government reports
    SEE: reports
Graduate Programs in Chemistry 312
GRAph command (STN) 148
graphics search options 145-146
Grateful Med front-end software 46, 47
Guidelines and Suggested Title List for Undergraduate Chemistry Libraries 278
guides 19, 21
    chemistry, general 20, 25-28
    directory 31
    features 21-22
    patents 113
    SEE ALSO: databases—guides

Handbook of Environmental Data on Organic Chemicals 257
Handbook of Reactive Chemical Hazards 257-258
handbooks 11, 168-169, 200-205, 257
Handbooks and Tables in Science and Technology 182
handbuchs 181, 185
Hayes modems 40
Hayes protocols
    SEE: protocols
Hazardous Substances Data Bank
    SEE: HSDB database
HCO command (STN) 151
Heading Parent 117, 133-134, 225
HEILBRON database 207-208
Hill System (formulas) 120, 121
history
    SEE: chemistry—history
hit charges 35-36

HODOC database 208
Houben-Weyl
    SEE: *Methoden der Organischen Chemie*
how-to-write books
    SEE: writing—guide
HSDB database 248, 249, 255
HTSS software 156, 157
hydrogen count command (STN)
    SEE: HCO command (STN)
hyphenated authors' names
    SEE: authors' names, hyphenated

IDE display format 134-135, 195
IFI/Plenum 107, 110
Index Chemicus 141, 228-230, 232
    codes 228
    databases 230, 232
    example 229
    indexes 230
Index Name (*CA*)
    SEE: chemicals, names (*CA*)
Index to Scientific and Technical Proceedings 67-68, 93
Index to Scientific Book Contents 276, 278
Index to Scientific Reviews 275
Index to Standard Interest Profiles in Science and Technology 274
indexes, online
    SEE: inverted dictionary files
indexing 6, 82-83
indexing journals
    SEE: journals—indexing
Individual Search Service (CAS) 272, 273
industrial chemistry
    SEE: chemical industry
InfoMaster 41, 46, 47
INFONET 41
information analysis centers 182-183
information gatekeeper 29
initials in authors' names
    SEE: authors' names, initials
inorganic chemistry 197-200
    reviews 237, 238
    treatises 219-220, 223, 277-278
Inorganic Reactions and Methods 218, 219
Inorganic Syntheses 237
INPADOC 107, 109-110
INSPEC database 209
interlibrary loan 279
International Standard Serial Number
    SEE: ISSN
International Union of Biochemistry 294

INDEX **347**

International Union of Pure and Applied Chemistry, SEE: IUPAC
Introduction to Chemical Nomenclature 294
invention 102, 103
inverted dictionary files 51-53
   examples 52,
invisible college 7
isomers 121
ISSN 292
IUPAC 293-294, 295

JANAF Thermochemical Tables 203, 207
   database 209
Journal Citation Reports 279
Journal of Chemical Education 258, 278, 309, 311, 312
Journal of Physical and Chemical Reference Data 207
Journal of Synthetic Methods 232-235, 240
   codes 234
   example 233
journals 8
   abbreviations 279, 291, 291-293
   abstracting 177
      guides 177
   citation format 289-290
   cover-to-cover translation 282-283
   current awareness 13
   historical 279
   indexing
   letters 8
   news 8
   older 279, 292
   peer review 8
   ranking 279

Kaye and Laby's Tables
   SEE: *Tables of Physical ...*
key
   SEE: search keys
key-word index 7, 75, 77, 79, 279
   SEE ALSO: uncontrolled vocabulary
      subject index
key words 75
Kirk-Othmer Encyclopedia of Chemical Technology 95, 255, 276-277
KWIC
   SEE: key-word index
KWOC
   SEE: key-word index

Lab Guide issue of *Analytical Chemistry* 278
laboratories 258
laboratory equipment
   SEE: equipment
laboratory information management systems
   SEE: LIMS
laboratory notebooks 285, 289
Landolt-Börnstein 183-185
Lange's Handbook of Chemistry 191, 202, 295
LC_{50} 256
LD_{50} 252, 256, 257
letters to the editor 252
Library of Congress call numbers 4
Library of Congress Subject Headings 30, 82
LIFO 35
LIMIT command (DIALOG) 77
LIMS 289
line notations 124-125
LINK logical operator 55, 97
locater fields
   SEE: databases—locater fields
logging on/logging off 43-45

MACCS 140-141, 158
magazines 8
main entry 293
MAPRN command (DIALOG) 129
marketing 307-308
Markush structure 144-145, 159
masking 133
material safety data sheets 209, 245, 250, 255, 256, 261
Materials Property Data Network 209
materials science 28
matrix 177
McGraw-Hill Dictionary of Chemistry 276, 277
McGraw-Hill Encyclopedia of Science and Technology 276
McGraw-Hill Modern Scientists and Engineers 302
medicinal chemistry
   treatise 222-223
MEDLINE database 95, 118, 174
memoirs
   SEE: biographical memoirs
menu-driven software 44
Merck Index 118, 119, 203-204, 295
   database 208, 250
MeSH 118
Methoden der Organischen Chemie 218-219
methods 161
Methods in Enzymology 166-167

Methods of Biochemical Analysis 175-176
Methods of Enzymatic Analysis 167
microcomputers 140
mixtures 133
molecular biology 169, 173
molecular formulas
   SEE: formulas
molecular sequence databank
   SEE: sequence databases
molecular weight
   SEE: formula weight
MOLKICK software 156
monograph (drugs) 203-204
monographs 11, 278
multigraphs 10, 278
mutagenic 245

name reactions 211, 220, 239-240
NASA 22
National Aeronautics and Space Administration
   SEE: NASA
National Library of Medicine
   SEE: NLM
National Referral Center 304
National Translations Center 282
natural-language index
   SEE: uncontrolled vocabulary subject
      index; key-word subject index
NEIGHBOR command
   SEE: expanding
nesting 54
newsletters 260
NLM 22, 130, 247
Nobel Prize 300, 302
NODe command (STN) 148, 152, 153
node specification command (STN)
   SEE: NSPEC command (STN)
non-bibliographic databases 26
normalized bond 150
NOT logical operator 35
novelty 102
NSPEC command (STN) 150-151
NTIS database 94, 95
numeric data indexes 206-207
numeric data compilations 11, 183-200
   arranged by property 204-205
   comprehensive indexes 184-185, 205-206
   guides 182
numeric databases
   SEE: databases—types—numeric
Numerica 22, 208, 250

Official Gazette of the United States Patent and Trademark Office 108
OHM/TADS database 250, 252, 253
Oil and Hazardous Materials Technical Assistance Database
   SEE: OHM/TADS database
OneSearch feature (DIALOG) 63, 80
online chemical dictionaries 129-136
   SEE ALSO: Registry File; SANSS, SYNDEX
online database searching
   SEE: databases—searching
online database producers
   SEE: database producers
online database vendors
   SEE: database vendors
online indexes
   SEE: inverted dictionary files
online searching 20
   command languages 37
   costs 38-39
   features 38-39
online searching commands
   SEE: databases—searching—commands
online searching equipment
   SEE: databases—searching—equipment
OPD Chemical Buyer's Directory 305
optical character readers 265
OR logical operator 35
ORAC software 241-242
ORBIT 22, 33, 37, 60, 63, 128, 130, 158
organic chemistry
   reviews 220, 221, 222, 236-237, 238
   treatises 220-223
Organic Syntheses 236-237, 239, 240

PARAGRAPH logical operator 97
parity bit 41
patent application 102
Patent Cooperation Treaty 104
patent assignee 69
patent family 103
patent specification 102
patentee 69, 102
patents 101-106, 174, 232
   application numbers 106
   chemistry 103, 105
   classification 105-106, 108-109
   current awareness 109, 110
   equivalent 103
   example 9
   features 106

infringement 105
numbering 106
translation 105, 283
value 105-106
patents (CA) 110-112
Patty's Industrial Hygiene and Toxicology 256-257
peer review
   SEE: journals—peer review
Pergamon 219
periodic groups 134
periodic table 199
periodical 8
personal information management system
   SEE: PIMS software
personal names
   SEE: authors' names
pesticides 255-256
pharmacopoeia 168
physical chemistry
   reviews 210
   treatises 209
Physical Methods of Chemistry 163
physical properties 181-210
   guides 182
   SEE ALSO: handbooks, numeric data compilations
Physics Abstracts 107, 209
PIMS software 286
plant equipment
   SEE: equipment
Poggendorff's Bio-Bibliographical Handbook of the Exact Sciences 301
polymers 27, 122, 132, 133
   nomenclature 222
   treatise 222
positional operator
   SEE: proximity logical operator
precision 38
preparation of chemical substances
   SEE: chemicals, preparation
preprints 285
primary literature 7
principle of last position 198
PRINT command (ORBIT) 36
PRINT command (STN) 36
PRINT SELECT RN command (ORBIT) 129
priority date 102
proceedings
   SEE: conference proceedings
processing 307
Pro-Cite software 288

producers
   SEE: database producers
Project SERAPHIM 309, 310
property hierarchy search (Beilstein) 196
Pro-Search front-end software 47
protocols 40-42
proximity logical operator 55-56, 96, 97, 226
Prudent Practices for Handling Hazardous Substances 258
publication date searching
   SEE: RANGE command (STN)
        LIMIT command (DIALOG)
punctuation marks 80, 133
Pure and Applied Chemistry 293, 295, 296

qualified substances (CA) 117, 120, 224
   subject searching 117, 120, 224-225
Questel 22, 37, 128, 139, 158

RANGE command (STN) 37, 71-72, 86
rapid communication journals
   SEE: journals—letters journals
REACCS software 232, 240-241, 243
reaction chemistry 3, 211-243
   abstracts 223-236
   compendia 238-239
   databases 213, 225, 232, 240-242
   guide 212-213, 243
   reviews 220, 235, 236-238
   treatises 219-223
reaction diagrams 228, 229, 231, 233, 236
reaction mechanism 211
reagents 168, 220, 239
recall 38
records 35, 46, 51
   example (CA) 48-49
Registry File online chemical dictionary database 46, 131-136, 145
   Basic Index 132
   chemical name searches 132, 133
   components, number of 134
   DISPLAY 134-136, 151, 152
      example 153-155
      example, graphics 135
   formula weight 133
   inverted dictionary files 132
   relationship to CA and CAOLD Files 46, 50, 89, 90, 131
   search sequence 152

structure building 147
structure searching 146
  bond codes 150
  closed substructure search 146, 150
  exact search 146, 154
  family search 146
  generic group symbols 149
  node symbols 148
  normalized bonds 150
  shortcut symbols 149
  substructure search 147,
registry numbers 46, 89, 110, 116, 126, 128, 129, 142
  index 202, 255, 256, 257
  examples 48, 49, 126
registry numbers, online searching 127-128, 134, 142, 177, 225, 226
Registry of Toxic Effects of Chemical Substances
  SEE: RTECS
regulations
  SEE: chemicals, regulations
relevance
  SEE: precision
reports 9, 94
  example 9
reprints 281
research front 80, 275
research in progress 263, 274
retrospective search 13
review articles 12, 175, 275-276, 283
  SciSearch 77
  searching, example 76
  SEE ALSO: biochemistry—reviews, organic chemistry—reviews, etc.; CA File—reviews; *Chemical Abstracts*— reviews
review serials 11, 175
ring indexes 122-123, 221
ring specification command (STN)
  SEE: RSPEC command (STN)
ring systems (REG) 147-148
Ring Systems Handbook 123
  example 123
ROSDAL structure codes 197
RSPEC command (STN) 151
RTECS 118, 119, 247, 248, 250, 252, 254 255, 256

Sadtler spectral compilations 171
safety
  SEE: chemicals, safety
salaries 313
salts 122, 198

SANDRA software 188-190
SANSS online chemical dictionary database 130-131, 142-144, 250
  example 143
Schaum's outline series 312
SCI CD-ROM database 18
Science Citation Index 60-63
  anonymous works in 63
  Citation Index 60, 61
    example 61-62
  Corporate Index 67
    example 68
  errors 63
  journals covered 61
  Permuterm Subject Index 60, 77-79, 82
  Source Index 60, 61, 78
  time periods covered 60
Scientific and Technical Books and Serials in Print 278
scientific papers 289, 291
SciSearch database 60, 72
  author search 64
  Basic Index 80
  cited reference search 64-66, 72
    example 65
  co-citation search 66
  corporate source search 66-67
    example 66
  errors 63-64
  journal coverage 60, 64
  subject search 78, 80
    example 81
  time periods covered 63
screens 124, 151
  SEE ALSO: fragment codes
SDI 13, 35, 271-272
Search for Data in the Physical and Chemical Sciences 182
search intermediary 38
search keys 75
search strategy 22, 29-32, 152
secondary literature 10
  selection 30-31
  time lag 13-15
section codes (CA) 91
  examples 48, 49
SEE reference
  SEE: cross reference
SELECT command (STN) 128, 136
SELECT STEPS command (DIALOG) 67
selective dissemination of information
  SEE: SDI
SENTENCE logical operator 97
sequence databases 173-175, 178

SERAPHIM
   SEE: Project SERAPHIM
serials 8
   guides 279
   standards
   SEE ALSO: review serials
SGML
   SEE: standard generalized mark-up
      language
software
   SEE: teaching—software; telecommunications
      software
specifications 308
spectral data collections 170-172, 179
spectroscopy 169
standard generalized mark-up language 96
standard interest profiles 13, 272-274
   SEE ALSO: current awareness journals;
      SDI
standard methods of analysis 167-168
standards 168, 291, 295-296, 308-309
statutory invention registration 105
stereochemistry 126, 141, 173
STN 22, 37, 38
   logon procedure 44-45
STN Express front-end software 40, 46, 47, 156
STN Mentor software 72
stop bit 41
stop list 77, 78
string searching 56
structure-activity relationships 26
structure searching 139, 213
style guide 289-290
subject indexes 6
subject searches 31, 73-74
   example (CA) 76, 86, 99
   example (*SCI*) 79
   example (SciSearch) 80, 81
subject term authority lists 82-83, 178
substrate 212
substructure searching
   SEE: structure searching
symbols 295, 296
SYNDEX online chemical dictionary database 251
SYNLIB software 242
synthesis
   SEE: reaction chemistry
system search software
   SEE: databases—searching—software
systematic nomenclature rules 294

3-D 135, 136, 287
table of contents journals
   SEE: journals—current awareness
table of contents service 264-265
Tables of Physical and Chemical Constants and
      Some Mathematical Functions 202
tautomers 150
teaching 309-312
   audiovisual materials 310
   computers 310-311, 314
   demonstrations 311-312
   methodology 309
   software 310
technical reports
   SEE: reports
Techniques of Chemistry 162, 163
telecommunications companies 35-36, 40-41
telecommunications problems 42
telecommunications software 39-40, 57
Telenet 41, 43, 44
templates 156
teratogenic 245-246
tertiary literature 19, 21, 26, 28
text databases
   SEE: databases—types
text modifications 120, 127, 224
   examples 48, 49, 224
Theilheimer's Synthetic Methods of Organic
      Chemistry 234-235, 240
   codes 235
   guide 234
Thermodynamics Research Center 171
thesaurus
   SEE: subject term authority list
theses
   SEE: dissertations
Thomas Register of American Manufacturers 306
three-dimensional display
   SEE: 3-D
Threshold Limit Values 252, 258
TOPFRAG software 158
toxicity 245
toxicology 245
   guide 246-247, 261
   SEE ALSO: chemicals, toxic
TOXLINE databases 248
TOXNET databases 247-249, 261
trade literature 204, 306
   guide 306
trade names 255
trademarks 112

translations 27, 94, 105, 281-283
    patents 283
Treatise on Analytical Chemistry 163-164
treatises 10, 163
    SEE ALSO: individual classes of chemistry, e.g., reaction chemistry—treatises
truncation 74, 81
truncation (DIALOG) 64, 80
truncation (STN) 72, 74-75
Tymnet 41
TYPE command (DIALOG) 37, 67

Ullmann's Encyclopedia of Industrial Chemistry 277
uncontrolled-vocabulary subject index 7, 77, 82, 87
Undergraduate Professional Education in Chemistry 312
union list of serials 279
United States Patent and Trademark Office 107-108
uniterms 110
user aids
    SEE: guides

Using CAS ONLINE 86, 131, 134, 148
utility 102, 103
utility patents 108

vendors
    SEE: database vendors

wild cards
    SEE: truncation
Wilson and Wilson's Comprehensive Analytical Chemistry 164-166
Wilsonline 22
Wiswesser Line Notation 123, 124-125
WITH logical operator 55, 56
wordprocessing 287, 288, 296
WPI database 108-109, 145
writing 285
    guides 289-291

X-ray crystallography
    SEE: crystallography